Magneto-Optics and Spectroscopy of Antiferromagnets

V.V. Eremenko N.F. Kharchenko
Yu.G. Litvinenko V.M. Naumenko

Magneto-Optics and Spectroscopy of Antiferromagnets

With 125 Illustrations

Springer-Verlag
New York Berlin Heidelberg London Paris
Tokyo Hong Kong Barcelona Budapest

V.V. Eremenko
Academy of Sciences of Ukraine
Lenin Avenue 47
Kharkov 310164
Ukraine

N.F. Kharchenko
Academy of Sciences of Ukraine
Lenin Avenue 47
Kharkov 310164
Ukraine

Yu.G. Litvinenko
Academy of Sciences of Ukraine
Lenin Avenue 47
Kharkov 310164
Ukraine

V.M. Naumenko
Academy of Sciences of Ukraine
Lenin Avenue 47
Kharkov 310164
Ukraine

This book is a translation of *Magnitooptika i Spektroskopiia͡ Antiferromagnetikov* by V.V. Eremenko et al., published by Naukova Dumka Publishers (the Ukrainian SSR Academy of Sciences), Kiev, 1989.

Library of Congress Cataloging-in-Publication Data
Magnitooptika i spektroskopiia͡ antiferromagnetikov. English
Magneto-Optics and spectroscopy of antiferromagnets / V.V. Eremenko ... [et al.].
p. cm.
Translation of: Magnitooptika i spektroskopiia͡ antiferromagnetikov.
Includes bibliographical references.
ISBN-13: 978-1-4612-7694-4 e-ISBN-13: 978-1-4612-2846-2
DOI: 10.1007/978-1-4612-2846-2
1. Magneto-optics. 2. Magnetic crystals. 3. Magnetic crystals—Spectra. 4. Antiferromagnetism. I. Eremenko, V.V. (Viktor Valentinovich) II. Title.
QC675.M2413 1992
538'.44—dc20 91-30896

Printed on acid-free paper.

Production coordinated by Brian Howe and managed by Francine Sikorski; manufacturing supervised by Vincent Scelta.
Typeset by Asco Trade Typesetting Ltd., Hong Kong.

9 8 7 6 5 4 3 2 1

Preface

The magneto-optics and spectroscopy of magnetically ordered crystals have actively been developing since the early 1960s. At present, not only a considerable body of experimental results is accumulated but also some concepts of the nature of the optical and magneto-optical properties of magnetic crystals are formulated. Moreover, it is high time to apply the optical and magneto-optical properties of magnetically ordered media in designing new devices, computer memory elements, etc.

The results of studies into the optical and magneto-optical properties of magnetically ordered crystals are scattered in numerous journal articles. The first attempts to summarize these results were made in 1975 when the monographs *The Introduction to the Optical Spectroscopy of Magnetic Crystals* by V.V. Eremenko and *The Theory of Magnetic Excitations* by E.G. Petrov were published by the Ukrainian SSR publishers "Naukova Dumka."

This new monograph is an attempt to summarize the results obtained between 1975 and 1986, and to give an account of them in terms of the modern concepts of the nature of optical and magneto-optical phenomena in magnetic crystals. Naturally, all the problems concerning the optics, magneto-optics, and spectroscopy of magnetic crystals cannot be dealt with in a single book. Thus, for example, the book is not concerned with problems of the optics of magnetic metals or magneto-optics applications. The authors prefer to place primary emphasis on the magneto-optics and spectroscopy of antiferromagnetic insulating crystals or, rather, on those aspects of the problem which their works deal with and in which they are well versed.

The first two chapters (by V.V. Eremenko and N.F. Kharchenko) are concerned with magneto-optical effects in antiferromagnetic crystals and their application to investigations of the structure under magnetic phase transitions induced by an external magnetic field, and the visualization of collinear antiferromagnetic domains and two-phase magnetic structures. The symmetry of the optical properties of magnetically ordered crystals is considered; in particular, a more comprehensive consideration is given to the symmetry of the effects found by the authors which are typical for noncentrally antisymmetric antiferromagnetic crystals, namely, the linear magneto-optical effect and the squared-in-a-field strength magnetic gyrotropy. The experimental data on the novel magneto-optical effects in antiferromagnetic insulating crystals are discussed. In particular, it is shown that the experimental studies of the linear magneto-optical effect, together with the symmetry analysis, permit the magnetic symmetry point group of complex antiferromagnets to be found that has only been feasible earlier with sophisticated and expensive neutron diffraction techniques.

The third and fourth chapters (by V.V. Eremenko and Yu.G. Litvinenko) deal with the physics of optical magnetic excitations (magnetic excitons) and their application to the study of the physical properties of antiferromagnets, namely, the magnetic configurations and their transformation in external magnetic fields, the parameters of the spin-wave spectrum, and the photo-induced properties.

And, finally, the fifth chapter (by V.V. Eremenko and V.M. Naumenko) discusses the problem of impurity magnetic excitations in antiferromagnetic insulators, as well as experimental and theoretical studies of the conditions for their delocalization.

Kharkov, Ukraine

V.V. Eremenko
N.F. Kharchenko
Yu.G. Litvinenko
V.M. Naumenko

Contents

CHAPTER 1

Magneto-Optic Effects in Non-Centroantisymmetrical Antiferromagnetic Crystals

The reduction of the space–temporal symmetry of the medium during magnetic ordering leads to qualitative changes in its physical properties. For example, there are well-known effects such as piezomagnetism and the magnetoelectric effect [1] and some galvano- and thermomagnetic effects [2], forbidden in magnetically nonordered crystals and allowed in ordered crystals with a definite magnetic symmetry. The majority of optical properties of magnetically ordered media, which appear during magnetic ordering, are similar to those of magnetically nonordered media magnetized by an external magnetic field. For example, the spontaneous magnetic rotation of the plane of polarization in a ferromagnet is proportional to the magnetic moment, in the same way as the Faraday rotation in a paramagnet, induced by a field, is proportional to its magnetization. Similar correspondence can be found between the spontaneous and field-induced quadratic effects of magnetic linear birefringence. Spontaneous magnetic gyrotropic birefringence is possible in antiferromagnetic crystals of some classes, i.e., the birefringence of linearly polarized light, which changes sign when the direction of light propagation or the directions of elementary magnetic moments of the crystal are reversed [3–9]. Similar to this is the effect of induced gyrotropic birefringence in nonmagnetic crystals found in non-centrosymmetric crystals in the presence of a magnetic field [9–11]. The sublattice structure of ferri- and antiferromagnets is not an obstacle to this analogy. The optical properties of a multisublattice magnet can often be represented as a simple superposition of the properties of mutually penetrating magnetic subsystems.

However, this correspondence between the properties of magnetically ordered and nonordered media may be incomplete. The magnetically ordered medium exposed to external influences can have a lower space–temporal symmetry than magnetized para- or diamagnets, and its optical properties may have no analogue among the properties of magnetically nonordered media. The optical effects allowed in magnetically ordered media are sensitive only to the magnetic symmetry of the medium, and the reversal of the directions of all elementary magnetic moments in the medium. The properties of these new effects make them interesting for studying the symmetry of magnetically ordered substances, and for visual investigation of the so-called time-inversed or collinear (180 degrees) antiferromagnetic domains.

The results of the symmetry analysis of the magneto-optic effects in magnetic crystals, with the symmetries described by the Shubnikov groups, are presented in this chapter, and the experimental results on the two magneto-optic effects are discussed; namely the linear magneto-optic effect and the quadratic as to magnetic field rotation of the light polarization plane. Primary attention is paid to the linear magneto-optic effect.

The spontaneous Faraday and Cotton–Mouton magneto-optic effects, as well as the effects induced by the magnetic field inherent both to magnetically ordered and magnetically nonordered crystals, are not discussed here. The spontaneous magneto-optic properties of ferro-, ferri-, and antiferromagnets were studied by many authors. The results of these investigations are summarized in a number of review articles, e.g., [12–22].

1.1. Symmetry of the Optical Properties of Magnetically Ordered Crystals

The optical properties of a physical medium can be described by the tensors of dielectric, $\varepsilon_{ij}(\omega, \mathbf{k})$, and magnetic, $\mu_{ij}(\omega, \mathbf{k})$, permeabilities which depend on the frequency and the wave vector of the light wave. The question of the physical content of the magnetic permeability tensor definition at optical frequencies has been discussed in detail in papers [23–30]. This question has no simple solution, as the selection of tensors ε_{ij} and μ_{ij} is to some extent arbitrary [29]. The reaction of the medium to the electromagnetic wave disturbance can be described by an effective conductivity tensor without introducing the tensor μ_{ij} [28, 31–33]. In most cases, the optical properties of a transparent nongyrotropic medium are sufficiently described by the ordinary dielectric permeability tensor $\varepsilon_{ij}(\omega)$. The tensor $\mu_{ij}(\omega)$ is required for the description of contributions to magnetic dipole electron transitions [34]. These contributions are usually small and may be noticeable only in the immediate proximity of the corresponding absorption bands of the optical spectrum. But, in paramagnets and magnetically ordered media, the magnetic dipole transitions, corresponding to the magnetic resonant absorption in the radio frequency and far IR regions, prove to be substantial in the visual region also [35–38]. Though in this case the magnetic dipole contributions can also be described by the effective conductivity tensor or the effective dielectric tensor depending on the wave vector $\mathbf{k}$, it is also convenient to use the magnetic permeability tensor $\mu_{ij}(\omega)$ [30, 39, 40]. At comparable values of $(1 - \varepsilon_{ii})$ and $(1 - \mu_{ii})$, $\mu_{i \neq j}$, qualitatively new optical properties may appear in anisotropic media [30, 41–46]. However, this situation is far from that actually observed in solid media for the visible part of the spectrum, where the inequality $\mu_{ij} \ll 1$ is always satisfied. Therefore, when studying the symmetry of the optical properties of magnetic crystals, it is sufficient to investigate the dependence of the dielectric tensor $\varepsilon_{ij}(\omega, \mathbf{H}, \mathbf{k})$ components on the medium symmetry.

To determine the symmetry properties of the dielectric tensor the methods of nonequilibrium thermodynamics are used. If at $\omega = 0$ the tensor ε_{ij} is always symmetrical, at $\omega \neq 0$ its components are subject to more complex symmetry restrictions. They should satisfy the Onsager symmetry relations for kinetic coefficients [23, 47–50]. The Onsager relations, written in the form

$$\varepsilon_{ij}(\mathbf{H}, \mathbf{k}) = \varepsilon_{ji}(-\mathbf{H}, -\mathbf{k}), \tag{1.1}$$

are the consequence of medium symmetry with respect to the time-inversion operation or motion reversal, and are applicable to magnetically nonordered media only—paramagnets and diamagnets placed in a magnetic field. Therefrom follows an even dependence on $\mathbf{H}$ for the symmetrical components of the tensor ε_{ij} describing the birefringence of linearly polarized light, and an odd dependence for the antisymmetrical components describing the Faraday rotation of the plane of polarization. But, for the magnetically ordered media the time-inversion operation $\underline{1}$ is not a symmetry operation, and relation (1.1) should be generalized. The generalized Onsager symmetry relations for kinetic coefficients include space transformation operations [50–55]. As the kinetic coefficients characterize the macroscopic properties, their symmetry is defined by the point magnetic group operations. Besides, the symmetry restrictions on the kinetic coefficients describing the centrosymmetrical properties are the same for all crystals composing one Laue class, the group of which is formed by the point magnetic group elements by substituting the identification operation and the anti-identification for all improper rotations $\bar{R} = \bar{1} \cdot R$ and improper antirotations $\bar{\underline{R}} = \bar{1} \cdot \underline{1} \cdot R$. When investigating spontaneous and induced optical effects caused by spatial dispersion, it is necessary to account for the dependence of the dielectric tensor components on the wave vector of the light wave $\mathbf{k}$. As the direction of $\mathbf{k}$ is reversed, both at the time inversion $\underline{1}$ and at the space coordinates inversion $\bar{1}$, the Onsager relations should be generalized for the non-centrosymmetrical properties also. They should be completed by the restrictions on improper rotations and antirotations. In the most general form these relations can be written as follows:

for rotations R

$$\varepsilon_{\mu\nu}(R\mathbf{H}, R\mathbf{k}) = R_{\mu m} R_{\nu n} \varepsilon_{mn}(\mathbf{H}, \mathbf{k}); \tag{1.2}$$

for antirotations R

$$\begin{aligned} &\varepsilon_{\mu\nu}(\underline{R}\mathbf{H}, \underline{R}\mathbf{k}) = R_{\mu m} R_{\nu n} \varepsilon_{nm}(\mathbf{H}, \mathbf{k}), \\ &\underline{R}\mathbf{H} = -R\mathbf{H}, \qquad \underline{R}\mathbf{k} = -R\mathbf{k}; \end{aligned} \tag{1.3}$$

for improper rotations $\bar{R}$

$$\begin{aligned} &\varepsilon_{\mu\nu}(\bar{R}\mathbf{H}, \bar{R}\mathbf{k}) = R_{\mu m} R_{\nu n} \varepsilon_{mn}(\mathbf{H}, \mathbf{k}), \\ &\bar{R}\mathbf{H} = R\mathbf{H}, \qquad \bar{R}\mathbf{k} = -R\mathbf{k}; \end{aligned} \tag{1.4}$$

for improper antirotations $\overline{\underline{R}}$

$$\begin{aligned} \varepsilon_{\mu\nu}(\overline{\underline{R}}\mathbf{H}, \overline{\underline{R}}\mathbf{k}) &= R_{\mu m} R_{\nu n} \varepsilon_{nm}(\mathbf{H}, \mathbf{k}), \\ \overline{\underline{R}}\mathbf{H} = -R\mathbf{H}, &\qquad \overline{\underline{R}}\mathbf{k} = R\mathbf{k}. \end{aligned} \tag{1.5}$$

The disturbances of the crystal dielectric properties caused by a magnetic field, space dispersion, or other external influences, as well as by spin ordering, are small. The expressions for the dielectric tensor components can be represented by expansion into a Taylor series as per the following parameters:

$$\varepsilon_{\mu\nu}(\mathbf{H}, \mathbf{k}) = \sum_{mn} \tau_{\mu\nu\alpha_1\ldots\alpha_m\delta_1\ldots\delta_n} H_{\alpha_1} \ldots H_{\alpha_m} k_{\delta_1} \ldots k_{\delta_n}. \tag{1.6}$$

In (1.6) the summing is carried out as to repeated indices α_i, and also δ_i. Using [54], the following general symmetry relations can be written for the coefficients $\tau_{\mu\nu\alpha_1\ldots\alpha_n\delta_1\ldots\delta_n}$:

for R

$$\tau_{\mu\nu\alpha_1\ldots\alpha_m\delta_1\ldots\delta_n} = R_{\mu\kappa} R_{\nu\lambda} R_{\alpha_1\beta_1} \ldots R_{\alpha_m\beta_m} R_{\delta_1\gamma_1} \ldots R_{\delta_n\gamma_n} \tau_{\kappa\lambda\beta_1\ldots\beta_m\gamma_1\ldots\gamma_n}, \tag{1.7}$$

for $\underline{R}$

$$\tau_{\mu\nu\alpha_1\ldots\alpha_m\delta_1\ldots\delta_n} = (-1)^{m+n} R_{\mu\kappa} R_{\nu\lambda} R_{\alpha_1\beta_1} \ldots R_{\alpha_m\beta_m} R_{\delta_1\gamma_1} \ldots R_{\delta_n\gamma_n} \ldots \tau_{\lambda\kappa\beta_1\ldots\beta_m\gamma_1\ldots\gamma_n}, \tag{1.8}$$

for $\overline{R}$

$$\tau_{\mu\nu\alpha_1\ldots\alpha_m\delta_1\ldots\delta_n} = (-1)^{n} R_{\mu\kappa} R_{\nu\lambda} R_{\alpha_1\beta_1} \ldots R_{\alpha_m\beta_m} R_{\delta_1\gamma_1} \ldots R_{\delta_n\gamma_n} \tau_{\lambda\kappa\beta_1\ldots\beta_m\gamma_1\ldots\gamma_n}, \tag{1.9}$$

for $\overline{\underline{R}}$

$$\tau_{\mu\nu\alpha_1\ldots\alpha_m\delta_1\ldots\delta_n} = (-1)^{m+2n} R_{\mu\kappa} R_{\nu\lambda} R_{\alpha_1\beta_1} \ldots R_{\alpha_m\beta_m} R_{\delta_1\gamma_1} \ldots R_{\delta_n\gamma_n} \tau_{\lambda\kappa\beta_1\ldots\beta_m\gamma_1\ldots\gamma_n}. \tag{1.10}$$

We are interested only in the magneto-optic effects caused by an external magnetic field. Let us rewrite (1.6) in expanded form restricting to quadratic in $\mathbf{H}$ additions to the dielectric tensor

$$\varepsilon_{ij} = \varepsilon_{ij0} + \Delta\varepsilon_{ij_0} + \tau_{ij\alpha} H_\alpha + \gamma_{ij\delta} k_\delta + \Gamma_{ij\alpha\delta} H_\alpha k_\delta + \tau_{ij\alpha\beta} H_\alpha H_\beta. \tag{1.11}$$

Here $\Delta\varepsilon_{ij}$ comprises all changes of the dielectric tensor caused by magnetic ordering. All tensor matrices in (1.7) contain symmetrical and antisymmetrical parts with respect to the permutation of indices i and j. We will introduce special notations for some of them, describing the known optical effects

$$\tau^s_{ij\alpha} = q_{ij\alpha}, \qquad \tau^a_{ij\alpha} = if_{ij\alpha}, \tag{1.12}$$

and

$$\tau^s_{ij\alpha\beta} = C_{ij\alpha\beta}, \qquad \tau^a_{ij\alpha\beta} = iB_{ij\alpha\beta}. \tag{1.13}$$

Tensor $f_{ij\alpha}$ describes the well-known Faraday effect—rotation of the plane of polarization caused by a magnetic field [33, 34, 56]. Tensor $C_{ij\alpha\beta}$, symmetrical in respect to the rearrangement of indices i and j, α and β, describes the quadratic as to **H** birefringence of linearly polarized light—Voigt (Cotton–Mouton) effect. Tensor $i\gamma^{a}_{ij\delta}$, antisymmetrical as to i and j, describes the natural spontaneous optical activity caused by the spatial dispersion of the refraction coefficient in non-centrosymmetrical crystals [42, 57]. Tensor $\Gamma^{s}_{ij\alpha\delta}$, symmetrical as to i and j, describes the magnetic-field-induced birefringence of linearly polarized light, the sign of which is reversed when the light propagation direction is reversed [9–11, 49]. This linear birefringence is called the gyrotropic or nonmutual birefringence, and has been discovered recently [10]. The effects mentioned above are symmetrically allowed both in magnetically ordered and magnetically nonordered crystals. Tensors $q_{ij\alpha}$, $\gamma^{s}_{ij\delta}$, $iB_{ij\alpha\beta}$, $i\Gamma^{a}_{ij\alpha\delta}$ describe the effects which are forbidden in magnetically nonordered media, and symmetrically allowed after magnetic ordering only, which breaks time-inversion symmetry. Tensors $q_{ij\alpha}$ and $\gamma^{s}_{ij\delta}$, symmetrical as to i and j, describe the birefringence of linearly polarized light. Tensor $q_{ij\alpha}$ characterizes the birefringence which is proportional to the first order of the magnetic field strength contrary to the Cotton–Mouton effect, and changes its sign when the field direction is reversed. This magneto-optic phenomenon is called the "linear magneto-optic effect," similar to the linear electro-optic effect [58–60]. Tensor $\gamma^{s}_{ij\delta}$ describes spontaneous gyrotropic birefringence [3–9, 61], which according to estimates [6] can be as low as 10^{-8}, and has not been observed yet. Tensor $B_{ij\alpha\beta}$ describes the magnetic wave vector independent rotation of the plane of polarization, which in contrast to the Faraday effect is quadratic as to the magnetic field strength [62–65]. Tensor $\Gamma^{a}_{ij\alpha\delta}$ describes linear in **H** variations of natural optical activity caused by spatial dispersion and depending on the light wave vector [66, 46]. This effect has not yet been observed.[1]

Relations (1.7)–(1.10) make it possible to find the conditions at which the tensor components included in expansion (1.11) are not identically equal to zero. The results of the most general analysis are given in Table 1.1. This table also indicates the symmetry of physical tensors with respect to the space- and time-inversion operations in terms proposed by Birss [51]: the ith tensor is invariant and the cth tensor is noninvariant with respect to the time-inversion operation. It is obvious that all effects described by c-tensors are forbidden in magnetically ordered media, as their components change sign under the action of operation $\underline{1}$, which is the symmetry operation.

[1] Experimental observations of this effect in magnetically nonordered crystals are described in [67], and in [68] an attempt was made to present the symmetry foundation of the feasibility of its appearance in nonmagnetic crystals by postulating new symmetry relations for the antisymmetrical kinetic coefficients. However, earlier conclusions were not confirmed by further experiments. The introduction of new symmetry relations requires a deeper foundation from our point of view.

1.2. Symmetry of the Linear Magneto-Optic Effect (LMOE)

Linear in the field strength variations of the symmetrical components of the dielectric tensor lead to specific changes of the optical properties of a magnetic crystal, which can be represented as various manifestations of the linear magneto-optic effect (LMOE). As follows from Table 1.1, the LMOE is described by the axial *c*-tensor of third order, $q_{ij\alpha}$ symmetrical as to the first pair of indices. It follows from the fact that the effect is described by the *c*-tensor that the birefringence of linearly polarized light caused by it should change its sign as the directions of all elementary magnetic moments in the crystal are reversed, only if the direction of the magnetic field remains unchanged.

It also follows from Table 1.1 that the components of $q_{ij\alpha}$ are not identically equal to zero in the crystals, the point groups of which do not belong to the Laue gray groups, i.e., $q_{ij\alpha}$ becomes equal to zero in all antiferromagnetics of the second type, the magnetic spatial groups of which contain only antitranslations $\underline{1} \cdot T$ as the antisymmetrical operators. The components of $q_{ij\alpha}$ also become identically equal to zero in all centroantisymmetrical crystals which are symmetrical as to the anti-inversion operation $\underline{\bar{1}}$. Indeed, for the symmetry of the macroscopic centrosymmetrical property, only the symmetry operations of the Laue groups are substantial, but in these groups the symmetry operations $\underline{T}$ and $\underline{\bar{1}}$ are represented by the time-inversion $\underline{1}$ operation. This is the operation which makes all the components of tensor $q_{ij\alpha}$ vanish. The components of tensor $q_{ij\alpha}$, which are not identically equal to zero, can be directly determined from the equations which follow from the general symmetry relations (1.7) and (1.8):

for R

$$q_{ij\alpha} = R_{ik} R_{jl} R_{\alpha\beta} q_{kl\beta}, \tag{1.14}$$

for $\underline{R}$

$$q_{ij\alpha} = -R_{ik} R_{jl} R_{\alpha\beta} q_{lk\beta} = -R_{ik} R_{jl} R_{\alpha\beta} q_{kl\beta}. \tag{1.15}$$

When determining the matrices of tensor $q_{ij\alpha}$, the fact can be taken into account that the matrix of the third-rank axial *c*-tensor of any magnetic crystal should be the same as the matrix of the polar *i*-tensor which it describes, e.g., the linear electro-optic effect in a nonmagnetic crystal, the symmetry of which is determined by the set of the Laue group operators of magnetic crystal by replacing the proper antirotation operations by the nonproper rotation operations. For example, matrix $q_{ij\alpha}$ for the crystals with point magnetic group $\underline{4}/mm\underline{m}$ (Laue group $\underline{4}2\underline{2}$, its symmetry operators $-1, 2_x, 2_y, 2_z, \underline{2}_{xy}, \underline{2}_{-xy}, \pm\underline{4}_z$) is the same as the matrix of the Pockels effect coefficients for the crystal with point group $\bar{4}2m \cdot \underline{1}$ (symmetry operators $- 1, 2_x, 2_y, 2_z, \bar{2}_{xy}, \bar{2}_{-xy}, \pm\bar{4}_z$). The matrices of magneto-optic coefficients of single-color non-centrosymmetrical crystals are the same as the matrices of electro-optic coefficients of a nonmagnetic crystal with the same spatial symmetry operators.

It should also be noted that, as the restrictions imposed by the space–

Table 1.1. Magneto-optic properties of magnetically ordered crystals.

Tensor of property	Space–temporal symmetry of tensor	Optical properties	Shubnikov's classes[a]					
			Gray		One- and bicolor			
					CS		NCS	
			CS	NCS	CAS	NCAS	CAS	NCAS
$\Delta\varepsilon^{s}_{ij0}$	polar *i*-tensor	birefringence of linearly polarized light (spontaneous Cotton–Mouton or Voigt effect)	+	+	+	+	+	+
$\Delta\varepsilon^{a}_{ij0}$	polar *c*-tensor	spontaneous rotation of plane of polarization (spontaneous Faraday effect)	−	−	−	+ for pyromagnetics	−	+ for pyromagnetics
$q_{ij\alpha}$	axial *c*-tensor	induced birefringence of linearly polarized light, proportional to the magnetic field strength (linear magneto-optic effect)	−	−	−	+	−	+
$f_{ij\alpha}$	axial *i*-tensor	magnetic-field-induced rotation of plane of polarization (Faraday effect)	+	+	+	+	+	+
$\gamma^{s}_{ij\delta}$	polar *c*-tensor	nonmutual spontaneous birefringence of linearly polarized light, proportional to wave vector **k** (spontaneous gyrotropic birefringence)	−	−	−	−	+	+
$\gamma^{a}_{ij\delta}$	polar *i*-tensor	spontaneous rotation of plane of polarization proportional to **k** (natural optical activity)	−	+	−	−	+	+

(continued)

Table 1.1 (*cont.*)

Tensor of property	Space–temporal symmetry of tensor	Optical properties	Shubnikov's classes[a]					
			Gray		One- and bicolor			
					CS		NCS	
			CS	NCS	CAS	NCAS	CAS	NCAS
$\Gamma^{s}_{ij\delta\alpha}$	axial *i*-tensor	magnetic-field-induced linear as to **k** and **H** birefringence of linearly polarized light (induced gyrotropic birefringence)	–	+	–	–	–	+
$\Gamma^{a}_{ij\delta\alpha}$	axial *c*-tensor	magnetic-field-induced rotation of plane of polarization linear as to **k** and **H** (magnetic-field-induced natural optical activity)	–	–	–	–	+	+
$C_{ij\alpha\beta}$	polar *i*-tensor	magnetic-field-induced birefringence of linearly polarized light (Cotton–Mouton effect, Voigt effect)	+	+	+	+	+	+
$B^{a}_{ij\alpha\beta}$	polar *c*-tensor	magnetic-field-induced rotation of plane of polarization quadratic as to **H** (quadratic magnetic rotation, even Faraday effect)	–	–	–	+	–	+

[a] CS and NCS—centrosymmetrical and non-centrosymmetrical classes; CAS and NCAS—centroantisymmetrical (containing operation $\underline{\bar{1}} = \bar{1}\cdot\underline{1}$) and non-centroantisymmetrical classes.

temporal symmetry on the symmetrical parts of the kinematic coefficient tensors are the same as the symmetry restrictions imposed on the symmetrical parts of the corresponding tensors of static properties [50–54], the axial magneto-optic *c*-tensor $q_{ij\alpha}$ has the same symmetry as the static axial *c*-tensor which describes the reverse piezomagnetic effect.

The matrices of tensor $q_{ij\alpha}$, for all magnetic crystals which are described by the Shubnikov groups, are given in Table 1.2 with the conventional reduction of symmetrical indices [69]. It should be noted that, as the conventional rules of reduction of the deformation tensor and the dielectric tensor differ [50, 69], some components of the matrices $q_{\lambda\alpha}$ differ by the coefficients from the components of the matrices of reverse piezomagnetic effect given in [70, 71]. In the crystals characterized by the magnetic point groups 432, $\bar{4}3m$, and $m3m$, the components of tensor $q_{ij\alpha}$ vanish although their groups do not contain the anti-inversion operation.

The LMOE can be described by tensor $q^M_{ij\alpha}$ connecting ε^s_{ij} with the projections of the crystal net magnetic moment, but not with the projections of the field. Such a description makes it possible to separate the spontaneous LMOE which can exist in pyromagnetic crystals, if the latter allow such projection of the magnetic moment which induces the LMOE. As can be seen from Table 1.2, all pyromagnetics allow spontaneous LMOE. But the spontaneous LMOE never reduces the optical symmetry of the crystal. In highly symmetrical pyromagnetics the LMOE cannot be observed at the light propagation along a high-order axis. The spontaneous LMOE, in contrast to the spontaneous Faraday rotation, is invariant as to the opposite directions of the spontaneous magnetic moment, because to remagnetize a pyromagnet all directions of the elementary magnetic moments should be reversed. This operation is equal to the action of the time-inversion operation and will change the signs of the magneto-optic coefficients $q^M_{ij\alpha}$. The simultaneous changing of signs $q^M_{ij\alpha}$ and M_α leaves the LMOE sign unchanged.

It should be noted that the LMOE can also be induced by mechanical deformations, which manifested itself by changing the photoelastic coefficients

$$\Delta\varepsilon^s_{ij} = P_{ijlm}u_{lm} + q^M_{ij\alpha}M_{\alpha\,\text{piezo}} = (P_{ijlm} + q^M_{ij\alpha}Q_{\alpha lm})u_{lm}. \tag{1.16}$$

Here $Q_{\alpha lm}$ and P_{ijlm} are the components of the piezomagnetic and photoelastic tensors.

Let us consider the changes of the crystal optical properties of magnetic crystals caused by the LMOE. Table 1.3 gives the additions to the components of tensor ε^s_{ij}, caused by the linear and quadratic magneto-optic effects at the magnetic field directions parallel to the crystallographic axes of coordinates. The coefficients of the optical indicatrix, determined by the equation

$$\eta_{ij}x_i x_j = 1, \tag{1.17}$$

where $\eta_{ij} = \varepsilon^{-1}_{ij}$, will have the same additions.

Table 1.3 shows that there are field directions at which the indicatrix varia-

Table 1.2. Symmetry of the linear magneto-optic effect tensor $q_{ij\alpha}$, and additions to the dielectric tensor components $\Delta\varepsilon_{ij}^s$ caused by the spontaneous linear magneto-optic effect.

Magnetic point group	Laue magnetic group and its generators	Tensor matrix $q_{ij\alpha}$	Components of spontaneous magnetic moment M_α	Components $\Delta\varepsilon_{ij}^s$ proportional to M_α
$1, \bar{1}$	$\bar{1}$	all $q_{\mu\alpha} \neq 0$	all $M_\alpha \neq 0$	all $\Delta\varepsilon_{ij}^s \neq 0$
$2, m, 2/m$	2 $(2\,\|\,X_3)$ $\{2_z\}$	$\begin{pmatrix} \cdot & \cdot & \bullet \\ \cdot & \cdot & \bullet \\ \cdot & \cdot & \bullet \end{pmatrix}$ $\begin{pmatrix} \bullet & \bullet & \cdot \\ \bullet & \bullet & \cdot \\ \cdot & \cdot & \bullet \end{pmatrix}$	M_z	xx, yy, zz, xy
$\underline{2}, \underline{m}, \underline{2}/\underline{m}$	$\underline{2}$ $(\underline{2}\,\|\,X_3)$ $\underline{2}_z$	$\begin{pmatrix} \bullet & \bullet & \cdot \\ \bullet & \bullet & \cdot \\ \bullet & \bullet & \cdot \end{pmatrix}$ $\begin{pmatrix} \cdot & \cdot & \bullet \\ \cdot & \cdot & \bullet \\ \bullet & \bullet & \cdot \end{pmatrix}$	M_x, M_y	xx, yy, zz, xy
$222, mm2$ mmm	222 $\{2_z, 2_x\}$	$\begin{pmatrix} \cdot & \cdot & \cdot \\ \cdot & \cdot & \cdot \\ \cdot & \cdot & \cdot \end{pmatrix}$ $\begin{pmatrix} \cdot & \bullet & \cdot \\ \cdot & \cdot & \cdot \\ \cdot & \cdot & \bullet \end{pmatrix}$	—	—
$\underline{2}\underline{2}2, \underline{m}\underline{m}2$ $\underline{m}m\underline{2}, \underline{m}\underline{m}\underline{m}$	$\underline{2}\underline{2}2$ $\{2_z, \underline{2}_x\}$	$\begin{pmatrix} \cdot & \cdot & \bullet \\ \cdot & \cdot & \bullet \\ \cdot & \cdot & \bullet \end{pmatrix}$ $\begin{pmatrix} \cdot & \bullet & \cdot \\ \bullet & \cdot & \cdot \\ \cdot & \cdot & \cdot \end{pmatrix}$	M_z	xx, yy, zz
$4, \bar{4}, 4/m$	4 $\{4_z\}$	$\begin{pmatrix} \cdot & \cdot & \bullet \\ \cdot & \cdot & \bullet \\ \cdot & \cdot & \bullet \end{pmatrix}$ $\begin{pmatrix} \bullet & \bullet & \cdot \\ \bullet & \bullet & \cdot \\ \cdot & \cdot & \cdot \end{pmatrix}$	M_z	$xx = yy, zz$

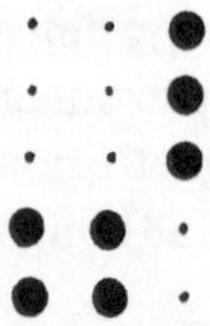

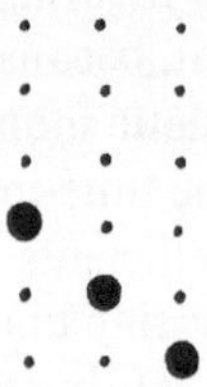

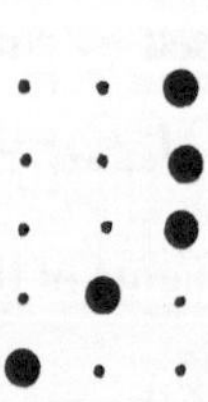

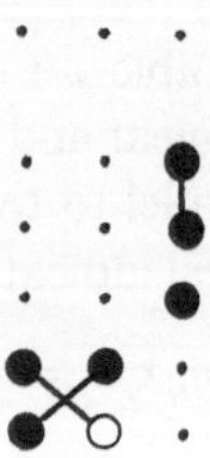

Table 1.2 (*cont.*)

Magnetic point group	Laue magnetic group and its generators	Tensor matrix $q_{ij\alpha}$	Components of spontaneous magnetic moment M_α	Components $\Delta\varepsilon^s_{ij}$ proportional to M_α
$\underline{4}, \bar{\underline{4}}, \underline{4}/m$	$\underline{4}$ $\{\underline{4}_z, 2_z\}$	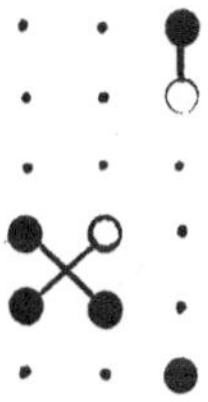	—	—
$422, 4mm$ $\bar{4}2m, 4/mmm$	422 $\{4_z, 2_x\}$	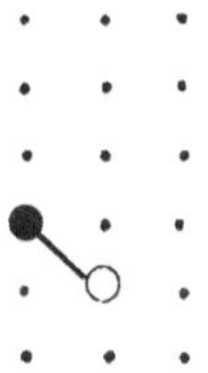	—	—
$\underline{4}2\underline{2}, \underline{4}m\underline{m},$ $\bar{\underline{4}}2\underline{m}, \bar{\underline{4}}\underline{2}m,$ $\underline{4}/mm\underline{m}$	$\underline{4}2\underline{2}$ $\{\underline{4}_z, 2_z, 2_y\}$		—	—
$4\underline{2}\underline{2}, 4\underline{m}\underline{m}$ $\bar{4}\underline{2}\underline{m}, 4/m\underline{m}\underline{m}$	$4\underline{2}\underline{2}$ $\{4_z, \underline{2}_z\}$	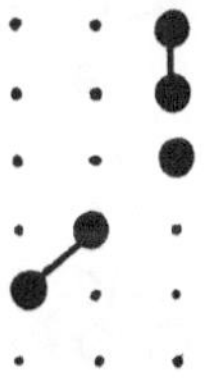	M_z	$xx = yy, zz$
$3, \bar{3}$	3 $\{3_z\}$	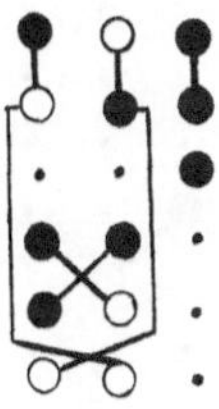	M_z	$xx = yy, zz$

Table 1.2 (*cont.*)

Magnetic point group	Laue magnetic group and its generators	Tensor matrix $q_{ij\alpha}$	Components of spontaneous magnetic moment M_α	Components $\Delta\varepsilon^s_{ij}$ proportional to M_α
$32, 3m, \bar{3}m$	32 $(2 \parallel X_1)$ $\{3_z, 2_x\}$	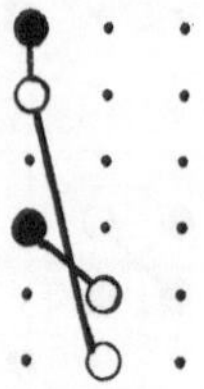	—	—
$3\underline{2}, 3\underline{m}, \bar{3}\underline{m}$	$3\underline{2}$ $(\underline{2} \parallel X_1)$ $\{3_z, 2_x\}$		M_z	$xx = yy, zz$
$6, \bar{6}, 6/m$	6 $\{3_z, 2_z\}$	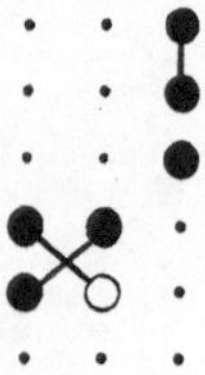	M_z	$xx = yy, zz$
$\underline{6}, \underline{\bar{6}}, \underline{6/m}$	$\underline{6}$ $\{3_z, 2_z\}$	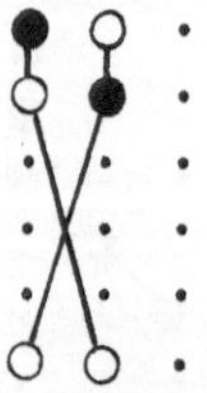	—	—
$622, 6mm,$ $\bar{6}m2, 6/mmm$	622 $\{3_z, 2_z, 2_x\}$	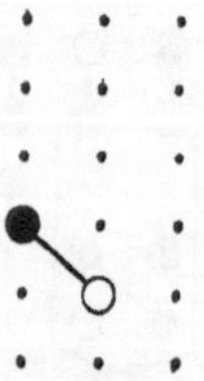	—	—

Table 1.2 (*cont.*)

Magnetic point group	Laue magnetic group and its generators	Tensor matrix $q_{ij\alpha}$	Components of spontaneous magnetic moment M_α	Components $\Delta\varepsilon_{ij}^{s}$ proportional to M_α
$6\underline{2}\underline{2}$, $6\underline{m}\underline{m}$, $\bar{6}\underline{m}\underline{2}$, $6/m\underline{m}\underline{m}$	$6\underline{2}\underline{2}$ $\{3_z, 2_z, \underline{2}_x\}$	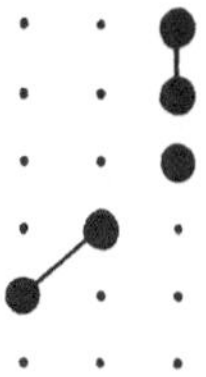	M_z	$xx = yy$, zz
$\underline{6}2\underline{2}$, $\underline{6}m\underline{m}$, $\underline{6}\underline{m}2$, $\underline{\bar{6}}m\underline{2}$, $\underline{6}/\underline{m}m\underline{m}$	$\underline{6}2\underline{2}$ $(2 \parallel X_1)$ $\{3_z, \underline{2}_z, 2_x\}$	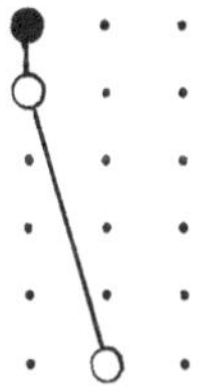	—	—
23, $m3$	23 $\{3_{xyz}, 2_z\}$	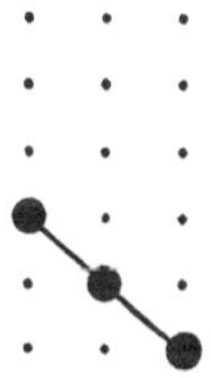	—	—
$\underline{4}3\underline{2}$, $\underline{\bar{4}}3m$, $m3\underline{m}$	$\underline{4}3\underline{2}$ $\{\underline{4}_z, 3_{xyz}, 2_z\}$			
432, $\bar{4}3m$, $m3m$	432 $\{4_z, 3_{xyz}\}$	all $q_{ij\alpha} = 0$	—	—

tions are the simplest ones. Let us study these cases in detail taking into account a small relative value of the induced effect.

In magnetic crystals with the rhombic system the LMOE, almost at all orientations $\mathbf{H} \parallel \mathbf{X}_i$, causes the rotation of the indicatrix. The quadratic effect does not rotate the indicatrix. In the crystals belonging to the Laue class 222 the indicatrix is rotated around the vector of the applied field. In the crystals of the Laue class $\underline{2}\underline{2}2$ the rotation takes place around the direction perpendicular to **H** and **Z**, only if **H** is not parallel to **Z**. When $\mathbf{H} \parallel \mathbf{Z}$ the indicatrix does not change its orientation but is only deformed. The angle of rotation linearly depends on the value of **H** and changes its sign when the direction of **H** is reversed.

Variations of the indicatrix in other crystal systems are more diverse. In tetragonal and trigonal crystals which belong to the Laue magnetic classes 422 and 32 and in the Laue hexagonal classes 622, $\underline{6}$, and $\underline{6}2\underline{2}$, the field orienta-

Table 1.3. Variation of the dielectric tensor components under the action of the linear and quadratic magneto-optic effects.

Laue magnetic class	Action of linear magneto-optic effect			Action of quadratic magneto-optic effect		
	H‖X	H‖Y	H‖Z	H‖X	H‖Y	H‖Z
1	all $\Delta\varepsilon_{ij}^{s} \neq 0$					
2	yz, xz	yz, xz	xx, yy, zz xy	xx, yy, zz, yx	xx, yy, zz, yx	xx, yy, zz,
$\underline{2}$	xx, yy, zz xy	xx, yy, zz xy	yz, xz			
222	yz	xz	xy	xx, yy, zz	xx, yy, zz	xx, yy, zz
$\underline{22}2$	xz	yz	xx, yy, zz			
4	yz, xz	yz, xz	xx = yy, zz	xx, yy, zz, xy	xx, yy, zz, xy	xx = yy, z
$\underline{4}$	yz, xz	yz, xz	xx = −yy, xy			
422	yz	xz	—			
$\underline{4}2\underline{2}$	yz	xz	xy	xx, yy, zz	xx, yy, zz	xx = yy, z
4$\underline{22}$	xz	yz	xx = yy, zz			
3	xx = −yy, yz xz, xy	xx = −yy, yz xz, xy	xx = yy, zz	xx, yy, zz xz, xy, yz	xx, yy, zz xz, xy, yz	xx = yy, z
32	xx = −yy, yz	xz, xy	—	xx, yy, zz, yz	xx, yy, zz, yz	xx = yy, z
3$\underline{2}$	xz, xy	xx = −yy, yz	xx = yy, zz			
6	yz, xz	yz, xz	xx = yy, zz	xx, yy, zz, xy	xx, yy, zz, xy	xx = yy, z
6$\underline{2}$	xx = −yy, xy	xx = −yy, xy	—			
622	yz	xz	—			
6$\underline{22}$	xz	yz	xx = yy, zz	xx, yy, zz	xx, yy, zz	xx = yy, z
$\underline{6}2\underline{2}$	xx = −yy	xy	—			
23	yz	xz	xy	xx, yy, zz	xx, yy, zz	xx, yy, zz
$\underline{4}3\underline{2}$	yz	xz	xy	xx, yy = zz	xx, yy, zz	xx = yy, z

tion along a high-order axis causes no change in the indicatrix. The same field orientation in crystals of Laue classes 4, 4$\underline{22}$, 3, 3$\underline{2}$, 6, and 6$\underline{22}$ only deforms the indicatrix without changing its symmetry and orientation. The field applied perpendicular to a high-order axis causes, due to the LMOE, the indicatrix rotations which do not appear under the action of the quadratic effect only. In crystals of Laue classes $\underline{6}$ and $\underline{6}2\underline{2}$, when the field is oriented along axes X or Y one of the principal axes of indicatrix X_3 remains directed along C_6. In crystals of class $\underline{6}2\underline{2}$ at **H**‖**Y** axes X_1 and X_2 are oriented along the diagonal directions and do not change their directions when the value of **H** changes. At

H || **X** the indicatrix axes are oriented along the crystallographic axes. In crystals of class $\underline{6}$ axes X_1 and X_2 are rotated linearly as to **H** around axis C_6 at both orientations of the field.

In crystals of Laue classes 422, $\underline{4}2\underline{2}$, $4\underline{2}\underline{2}$, 622, and $6\underline{2}\underline{2}$, at the transverse direction of the field one of the indicatrix axes X_1 or X_2, but not necessarily coinciding with the direction of **H**, preserves its direction along the axis of coordinates. The directions of the other axes depend on the field strength. But, if $\Delta\varepsilon_{ij} = q_{ij\alpha}H_\alpha \ll |\varepsilon_{110} - \varepsilon_{330}|$, as usually takes place, the angle of deflection of the indicatrix axes from the crystallographic axes is small.

In crystals of classes 4, $\underline{4}$, 3, and 6, at both transverse orientations of the field, all three axes do not coincide with the crystallographic axes. But in the measure of inequality $q_{ij\alpha}H_\alpha \ll |\varepsilon_{110} - \varepsilon_{330}|$, axis X_3 is close to axis Z. But in crystals of classes 32 and $3\underline{2}$ the same situation takes place at **H** || **Y** for class 32 and at **H** || **X** for $3\underline{2}$, but the angles of rotation of the indicatrix axes can be close to 45 degrees.

Unusual at first appearance, manifestation of the LMOE should be observed in tetragonal crystals of magnetic classes $\underline{4}$, $\underline{\bar{4}}$, $\underline{4}/m$ (Laue class $\underline{4}$), $\underline{4}2\underline{2}$, $\underline{4}m\underline{m}$, $\underline{\bar{4}}2\underline{m}$, $\underline{4}/mm\underline{m}$ (Laue class $\underline{4}2\underline{2}$), and in cubic magnetic crystals of classes $\underline{4}32$, $\underline{\bar{4}}3m$, $m3\underline{m}$ (Laue class $\underline{4}32$). In these crystals the effect is allowed for the light propagating along the high-order axis in the magnetic field to be also directed along the high-order axis. As a result of LMOE, the crystal in a magnetic field reduces its optical symmetry, and becomes optically biaxial in the same way as at the Pockels longitudinal effect. It should be noted that the magnetic field does not thereat disturb the arrangement of the crystal elementary magnetic moments directed along the high-order axis of symmetry if, in the absence of the field, the structure is collinear. Note that in contrast to the longitudinal electro-optic effect the longitudinal LMOE is accompanied by the magnetic circular birefringence, and at the longitudinal geometry of the experiment the crystal becomes elliptically birefringent.

In the crystals where the LMOE is allowed, the reverse piezomagnetic effect is also allowed. Therefore, in the same way as both the electro-optical effect and the quadratic magneto-optic effect, the LMOE can be represented as the superposition of the primary and secondary effects. The primary effect is caused by direct influence of the magnetic field upon the energy structure of the crystal ions and their polarizability. The secondary effect is connected via the photoelastic effect to the crystal deformations resulting from the action of the reverse piezomagnetic effect. It is obvious that the measured value of the effect depends on the experimental conditions.

If the crystal is squeezed from all sides, or a free sample is measured in an alternating magnetic field the frequency of which exceeds the mechanical resonance frequencies of the sample, the birefringence is determined by the primary effect only. But when measuring a free sample at low frequencies of the magnetic field, both mechanisms contribute to the measurement results. The secondary effect contribution is highest near the resonance frequencies. In the general case the expression for variation of the indicatrix coefficients η_{ij} under the action of the magnetic field and deformations u_{kl} can be written as follows:

$$\eta_{ij} = \eta_{ij0} + q'_{ij\alpha} H_\alpha + P_{ijkl} u_{kl}. \qquad (1.18)$$

Here $q'_{ij\alpha}$ are the LMOE magneto-optic coefficients connecting the dielectric impermeability tensor η_{ij} with the field projections H_α, and P_{ijkl} are the photoelastic tensor components. If the elastic deformations, determined by the minimum of the thermodynamic potential including elastic and magnetoelastic interactions, are expressed as follows:

$$u_{kl} = S_{klmn} \sigma_{mn} + \Lambda_{kl\alpha} H_\alpha, \qquad (1.19)$$

where S_{klmn} is the compliance tensor, $\Lambda_{kl\alpha}$ is the linear magnetostriction tensor, and σ_{mn} is the elastic stress tensor, the variations of the indicatrix coefficients will be equal to

$$\eta_{ij} - \eta_{ij0} = \Delta\eta_{ij} = (q'_{ij\alpha} + P_{ijkl}\Lambda_{kl\alpha}) H_\alpha + P_{ijkl} S_{klmn} \sigma_{mn}. \qquad (1.20)$$

If the specimen is free, $\sigma_{mn} = 0$, and the alternating magnetic field frequencies are low, we can write

$$\Delta\eta_{ij}^{\text{free}} = (q'_{ij\alpha} + P_{ijkl}\Lambda_{kl\alpha}) H_\alpha = \tilde{q}'_{ij\alpha} H_\alpha. \qquad (1.21)$$

If the specimen is fixed from all sides or at high frequencies of the magnetic field variation and free specimen, $u_{kl} = 0$, $S_{klmn}\sigma_{mn} = -\Lambda_{kl\alpha} H_\alpha$, and

$$\Delta\eta^{\text{fix}} = q'_{ij\alpha} H_\alpha. \qquad (1.22)$$

According to (1.21) and (1.22) the value of the secondary effect can be determined from the difference of measurements of the free and fixed specimens, or calculated from the known photoelastic and magnetostriction constants

$$q_{ij\alpha}^{\text{sec}} = \tilde{q}'_{ij\alpha} - q'_{ij\alpha} = P_{ijkl}\Lambda_{kl\alpha}. \qquad (1.23)$$

From (1.23) follows the secondary effect compensation condition by means of the specimen elastic tension. If $\sigma_{mn} = \sigma_{mn}^{\text{comp}}$ is such that the equation

$$S_{klmn} \sigma_{mn}^{\text{comp}} + \Lambda_{kl\alpha} H_\alpha = 0 \qquad (1.24)$$

is satisfied, the variations of the indicatrix coefficients $\Delta\eta_{ij}$ are determined by the primary effect only.

Usually the magnetic ordering symmetry is more susceptible to changes caused by external influences, than by the atomic ordering symmetry. The above-mentioned symmetry investigation of the LMOE does not answer the question as to how the magneto-optic effect changes in one crystal lattice when the magnetic order is varied, and does not permit us to evaluate temperature variations of the effect, as well as predicting its behavior in a high-intensity magnetic field and under other external influences. To answer these questions, the dielectric tensor component variations should be expressed as the function of the parameters characterizing the magnetic subsystem ordering in the crystal lattice, and the variations of these parameters under external actions should be accounted for. The magnetic ordering should therefore be considered as an external action upon the crystal with the unchanged symmetry of its ion subsystem [50, 72, 73]. It is rational to choose as the magnetic characteristics magnetic vectors, $\mathscr{L}^{\lambda}$, which are linear combinations of the sublattice magnetic moments and converted by irreducible representations λ of the crystal space group [72, 73]. This approach was used for the description of galvanomagnetic [2, 74] and optical [75, 76] properties of two-sublattice magnetic crystals. As the disturbance of the crystal electronic subsystem responsible for the optical properties caused by the magnetic ordering is small, the corresponding additions to the dielectric tensor components can be expressed in the form of a series

$$\varepsilon_{ij}^{s} = \varepsilon_{ij0}^{s} + Q_{ijrs}^{kl}\mathscr{L}_{r}^{k}\mathscr{L}_{s}^{l} + Q_{ijrstu}^{klmn}\mathscr{L}_{r}^{k}\mathscr{L}_{s}^{l}\mathscr{L}_{t}^{m}\mathscr{L}_{u}^{n} + \dots. \tag{1.25}$$

In (1.25), $k, l, m, \dots$ run through all numbers of the irreducible representations and the summing is carried out over all repeated indices. From this point of view the crystal magnetic ordering is only the disturbance of a nonmagnetic crystal described by a nonmagnetic point group, therefore only the i-tensor invariant as to transformation $\underline{1}$ can be the tensor Q not equal to zero. And, as ε_{ij}^{s} is an i-tensor, and $\mathscr{L}^{k}$ vectors are c-vectors, each of the expansions (1.25) may contain the product of only an even number of projections $\mathscr{L}^{k}$. Confining (1.25) to the LMOE, only those terms should be considered which, together with any other, contain the projection of the magnetic vector $\tilde{\mathscr{L}}_{r}^{k}$ transformed as an axial vector projection. Considering an expansion of $\tilde{\mathscr{L}}_{r}^{k}$ only the linear terms in $\tilde{\mathscr{L}}_{r}^{k} = \tilde{\mathscr{L}}_{r0}^{k} + \chi_{r\alpha}^{k}H_{\alpha}$, and confining to the zero approximation for the remaining projections $\mathscr{L}^{k}$, we can express the magneto-optic coefficients $q_{ij\alpha}$ via the projections of the crystal magnetic vectors

$$q_{ij\alpha} = Q_{ijrs}^{kl}\chi_{r\alpha}^{k}\mathscr{L}_{s0}^{l} + Q_{ijrstu}^{klmn}\chi_{r\alpha}^{k}\mathscr{L}_{s0}^{l}\mathscr{L}_{t0}^{m0}\mathscr{L}_{u0}^{n} + \dots. \tag{1.26}$$

Here $\chi_{r\alpha}^{k}$ is the susceptibility as to the magnetic vector $\mathscr{L}^{k}$. It characterizes the induction of the rth projection of the vector $\mathscr{L}^{k}$ by the magnetic field $\mathbf{H}_{\alpha}$.

For a two-sublattice magnetic crystal, vectors $\mathscr{L}^{k}$ are *in situ* the antiferromagnetism (AFM) vector $\mathbf{L} = \mathbf{M}_1 - \mathbf{M}_2$ and ferromagnetism (FM) vector $\mathbf{M} = \mathbf{M}_1 + \mathbf{M}_2$. Confining ourselves to the terms with the product of not more than two projections of the magnetic vectors, we can write (1.25) as follows:

$$\varepsilon_{ij}^{s} = \varepsilon_{ij0}^{s} + Q_{ijrs}^{LL}L_{r}L_{s} + Q_{ijrs}^{ML}M_{r}L_{s} + Q_{ijrs}^{MM}M_{r}M_{s}. \tag{1.27}$$

The AFM vector projections which are transformed as the axial vector projections, can differ from zero in only antiferromagnets whose structure allows weak ferromagnetism (WFM). Let us denote by $\chi_{r\alpha}^{L}$ the "antiferromagnetic susceptibility" which characterizes the inducing of AFM vector r-projections (i.e., the inducing of the internal effective magnetic field which changes its sign from ion to ion and is directed along axis r). Then, (1.27) can be written as follows:

$$\begin{aligned}\Delta\varepsilon_{ij}^{s} &= Q_{ijrs}^{LL}\tilde{L}_{r0}L_{s0} + Q_{ijrs}^{ML}M_{r0}M_{s0} \\ &\quad + (Q_{ijrs}^{ML}\chi_{r\alpha}^{M}L_{s0} + Q_{ijrs}^{LL}\chi_{r\alpha}^{L}L_{s0} + Q_{ijrs}^{MM}\chi_{r\alpha}^{M}M_{s0})H_{\alpha}.\end{aligned} \tag{1.28}$$

The expression in the parentheses in the considered approximation is equal to $q_{ij\alpha}$. Here the AFM susceptibility $\chi^L_{r\alpha}$ is related to the ordinary magnetic susceptibility $\chi^M_{r\alpha}$ by the expression

$$\chi^L_{r\alpha} = D_{rj}\chi^M_{j\alpha}, \tag{1.29}$$

where D_{ij} are the constants proportional to the Dzyaloshinsky–Moria constants which combine the FM and AFM vector projections in a thermodynamic potential [73].

The matrices of magneto-optic coefficients in (1.28) can be obtained taking into account the properties of the AFM and FM vector transformations. The internal magnetic property can be changed under the action of the non-magnetic group symmetry operation. For example, if operation R changes the sublattice numbering, the direction of the AMF vector is changed. As in this case the internal magnetic properties play the role of external action, this feature of some symmetry operators should be taken into account. According to [73] the transformation properties of magnetic vectors can be accounted for by introducing even and odd operations. The odd operations permutate the AFM sublattice numbers. As the crystal property should not change after the symmetry transformation, the following general relations can be written for the transformation of the magneto-optic coefficients symmetrical as to the first two indices:

R, Laue group operator:

$$Q^{\alpha_1\alpha_2\ldots\alpha_n}_{mlrs\ldots w} = (-1)^{k\le n} R_{m\mu}R_{l\lambda}R_{z\kappa}R_{s\theta}\ldots R_{w\xi}Q^{\alpha_1\alpha_2\ldots\alpha_n}_{\mu\lambda\kappa\theta\ldots\xi}. \tag{1.30}$$

$\underline{1}$:

$$Q^{\alpha_1\alpha_2\ldots\alpha_n}_{mlrs\ldots w} = (-1)^n Q^{\alpha_1\alpha_2\ldots\alpha_n}_{mlrs\ldots w}. \tag{1.31}$$

Here k is the number of magnetic vectors connected with the coefficient being considered, for which operation R is an odd one. Relation (1.31) is always valid too as, undisturbed by the magnetic ordering crystal, it is symmetrical as to operation $\underline{1}$. It also should be remembered and taken into account the fact that the AMF vector is transformed as the axial vector $\mathbf{A}^k$ (taking into account the evenness of operations) which not always coincides with the AMF vector itself, although its projections are composed of the AMF vector projections. This fact changes the indices of Q in (1.30) when we proceed to (1.25), as the indices at Q in (1.30) correspond to the indices of the vector $\mathbf{A}^k$ but not $\mathscr{L}^k$.

Let us note that, as the product of components of only the FM vector or only the AFM vector is always transformed in the same way and similar to the components of the polar i-tensor, the symmetry of Q^{MM}_{mlrs} and Q^{LL}_{mlrs} is the same and similar to the symmetry of the polar i-tensor of the fourth rank. If the indices of dyads $(\mathbf{M}, \mathbf{M})$ and $(\mathbf{L}, \mathbf{L})$ are canceled according to the covariant type, at which $(\mathbf{M}, \mathbf{M})_\nu = M_iM_j$ for $i = j$ and $(\mathbf{M}, \mathbf{M})_\nu = 2M_iM_j$ for $i \neq j$, and the indices of the dielectric tensor ε_{ij} are canceled, as usual, according to the contravariant type ($\varepsilon_{ij} = \varepsilon_\mu$ for all indices i and j), the matrices do not completely coincide with the photoelastic tensor matrices, because the indices of the deformation tensor components are canceled according to the covariant type. Matrices $Q_{\mu\nu}$ are listed in Table 1.4.

The coefficients of Q_{ijrs} can be found from (1.30) by rewriting the latter for the two-sublattice antiferromagnets

$$Q_{ijrs}^{ML} = \pm R_{i\mu} R_{j\nu} R_{r\kappa} R_{s\theta} Q_{\mu\nu s\theta}^{ML}, \tag{1.32}$$

where the "+" sign corresponds to the even R^+, and the sign "−" to the odd R^- operations of symmetry. The matrices composed by means of (1.32) are listed in Table 1.5 where the reduced notation $Q_{ijrs}^{ML} \leftrightarrow Q_{\mu\nu}^{ML}$ is used in which the symmetrical indices i and j are conventionally canceled

$$i, j = 11 \quad 22 \quad 33 \quad 23 \quad 13 \quad 12$$

$$\mu = 1 \quad 2 \quad 3 \quad 4 \quad 5 \quad 6$$

and the second pair is canceled according to the rule

$$r, s = 11 \quad 22 \quad 33 \quad 23 \quad 13 \quad 12 \quad 32 \quad 31 \quad 21$$

$$\nu = 1, \quad 2, \quad 3, \quad 4, \quad 5, \quad 6, \quad 7, \quad 8, \quad 9.$$

The components of Q_{ijrs}^{ML} have the same identically not-equal-to-zero indices as the components of the reverse linear piezomagnetic effect. The obtained matrices $Q_{\mu\nu}^{ML}$ agree with the matrices of the tensor connecting the crystal deformation with projections **M** and **L** [77].

The determined relation between the magneto-optic coefficients Q and the projections of magnetic vectors **M** and **L** makes it possible to predict the LMOE behavior at spin-orientation transitions, and to calculate the temperature dependence of the effect within the framework of one or the other model. The matrices of the magneto-optic coefficients and for multi-sublattice anti- and ferromagnetic crystals can be obtained in a similar way.

It is interesting to describe the LMOE microscopically and reveal its mechanisms. It is obvious that these questions are connected to the questions concerning the induction by the magnetic field of the sublattice nonequivalences in a crystal in respect to each other and to the states in fields of two opposite directions. This nonequivalence can appear only under the simultaneous action of the magnetic field, the intracrystal field, and the spin-orbital interaction.

The sublattice nonequivalence should lead to the nonequivalence of orientation and the value of the polarizability tensor of magnetic ions together with their ligand surroundings. These mechanisms should be to some extent kindred to the reverse piezomagnetic effect mechanisms. The LMOE is an optical detector of changes in the quantum mechanical states of the ions, a part of which leads to changes in the crystal sizes. If the reverse piezomagnetic effect reflects changes in the ground state of a magnetic ion or system of ions, the LMOE reflects changes in other states as well. In papers [78–81] the first steps were made towards the creation of the LMOE microscopic theory in tetragonal and rhombohedral AFM.

Table 1.4. Symmetry of tensors Q_{ijrs}^{MM} and Q_{ijrs}^{LL}.

Crystal point group	Tensor matrix $Q_{ijrs} \leftrightarrow Q_{\mu\nu}$	Crystal point group	Tensor matrix $Q_{ijrs} \leftrightarrow Q_{\mu\nu}$
$1, \bar{1}$	all $Q_{\mu\nu} \neq 0$		
$2, \bar{2}, m,$ $\overline{m}, 2/m$	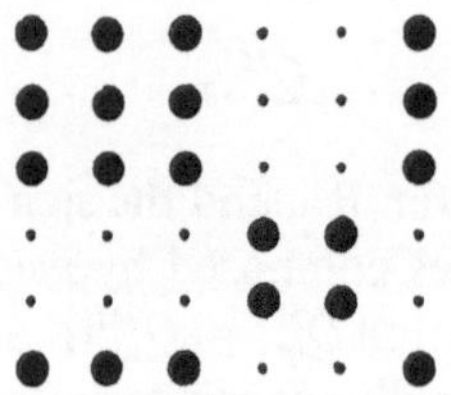	$3, \bar{3}$	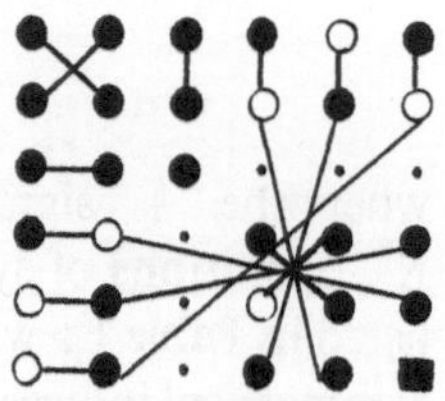
$222, mmm,$ $mm2$	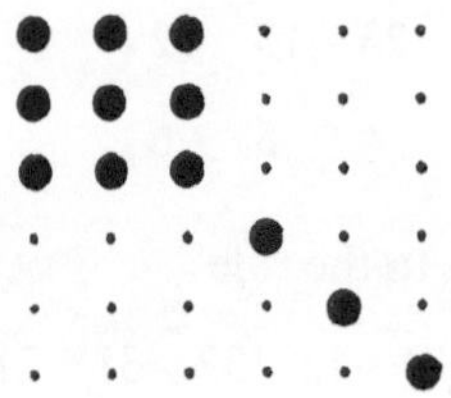	$32, 3m,$ $\bar{3}m$	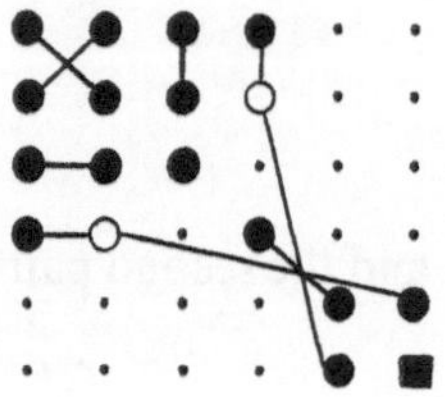
$4, \bar{4}, 4/m$	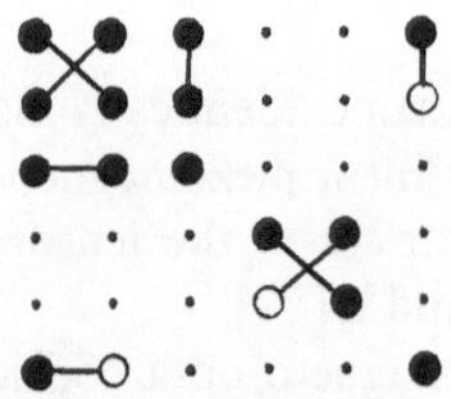	$6, \bar{6}, 6/m$	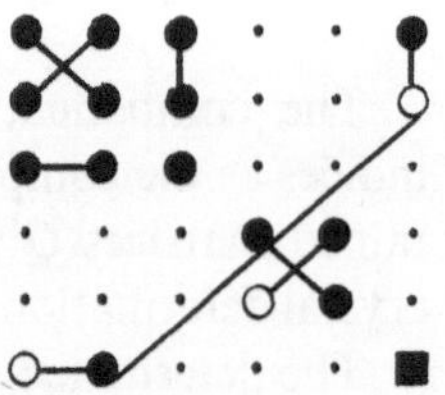
$4mm, \bar{4}2m$ $422, 4/mmm$	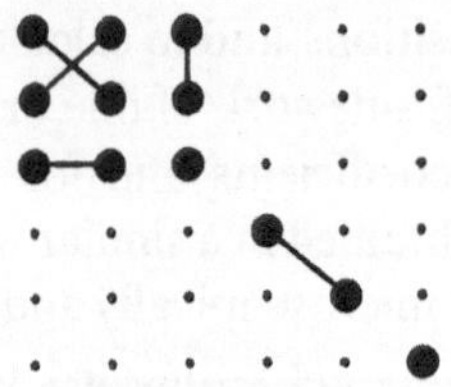	$622, \bar{6}m2,$ $6mm, 6/mmm$	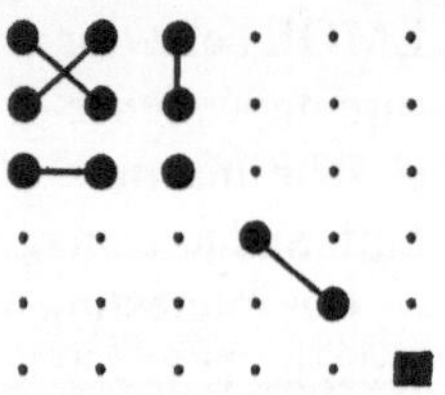
$23, m3$	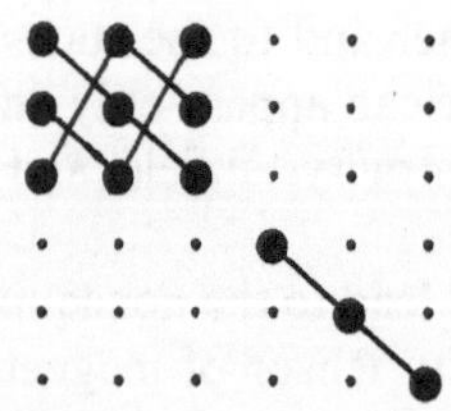	$432, \bar{4}3m,$ $m3m$	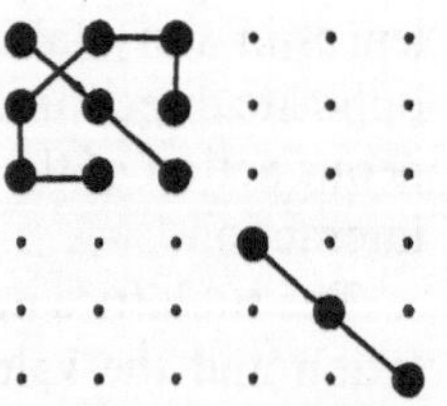

- • Zero component.
- ● Nonzero component.
- ●—● Equal components.
- ●—○ Components differing only by sign.
- ■ Component is equal to $\frac{1}{2}(Q_{11} - Q_{12})$.

Table 1.5. Symmetry of tensors Q^{ML}_{ijrs} in antiferromagnetic crystals.

Symmetry of antiferro-magnetic ordering	Tensor matrix $Q^{ML}_{ijrs} \leftrightarrow Q^{ML}_{\mu\nu}$	Symmetry of antiferro-magnetic ordering	Tensor matrix $Q^{ML}_{\mu\nu} \leftrightarrow Q^{ML}_{ijrs}$
2_z^-	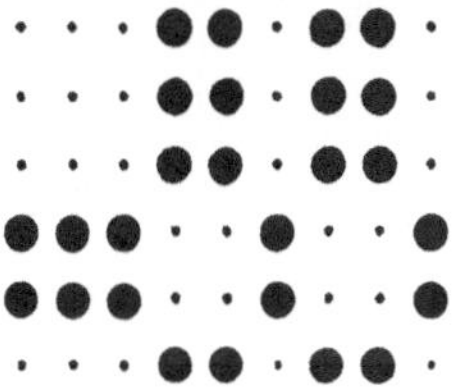	$4_z^+ 2_{xy}^-$	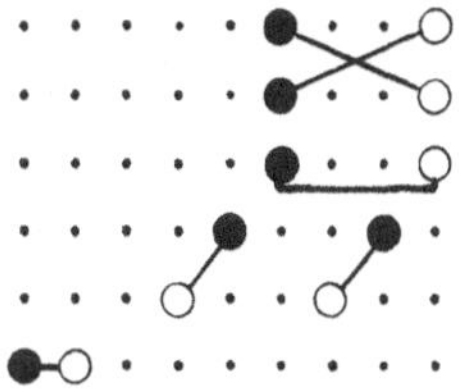
$2_x^+ 2_y^-$	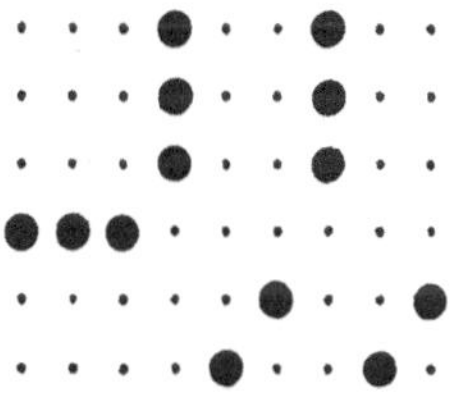	$4_z^- 2_{xy}^+$	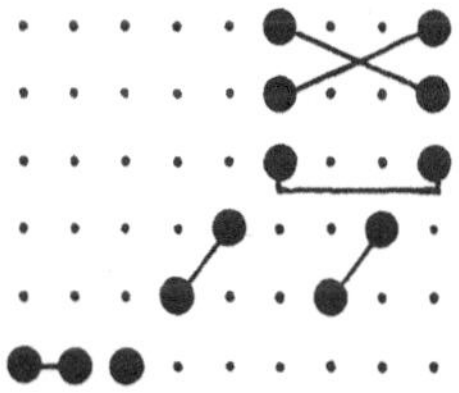
$2_x^- 2_y^+$	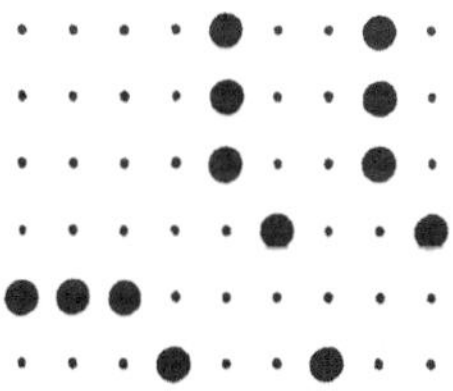	$4_z^- 2_{xy}^-$	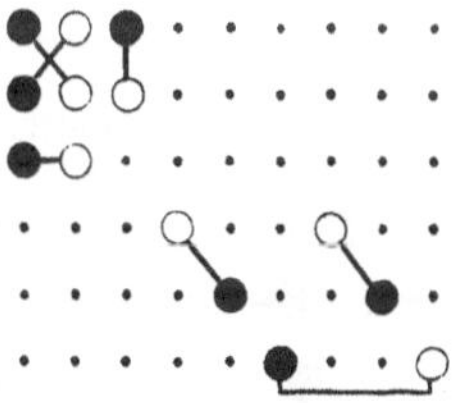
$2_x^- 2_y^-$	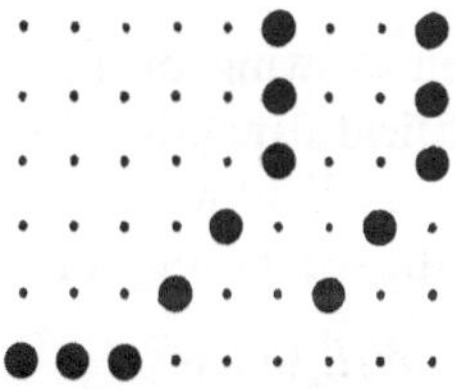	$3_z^+ 2_x^-$	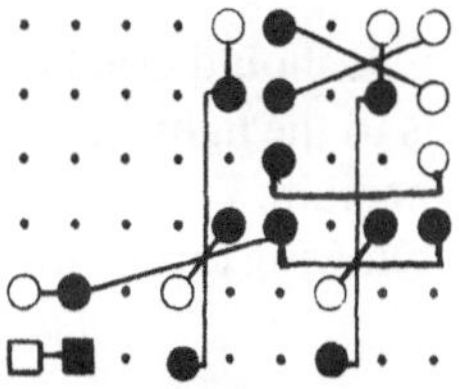
4_z^-	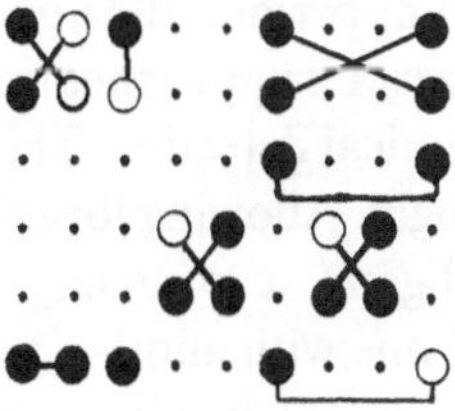	6_z^-	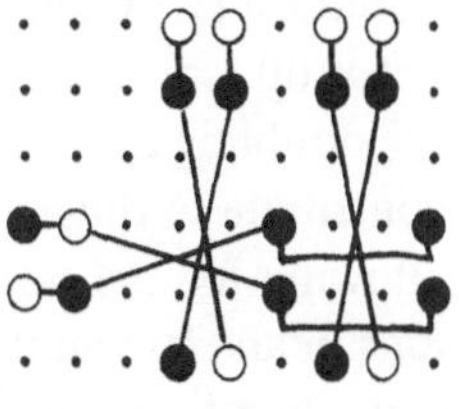

Table 1.5 (*cont.*)

Symmetry of antiferro-magnetic ordering	Tensor matrix $Q^{ML}_{ijrs} \leftrightarrow Q^{ML}_{\mu\nu}$	Symmetry of antiferro-magnetic ordering	Tensor matrix $Q^{ML}_{\mu\nu} \leftrightarrow Q^{ML}_{ijrs}$
$6_z^+ 2_x^-$	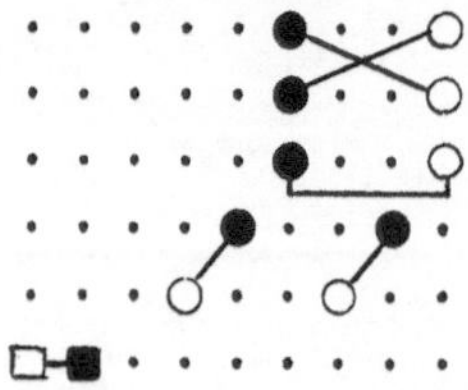	$6_z^- 2_x^+$	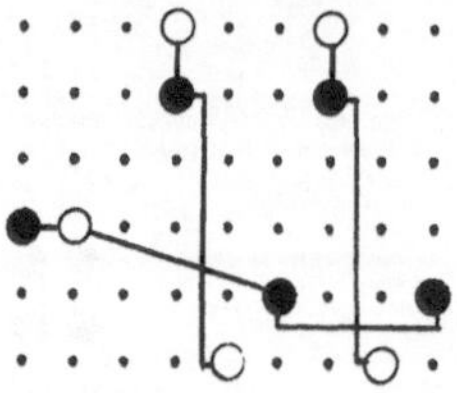
$6_z^- 2_x^-$	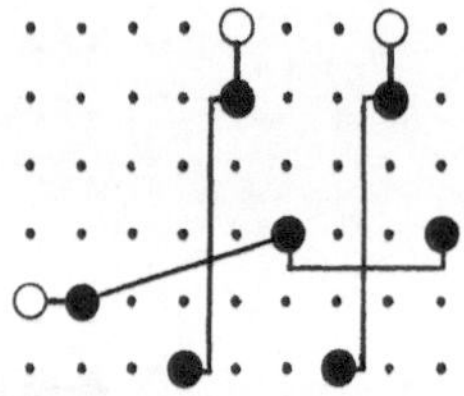	$4_z^- 3^+$	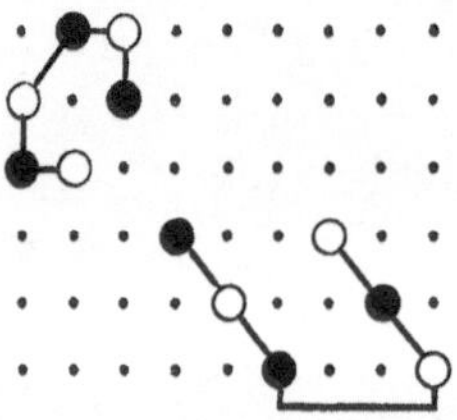

•	Zero component.
●	Nonzero component.
●–●	Equal components.
●–○, ■–□	Components differing only by sign.
■	Component is equal to $\frac{1}{2}(Q_{16} - Q_{26})$.

It should also be noted that the medium allowing for the LMOE is bilinear as to the light wave electric and magnetic field strengths, **E** and **H**, the optical medium. The polarization of this medium in a light wave can be written in the following form with an accuracy of up to the second-order members:

$$P_i = \alpha_{ik} E_k(\omega) + \alpha_{ijk} H_j(\omega) E_k(\omega) + \cdots . \tag{1.33}$$

Equation (1.33) shows that such a medium permits the appearance of nonlinear optical effects which appear in the second approximation, in particular, the doubling of optical frequencies and optical detection. The LMOE should contribute to the magnetic scattering of light. Allowing for this is the same as allowing for the nonequivalence of the ligand surrounding of the magnetic ions in an elementary cell, not connected one with another by the anti-inversion operation. The LMOE is most substantial when light is scattered at the magnetic excitations, characterized by the appearance of such projections of the magnetic moment, which cause the LMOE.

Also interesting is the manifestationof the LMOE in the dichroism spectra of linearly polarized light. The magnetic dichroism connected with the LMOE contains the information on the interaction of excitons and magnetic excitations in a crystal with magnetic field.

1.3. Light Polarization Transformation at Magneto-Optic Effects, and Specific Features of Experimental Methods

When studying the magneto-optic effects in crystals experimentally, one inevitably meets with the simultaneous manifestation of the birefringence of linearly and circularly polarized light [41, 82–84]. As the longitudinal LMOE is allowed symmetrically in many magnetically ordered crystals, the utilization of the longitudinal geometry of experiment, at which light propagates along the magnetic field vector directed along a high-order symmetry axis, does not guarantee the absence of the birefringence of linearly polarized light. Therefore the task of the separation of the effects described by the symmetrical and antisymmetrical components of tensor ε_{ij} is especially important in the experimental magneto-optics of magnetic crystals. This section birefly describes the questions concerning the light polarization transformation in the media, the dielectric tensor of which contains symmetrical and antisymmetrical components. It also deals with specific features of the experimental methods of separation of magneto-optic effects.

The optical properties of a crystal with nonzero antisymmetrical components of the dielectric tensor can be described by the generalized Fresnel equation. According to the compatibility condition of the plane wave equation

$$D_i = n^2[E_i - m_i(m_j E_j)], \tag{1.34}$$

and the equation coupling vectors **E** and **D** in the medium

$$D_i = \varepsilon_{ij}^s E_j + \varepsilon_{ij}^a E_j, \tag{1.35}$$

it follows that the determinant of the system of linear equations for components of **E** is equal to zero. For an arbitrary direction of the wave vector $\mathbf{k} = k\mathbf{m}$, in the system of coordinates built on the proper vectors of the dielectric tensor, this condition can be written as follows:

$$\begin{vmatrix} \varepsilon_{11}^s - n^2(1 - m_1^2) & \varepsilon_{12}^a + m_1 m_2 n^2 & \varepsilon_{13}^a + m_1 m_2 n^2 \\ -\varepsilon_{12}^a + m_1 m_2 n^2 & \varepsilon_{22}^s - n^2(1 - m_2^2) & \varepsilon_{23}^a + m_2 m_3 n^2 \\ -\varepsilon_{13}^a + m_1 m_3 n^2 & -\varepsilon_{23}^a + m_2 m_3 n^2 & \varepsilon_{33}^s - n^2(1 - m_3^2) \end{vmatrix} = 0. \tag{1.36}$$

By designating the gyration vector as $G_k = -i\varepsilon_{ij}^a e_{ijk}$, where e_{ijk} is the unit antisymmetrical tensor, we can obtain from (1.36) the generalized Fresnel equation

$$\begin{aligned} n^4(\varepsilon_{xx}m_x^2 + \varepsilon_{yy}m_y^2 + \varepsilon_{zz}m_z^2) - n^2[\varepsilon_{yy}\varepsilon_{zz}(m_y^2 + m_z^2) \\ + \varepsilon_{xx}\varepsilon_{zz}(m_x^2 + m_z^2) + \varepsilon_{xx}\varepsilon_{yy}(m_x^2 + m_y^2) - [\mathbf{mG}]^2] \\ + \varepsilon_{xx}\varepsilon_{yy}\varepsilon_{zz} - (\varepsilon_{xx}G_x^2 + \varepsilon_{yy}G_y^2 + \varepsilon_{zz}G_z^2) = 0. \end{aligned} \tag{1.37}$$

According to Born [85], it can be rewritten in the compact form

$$(n^2 - n_{10}^2)(n^2 - n_{20}^2) = \tilde{G}^2, \tag{1.38}$$

where n_{10} and n_{20} are the refraction indices for the normal light modes propagating in the crystal in direction **m** at $\varepsilon_{ij}^a = 0$, and $\tilde{G}$ is the scalar parameter of gyration equal to

$$\tilde{G} = [(\varepsilon_{ii}^s G_i^2 - n^2[\mathbf{mG}]^2)/\varepsilon_{ii}^s m_i^2]^{1/2} \tag{1.39}$$

and complexly depending on direction **m** and the refraction index. A simple solution of (1.38) can be obtained by the Voigt approximation [85] in which the influence of the birefringence on the gyration parameter, as well as the influence of the antisymmetrical components of the dielectric tensor or gyration vector on the linear birefringence, are neglected. In this approximation equation (1.39) is substantially simplified

$$\tilde{G} \approx (\mathbf{mG}) = -i(m_1 \varepsilon_{23}^a + m_2 \varepsilon_{13}^a + m_3 \varepsilon_{12}^a) = G_0, \tag{1.40}$$

and we can write for the birefringence of light propagating in the direction **m**

$$(n_2 - n_1)^2 = n_{10}^2 + n_{20}^2 - 2\sqrt{n_{10}^2 n_{20}^2 - G_0^2}. \tag{1.41}$$

The ellipticity of normal wave polarization in this approximation is equal to

$$\frac{a}{b} = 2G_0 \frac{1}{n_{20}^2 - n_{10}^2 + \sqrt{(n_{20}^2 - n_{10}^2)^2 + 4G_0^2}}. \tag{1.42}$$

Taking into account a small value of antisymmetrical components $|\varepsilon_{ij}^a| \ll \varepsilon_{ii}^s$, we obtain from (1.41)

$$n_2 - n_1 = \sqrt{(n_{20} - n_{10})^2 + G_0^2/n_{10}n_{20} + G_0^4/n_{10}^3 n_{20}^3 + \ldots}. \tag{1.43}$$

Neglecting the small terms higher than the second order in the Voigt approximation for the phase difference Δ between the normal light modes after they have passed through a plate of thickness t, we obtain the expression

$$\Delta = (\delta^2 + 4\rho^2)^{1/2}. \tag{1.44}$$

There $\delta = 2\pi(n_{20} - n_{10})t/\lambda$ is the phase difference caused by the symmetrical components of the dielectric tensor only, and $2\rho = 2\pi G_0 t/\lambda\sqrt{n_{10}n_{20}}$ is the phase difference caused only by the antisymmetrical components. These approximations are sufficient in almost all cases. If (1.44) is satisfied, the superposition principle [41, 69] is valid, according to which the action of an elliptically birefringent plate on the light polarization is equivalent to the action of the set of the same thickness of infinitely thin alternating linearly and circularly refracting plates, the birefringence of which is described by only symmetrical and only antisymmetrical components of the dielectric tensor. The validity of this principle makes it possible to consider separately the linear or circular birefringence of elliptically birefringent media.

To describe the variations of polarization of the light passed through an elliptically birefringent plate within the framework of the above-mentioned approximations, it is convenient to use the Poincaré sphere representation [86–88]. According to this representation, the light polarization transforma-

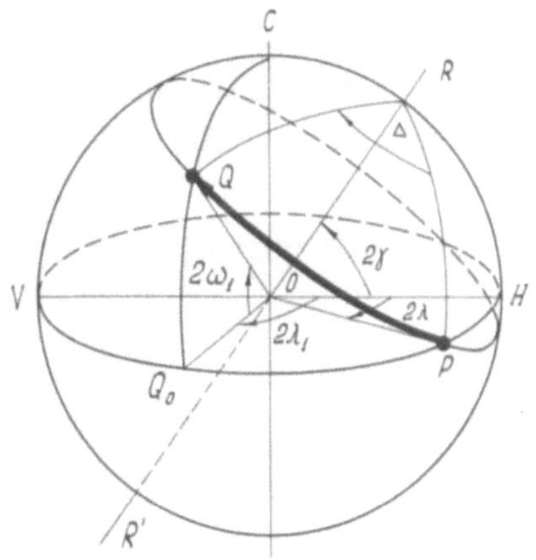

Fig. 1.1. Transformation of light polarization by an elliptically birefringent plate represented by a Poincaré sphere.

tion is represented by the motion of a point over the sphere around the axis, which is determined by the relation of the values and signs of the circular and linear birefringence. The angle between the axis and equatorial plane is equal to (see Fig. 1.1)

$$\tan 2\gamma = 2\rho/\delta = G_0/(n_{10}n_{20})^{1/2}(n_{20} - n_{10}). \tag{1.45}$$

The azimuth of the plane passing through the sphere poles and the axis of rotation coincides with the polarization azimuth of the fast light mode. The points of intersection of the axes with the sphere surface correspond to the orthogonal polarizations of the normal light modes propagating in the crystal. The ellipticity of modes (the relation of axes of the polarization ellipse, a/b) and angle γ have the following relation: $\tan \gamma = a/b$. The spherical trigonometry formulas make it possible to find all the necessary relations between the values of ellipticity and the azimuthes of axes of the polarization ellipses of the light incident on the plate and leaving it. The relations for the most frequently encountered cases of the polarization measurements appear below. The elliptically birefringent plate is characterized by phase shifts δ and 2ρ or Δ and the angle 2γ. Their relation can be seen from (1.44) and (1.45)

$$\delta = \Delta \cos 2\gamma, \qquad 2\rho = \Delta \sin 2\gamma. \tag{1.46}$$

The azimuth of the principal axis of the polarization ellipse of one (faster) of the normal light modes propagating in the plate is usually taken as the reference point.

(1) The incident light is linearly polarized, the azimuth of its polarization axis is λ. The coordinates of the corresponding point on the Poincaré sphere are $(2\lambda, 0)$. The azimuth of the polarization ellipse axis of passing light λ_1 is

determined by the expression

$$\tan 2\lambda_1 = \frac{\tan 2\lambda \cos \Delta + \sin 2\gamma \sin \Delta}{1 - \sin^2 2\gamma(1 - \cos \Delta) - \sin 2\gamma \sin \Delta \tan 2\lambda}. \quad (1.47)$$

The ellipticity is equal to

$$a/\ell = \tan \omega_1,$$

where

$$\sin 2\omega_1 = \cos 2\gamma \cos 2\lambda\{\sin 2\gamma(1 - \cos \Delta) + \tan 2\lambda \sin \Delta\}. \quad (1.48)$$

The variation of azimuth $(\lambda_1 - \lambda)$ is found from the expression

$$\tan 2(\lambda_1 - \lambda) = \frac{\sin 2\gamma \sin \Delta(1 + \tan^2 2\lambda) - (1 - \cos \Delta) \cos^2 2\gamma \tan 2\lambda}{\cos^2 2\gamma + \sin^2 2\gamma \cos \Delta + \tan^2 2\lambda \cos \Delta}. \quad (1.49)$$

If the polarization axis of the incident light lies in the principal plane of the plate, the plane of oscillations of the electric vector of the normal light mode in the absence of the circular birefringence, then $2\lambda = 0$ and (1.47), (1.48), and (1.49) are substantially simplified

$$\tan_{0^\circ} 2(\lambda_1 - \lambda) = \frac{\sin 2\gamma \sin \Delta}{1 - (1 - \cos \Delta) \sin^2 2\gamma} = \frac{2\rho\Delta \sin \Delta}{\delta^2 + 4\rho^2 \cos \Delta}, \quad (1.50)$$

$$\sin 2\omega_1 = \sin 2\gamma \cos 2\gamma(1 - \cos \Delta) = 2\rho\delta(1 - \cos \Delta)/\Delta^2. \quad (1.51)$$

At small $\Delta \ll 1, 4\rho^2 \ll \delta^2$,

$$(\lambda_1 - \lambda)_{0^\circ} = \lambda_1 \approx \rho(1 - \tfrac{1}{6}\delta^2), \quad (1.52)$$

$$\left(\frac{a}{\ell}\right)_{0^\circ} \approx \tfrac{1}{2}\rho\delta\left(1 - \frac{\Delta^2}{12}\right). \quad (1.53)$$

If the angle between the polarization plane of the incident light and the principal plane is 45 degrees, $2\lambda = \pi/2$, then

$$\tan_{45^\circ} 2(\lambda_1 - \lambda) = \sin 2\gamma \tan \Delta = 2\rho \tan \Delta/\Delta, \quad (1.54)$$

$$\sin_{45^\circ} 2\omega_1 = \cos 2\gamma \sin \Delta = \delta \sin \Delta/\Delta. \quad (1.55)$$

At small $\Delta \ll 1, 4\rho^2 \ll \delta^2$,

$$(\lambda_1 - \lambda)_{45^\circ} \approx \rho(1 + \tfrac{1}{3}\delta^2), \quad (1.56)$$

$$\left(\frac{a}{\ell}\right)_{45^\circ} \approx \tfrac{1}{2}\delta\left(1 + \frac{\Delta^2}{12} - \rho^2\right). \quad (1.57)$$

(2) The incident light is circularly polarized

$$\tan 2\lambda_1 = \frac{\sin \Delta}{\sin 2\gamma(1 - \cos \Delta)} \quad \text{or} \quad \lambda_1 = \tfrac{1}{2} \arctan\left\{\frac{\Delta}{2\rho} \cot \frac{\Delta}{2}\right\}, \quad (1.58)$$

$$\sin 2\omega_1 = \sin^2 2\gamma + \cos^2 2\gamma \cos \Delta \quad \text{or} \quad \frac{a}{\ell} = \tan\left\{\frac{\pi}{4} - \arcsin\left(\frac{\delta}{4} \sin \frac{\Delta}{2}\right)\right\}. \quad (1.59)$$

At small $\Delta \ll 1$

$$\tan \lambda_1 \approx -1 + \rho - \rho^2, \tag{1.60}$$

$$a/b \approx 1 - \delta(1 + \delta^2/24 - \rho^2/6). \tag{1.61}$$

To analyze the elliptically polarized light it is convenient to use a linear $\lambda/4$-plate. If the azimuth of its axis is equal to Λ and the incident light has an arbitrary polarization $P(2\lambda, 2\omega)$, then for the polarization of the light passed through the plate $Q(2\lambda_1, 2\omega_1)$, the following expressions are valid:

$$\tan 2(\lambda_1 - \Lambda) = \tan 2\omega/\cos 2(\Lambda - \lambda), \tag{1.62}$$

$$\sin 2\omega_1 = \cos 2\omega \sin 2(\Lambda - \lambda). \tag{1.63}$$

If there is a linear analyzer in the optical system, the passing light intensity at the elliptical polarization of the light incident on the analyzer is equal to

$$\mathcal{I} = \tfrac{1}{2}[1 + \cos 2\omega_1 \cos 2(\lambda_1 - \lambda_A)]. \tag{1.64}$$

Here $2\omega_1$ and $2\lambda_1$ are the coordinates on the Poincaré sphere corresponding to the incident light polarization and λ_A is the azimuth of the analyzer axis. Using (1.64) one can obtain the expressions for the passing light intensities in the cases of interest.

Many tasks connected with light propagation in elliptically birefringent crystals were investigated by various methods in [82–87].

As in the presence of circular and linear birefringence the polarization axis azimuth of the passing light λ_1 does not remain constant when the field or temperature is changed, the utilization of compensation methods for ellipticity measuring requires the measuring of azimuth λ_1 and correction of the compensator position. This hampers the measurement and reduces its accuracy.

A convenient method for studying the crystal optical properties is the conoscopic method. Although its accuracy of the measurement of birefringence is lower than the accuracy of other methods, it far exceeds all other methods by the volume of information, its visualizability, and the rapidity of its collection. The conoscopic method is an attractive express method for determining the directions of the proper axes of the symmetrical part of the dielectric tensor, and for obtaining the visual and unambiguous information on the qualitative changes of the tensor symmetry. In this method a conical beam of parallel-polarized rays is sent onto the specimen (plane-parallel crystal plate). A lens and an analyzer are placed behind the plate. All the previously overlapping parallel rays are spatially separated in the lens focal plane. Each point of the focal plane corresponds to only one direction of the parallel rays. As the polarization of the light passing through the crystal depends on the direction of light propagation in the crystal in respect to the optical axes, the path length, and the polarization azimuth axis in respect to the principal plane (the principal plane is not rigidly connected with the crystallographic axes but depends on the propagation direction), the light in various points of the focal plane has different polarization. The hollow cones of rays, for which the phase shift Δ acquired in the crystal between the normal light modes is $2\pi\nu$ radians, do not change their polarization. The analyzer makes it possible to visualize the corresponding closed lines in the focal plane and are called isochromates. The isochromates with $\nu = 0$ represent the points which correspond to the directions of the crystal optical axes. The dimensions, shape, and orientation of the isochromates carry information on the dielectric tensor components. The isochromate equation can be obtained from the intersection of the surface of equal phase difference with the corresponding planes [85, 89]. The investigation of conoscopic figures is limited to linear birefringent crystals, as the general case of an elliptically birefringent crystal is very awkward. Concerning the cases of elliptical birefringence studied in this paper, it is advantageous to write the expression for the phase difference in the system of coordinates connected with the directions of optical axes in the crystal. According to (1.44) for a biaxial crystal it has the following form

$$\Delta^2 = \left(\frac{2\pi t}{\lambda \cos\psi}\right)^2 \left\{\left(\frac{(n_m - n_p)\sin\theta_1 \sin\theta_2}{\sin^2 V}\right)^2 + (n_+ - n_-)^2 \cos^2\xi\right\}. \quad (1.65)$$

Here ψ is the average angle between the ray direction in the crystal and normal to the plate surface z. It is equal to

$$\cos\psi = \cos\theta_1 \cos V + \frac{\sin\theta_1 \sin V(\cos\theta_2 - \cos\theta_1 \cos 2V)}{\sin 2V \sin\theta_1}. \quad (1.66)$$

θ_1 and θ_2 are the angles between the light wave direction and the optical axes, $2V$ is the acute angle between the optical axes, ξ is the angle between the gyration vector $\mathbf{G}$ and the light wave direction $\mathbf{m}$, $(n_m - n_p) = \Delta n_{mp}$ is the value of linear birefringence for the light propagating along the acute bisectrix between the optical axes, and $(n_+ - n_-) = \Delta n_\pm$ is the value of circular birefringence for the light propagating along $\mathbf{G}$. The hollow cone of the beams which form the νth isochromate is determined by expression (1.65), where $\Delta = 2\pi\nu$. At $\mathbf{G} \parallel \mathbf{z}$ in the approximation of small angles θ_1, θ_2, and V from (1.65), the birefringence value Δn_{mp} can be expressed in terms of the angles characterizing the conoscopic figure

$$\Delta n_{mp} = \frac{V^2}{\theta_1 \theta_2}\left[\left(\frac{\lambda\nu}{t}\right)^2 - \Delta n_\pm^2\right]^{1/2}. \quad (1.67)$$

Here ν is the isochromate number. At the circular birefringence which is sufficiently small to make condition $\Delta_\pm \ll \nu\lambda/t$ or $\rho \ll \nu\pi$ valid, formula (1.67) is simplified. As (1.67) contains the relations of small angles, the angles can be replaced by the corresponding distances between the conoscopic figure points (see Fig. 1.2) as follows:

$$\Delta n_{mp} = \frac{v^2 \lambda\nu}{a_1 a_2 t}\left[1 - \left(\frac{\rho}{\pi k}\right)^2\right]^{1/2}. \quad (1.68)$$

At $\rho < \pi$, it is convenient to use the isochromate with $\nu = 1$. And if $\rho \gtrsim \pi$, the

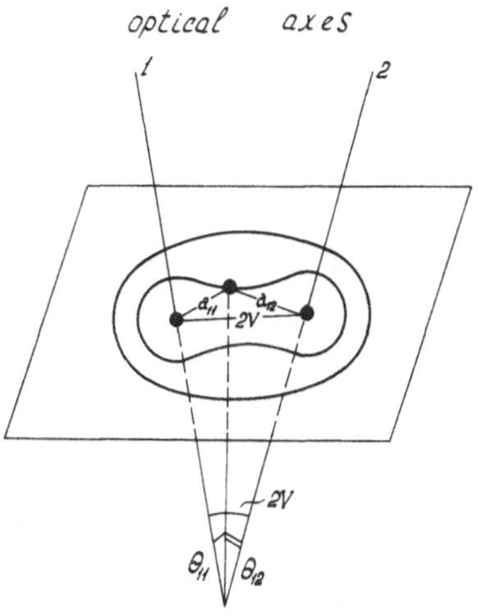

Fig. 1.2. Parameters for a conoscopic figure.

isochromates with $\nu > 1$ should be used, but in this case the measurement error becomes greater. As the magnetic linear birefringence is usually small compared with the natural one, the birefringence $\Delta n_{xy} = \Delta n_{mp} \ll \Delta n_{gm}$ which often appears in magnetic crystals can be approximately determined without measuring the Faraday rotation, only if the refraction coefficients of the undisturbed crystal are known with a not excessively high accuracy. To do this, it is necessary to measure the angle between the optical axes in the crystal $2V$ and use the well-known relation

$$V = \left[\frac{n_z^2(n_x^2 - n_y^2)}{n_x^2(n_z^2 - n_y^2)}\right]^{1/2}. \tag{1.69}$$

If the induced linear birefringence is small compared with the natural one, the approximate expression for determining the value of circular birefringence can be obtained from the parameters of the conoscopic figures. It is possible to show the validity of the expression

$$\Delta n_\pm^2 = \frac{\lambda}{t}\left[\nu^2 - \tilde{\nu}^2\left(\frac{\sin\theta_{\nu_1}\sin\theta_{\nu_2}}{\sin\tilde{\theta}_{\nu_1}\sin\tilde{\theta}_{\nu_2}}\right)^2\right], \tag{1.70}$$

where the "$\sim$" sign is related to conoscopic figure isochromates of the crystal, in which the magnetic gyration vector has vanished and the symmetrical components of the dielectric tensor have remained the same. At weak influence of the magnetic field on the natural birefringence $n_g - n_m = \Delta n_{gm}(H) = \Delta n_{gm}(H = 0) + \zeta(H)$, $\zeta(H) \ll n_{gm0}$, we can write for the νth isochromate angles

$$\sin\tilde{\theta}_{\nu_1}\sin\tilde{\theta}_{\nu_2} = \sin\theta_{\nu_{10}}\sin\theta_{\nu_{20}}\left(1 - \frac{2\zeta(H)}{\Delta n_{gm0}}\right). \tag{1.71}$$

Finally, taking into account the small value of the angles and going from the angular dimensions to the linear ones characterizing the isochromate, we can write

$$\Delta n_{\pm} = \frac{\lambda v}{t}\left\{1 - \frac{v_0^2}{v^2}\left[\frac{a_1 a_2}{a_{10} a_{20}}\left(1 + \frac{2\zeta(H)}{\Delta n_{\mathrm{gm0}}}\right)\right]^2\right\}^{1/2}. \tag{1.72}$$

Index "0" refers to the birefringence and isochromates of the same crystal in the absence of the field. Usually, $\zeta(H)/\Delta n_{\mathrm{gm0}} < 10^{-2}$ and the error caused by ignoring this addition does not exceed 2%.

From (1.65) and (1.72) there follows a simple approximate formula for determining the linear birefringence valid at $\zeta(H)/\Delta n_{\mathrm{gm}} \to 0$:

$$\Delta n_{\mathrm{mp}} = \frac{\lambda v V^2}{t\theta_{10}\theta_{20}} = \frac{\lambda v v^2}{t a_{10} a_{20}}. \tag{1.73}$$

For the initial uniaxial crystal $\theta_{10} = \theta_{20} = \theta_0$, and

$$\Delta n_{\mathrm{mp}} = \frac{\lambda v}{t}\left(\frac{V}{\theta_0}\right)^2 = \frac{\lambda v}{t}\left(\frac{v}{a_0}\right)^2. \tag{1.74}$$

Although the accuracy of determining the linear and especially circular birefringence by means of formulas (1.68), (1.72), and (1.73) cannot be high due to the errors of determining the distances a_i, v, these expressions can often be useful.

The conoscopic method was used for investigation of the LMOE in cobalt carbonate and fluoride AFM crystals [90–92] both in stationary and pulse magnetic fields. Figure 1.3 shows the diagram of the experimental equipment for photographing the conoscopic figures in pulsed magnetic fields. The lens L_1, which forms the light ray cone, is placed in the vicinity of the specimen inside the solenoid working, opening the diameter which was equal to 4 mm. The second lens L_2, which forms the conoscopic picture, was also located inside the solenoid behind the specimen. The focal planes of these lenses were aligned in the place of specimen location. The objective L_4 transmitted the

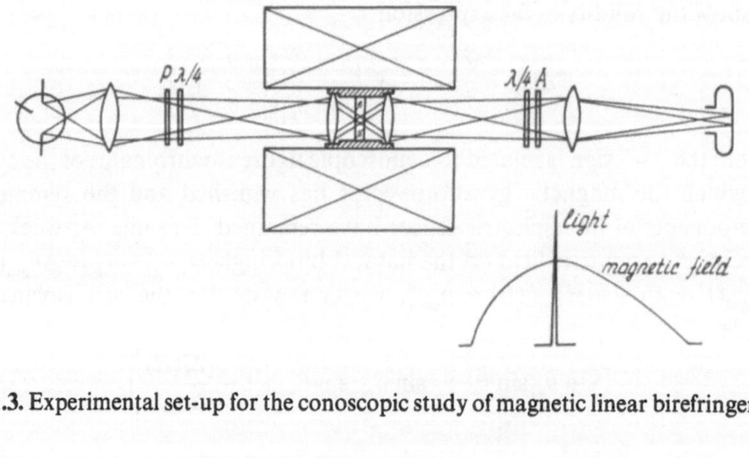

Fig. 1.3. Experimental set-up for the conoscopic study of magnetic linear birefringence.

conoscopic figure image into the plane where photographic film was located. A light flash of duration about $10^{-}5$ s was synchronized with the pulse magnetic field maximum, the rise time of which was equal to about $3 \cdot 10^{-3}$ s. During the flash the magnetic field varied with $5 \cdot 10^{-4}\ H_{ampl}$. One flash was sufficient for photographing. To reduce the Faraday effect influence, the circularly polarized light was used. In the circularly polarized light the conoscopic figure is the most simple. Isolated dark circles with easily determined coordinates of their centers correspond to the exits of the optical axes. Their position is independent from the Faraday effect, but as the Faraday rotation increases the contrast decreases and the circle dimensions increase. In the linearly polarized light the conoscopic picture varies in a complex way depending on the Faraday effect value. The position determining error of the points corresponding to the exists of the axes becomes larger as the Faraday rotation angle increases, because the kind of dependence of the phase shift between the normal light modes on the angle of light propagation in the crystal changes as the circular birefringence increases, and because the elliptical polarization of the light passing through the crystal in the direction close to the optical axis increases.

To reduce the influence of circular birefringence to more accurate measurements of the linear birefringence, the method was used in which circularly polarized light was sent on the specimen, and a linear polarizer rotating with frequency Ω served as the analyzer (Fig. 1.4) [91]. The intensity of the light passing through the specimen and analyzer varies according to the law

$$\mathscr{I} = \tfrac{1}{2}\mathscr{I}_0(1 + \sin 2\eta \cos 2\Omega t) \qquad \text{where} \quad \sin \eta = \frac{\delta}{\Delta} \sin \frac{\Delta}{2}. \tag{1.75}$$

At small $\Delta \ll 1$, $\sin \eta \approx \delta/2$, and modulation depth $\mathscr{I}_\sim/\mathscr{I}_-$ depends only on the linear birefringence value

$$\mathscr{I}_\sim/\mathscr{I}_- = \sin 2\eta \approx \delta. \tag{1.76}$$

This method is especially convenient when the position of the plane of optical axes changes during the experiment, because it does not require the information on the azimuth of the polarization ellipse axis of the passing light. And, if we determine the ellipse azimuth λ by measuring the alternating signal phase, we can find the linear birefringence and Faraday rotation by means of the

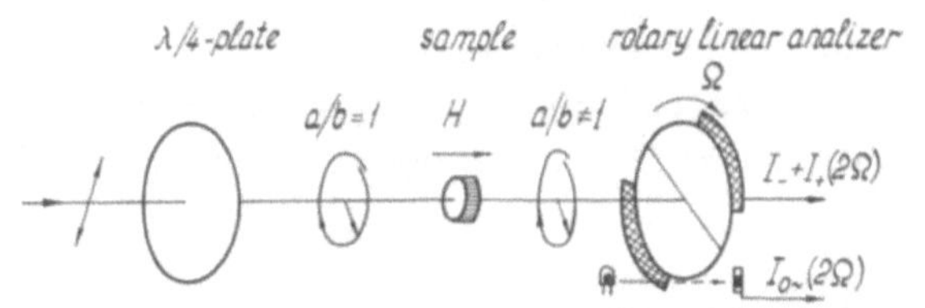

Fig. 1.4. Experimental set-up diagram for measuring linear birefringence in a gyrotropic crystal.

relations obtained by the investigation of constructions on the Poincaré sphere

$$\cos \Delta = \cos \eta \sin 2\lambda,$$

$$\delta = \frac{\Delta}{\sin \Delta} \sin \eta \sin 2\lambda, \tag{1.77}$$

$$\rho = \tfrac{1}{2} \delta \cos 2\lambda \cot \eta.$$

It is easy to obtain the required degree of circular polarization for a selected part of the spectrum by means of two linear phase plates.

At elliptical birefringence, the azimuth of the polarization ellipse axis of the light passing through the plate does not remain constant, as the value of birefringence varies at any azimuth of the incident light polarization. Besides, in magneto-optical experiments, often both the ellipticity and the azimuth of normal light modes vary. Therefore, the widely used Sénarmont method, which is very convenient for measuring magnetic birefringence, requires continuous adjustment of the $\lambda/4$-plate axis azimuth. This method can be used in its usual form only when either the spontaneous birefringence far exceeds the induced linear and circular ones, or when the resulting phase shift is small. The latter requirement can be satisfied by reducing the specimen thickness. The systematic errors of measurement are similar to the errors which occur at incorrect orientation of the $\lambda/4$-plate. They can be investigated by means of expressions (1.54), (1.55) and (1.62), (1.63) for each particular case.

In the cases when the nondiagonal symmetrical components of ε^s_{ij} play the main role in the magneto-optic effects, the magnetic rotation of the plane of polarization can cause great disturbance during determination of the rotation angles of the indicatrix optical axes. To eliminate the effect of magnetic rotation during the investigation of the LMOE, it is convenient to measure in two AFM domains in sequence in the same direction of field **H**. The Faraday rotation in this case remains constant (it is described by the axial *i*-tensor), and the indicatrix rotation angle changes its sign. If, in addition, the natural birefringence far exceeds the induced linear and circular ones, the rotation of the polarization ellipse axis, caused by the LMOE, is especially easily determined at small azimuth angles of the incident light polarization axis λ_0. According to formula (1.47), at $\delta \gg \rho$, $\Delta \approx \beta$, and $\lambda_0 \ll 1$, we can write

$$\lambda^{\pm} = \lambda_0^{\pm} \cos \delta^{\pm} + \gamma^{\pm} \sin \delta^{\pm}, \tag{1.78}$$

where the "+" and "−" signs refer to domains AFM^+ and AFM^-. Taking into account that $\delta^+ \approx \delta^-$, $\rho^+ = \rho^-$, $\lambda_0^+ = \lambda_0 + \theta$, $\lambda_0^- = \lambda_0 - \theta$, where θ is the angle of the indicatrix axes rotation, we obtain $\gamma^+ = \gamma^-$, and

$$\lambda^+ - \lambda^- = 2\theta \cos \delta. \tag{1.79}$$

The usual modulation technique of measuring the rotation of the light plane of polarization can be used to measure azimuthes $\lambda^{\pm}$ (Fig. 1.5). This method was used to study the LMOE in orthorhombic AFM dysprosium orthoferrite [93].

All experiments on study the new magneto-optic effects which are symmetrically allowed in magnetically ordered crystals should only be carried out on specimens with a uniformly ordered magnetic subsystem. Therefore, these experiments always involve the complex task of monodomainization of AFM samples. It is simplified due to the fact that in the investigated cases the combined action on the AFM crystal can lead to an energy nonequivalence of 180-degree or time-inverted states, which differ by the spin moments of the ions belonging to the same sublattices. So, the LMOE and QMR (quadratic magnetic rotation) are allowed in piezomagnetic AFM only. Therefore, the elastic compression of a crystal placed in a magnetic field of required direction and sufficient strength will move the domain wall between the AFM domains and monodomainize the specimen. This method of AFM monodomainization was used to study the piezomagnetic effect [94]; it was also acceptable for magneto-optic investigations [92, 95]. The other effect, also sensitive to AFM ordering, which is successfully used for AFM sample monodomainization, is the quadratic in field AFM magnetization [96, 97]. At a specially chosen orientation of the magnetic field, it is possible to obtain in most piezomagnetic AFM the energy nonequivalence of the AFM^+ and AFM^- states. This phenomenon also makes it possible to mondomainize AFM samples [98, 99]. It is remarkable than when studying the optical properties, the domain state of the sample can often be visually observed.

It is different to determine the odd character of the induced linear birefringence in AFM with weak ferromagnetism (WFM), because a quite definite AFM state is always realized in them in a magnetic field due to the Dzyaloshinskiy–Moriya-type interaction. Already at a small field strength the reversal of the field remagnetizes the WFM moment, which is inevitably accompanied by the remagnetization of all magnetic sublattices. As a result the LMOE preserves its sign. To fix the AFM ordering, external influences are necessary. In the described experiments on studying the cobalt carbonate AFM with WFM, the independent source of the transverse magnetic field component ($\mathbf{H}_\perp \perp C_3$) was used as the external influence. Interacting with the spontaneous magnetic moment, this projection delayed the process of the AFM remagnetization by field H_z, and made it possible to confirm the odd character of the induced birefringence.

1.4. Experimental Investigations of the Linear Magneto-Optic Effect

The number of magnetic crystals in which the LMOE was observed is still small. First, the "abnormal" behavior of the birefringence of linearly polarized light in a magnetic field was observed in the cobalt carbonate AFM [90] and then in dysprosium aluminum garnet [100]. In both cases the linear birefringence varied, obviously not proportionally, to the square of the field strength, increasing twofold at $T \ll T_N$ in the field of about 0.2 $H_{\mathbf{exch}}$ in $CoCO_3$, and varying in different ways in the stable and metastable AFM states of the complex multi-sublattice AFM DyAlG. Nearly linear regions can also be seen on the dependences of the linear birefringence on the haematite magnetic field strength obtained during the Morin magnetic phase transition [10]. Then the variation of birefringence in the field was studied in detail in $CoCO_3$, CoF_2, MnGeG, $DyFeO_3$, and α-Fe_2O_3. These investigations have shown that the observed specific features of the birefringence behavior in these crystals can be described as various manifestations of the LMOE.

1.4.1. The Linear Magneto-Optic Effect in Cobalt Carbonate

Cobalt carbonate is a two-sublattice AFM with a rhombic crystallographic structure (spatial group $P\bar{3}c$), and magnetic anisotrophy of the "easy plane" type. It is one of the most fully investigated weak ferromagnets [102–107]. The AFM ordering symmetry of cobalt carbonate $3^+2_X^-$ permits two independent components of the WFM moment. One of them, $\mathbf{M}_\perp$, is perpendicular to the trigonal axis and does not depend on the AFM vector projection orientation in the basal plane. The second, $\mathbf{M}_\parallel$, is parallel to C_3 and perpendicular to the basal plane. Its value depends on the orientation of $\mathbf{L}_\perp$ in the basal plane. At $\mathbf{L}_\perp \parallel C_2$, component $\mathbf{M}_\parallel$ is the greatest. It vanishes when L is located in the symmetry plane, when $\mathbf{L}_\perp \perp C_2$. Both components of the WFM moment are caused by the canting of the sublattice moments. The presence of a sufficiently high moment, $\mathbf{M}_\perp = 48$ emu/cm^3, makes it possible to vary easily the orientation of the sublattice moments by means of an external magnetic field $\mathbf{H} \perp C_3$, and thus vary the crystal magnetic symmetry. In this way, it is possible to obtain two symmetrical states described by the magnetic point groups $2/m$ and $\underline{2}/\underline{m}$. The first state is a spontaneous one. In this state, the AFM vector lies in the symmetry plane, does not tilt from the basal plane, is perpendicular to axis $C_2 \parallel X$, and the WFM moment is directed along C_2. The second state is characterized by the presence, besides $\mathbf{M}_\perp$, of the WFM moment component $\mathbf{M}_\parallel$. The AFM vector $\mathbf{L}$ in the case should be parallel to C_2. The phenomenological potential of the two-sublattice AFM with symmetry $3^+2_X^-$ has the following form [72, 102]:

$$\Phi = \frac{B}{2}M^2 + \frac{a}{2}L_z^2 + \frac{b}{2}M_z^2 + d(L_xM_y - L_yM_x)$$
$$+ \tfrac{1}{3}f[(L_x + iL_y)^3 + (L_x - iL_y)^3]M_z + \tfrac{1}{6}e[(L_x + iL_y)^6$$
$$+ (L_x - iL_y)^6] - \mathbf{MH}. \tag{1.80}$$

Equation (1.80) shows that state $\underline{2}/\underline{m}$, in which $L_y = 0$ and $L_x, M_y, M_z \neq 0$, can be induced by the magnetic field $\mathbf{H} \parallel Y \perp C_2$ or the field $\mathbf{H} \parallel Z \parallel C_3$.

From the optical point of view cobalt carbonate above T_N is a uniaxial negative crystal. Below T_N the crystal becomes a biaxial one [108]. The plane containing the optical axes passes through the AFM vector $\mathbf{L}$ and axis C_3 at any orientations of $\mathbf{L}_\perp$ in the basal plane. At temperature 4.2 K the angle between the optical axes is about 3.2° for light with a wavelength of about 590 nm, and the birefringence value for the light propagating along C_3 is close to 2.5×10^{-4}. It is sufficient for a description of the spontaneous magnetic birefringence to restrict ourselves to the dependence of the birefringence on the AFM vector transverse components [108–109].

The spontaneous magnetic birefringence makes it possible to observe the WFM domain structure of cobalt carbonate [108]. To monodomainize the specimen, a weak magnetic field $\mathbf{H} \perp C_3$ of about 0.02 T is sufficient. The specimen cannot be monodomainized in field $\mathbf{H} \parallel C_3$, as three orientations of vector $\mathbf{L}$ correspond to the same value of $M_{\parallel}$. The magnetic twinned state should be preserved in strong fields $H \simeq H_{\text{exch}}$. But a small error in the field orientation makes one of the magnetic twins energetically more advantageous. At a sufficiently small component $\mathbf{H}_{\perp}$, orientation $\mathbf{L} \perp C_3$ is determined by component $\mathbf{H}_{\parallel}$.

The magnetic field $\mathbf{H} \perp C_3$ has had small effect on the birefringence [109], but the action of the magnetic field $\mathbf{H} \parallel C_3$ suddenly proved to be abnormally strong [90, 91, 110]. At low temperatures the birefringence is doubled in the field to about 7.5 T, but at 30 T it was three times greater than the spontaneous one, in spite of the reduction of the transverse projection of the AFM vector perpendicular to the light propagation vector. At temperatures close to T_N the relative influence of the field is even stronger. The photographs of the conoscopic figures shown in Fig. 1.5. qualitatively illustrate the rise of birefringence, and the diagrams, Figs. 1.6 and 1.7, show its dependence on the field strength. Fast variations of the birefringence in weak fields are connected with the specimen monodomainization which ended in fields of about 1 T. The optical axes of a homogenous AFM sample ($H \gtrsim 1T$) are oriented in one of the planes containing the crystallographic axes C_2 and C_3. They preserve their position in this plane when the field is reversed. As the position of the plane of optical axes is determined by the AFM vector direction, their orientation in the plane $(C_2 C_3)$ indicates the rotation of $\mathbf{L}$ from the symmetry plane to the plane $(C_2 C_3)$. This behavior of the AFM vector agrees with the models of magnetic interactions $CoCO_3$ [72, 102].

At first sight, the increase of birefringence in a field is unexpected, as the transverse projections of the magnetic sublattice moments cannot vary so dramatically. Besides, these variations are quadratic as to the field and, even

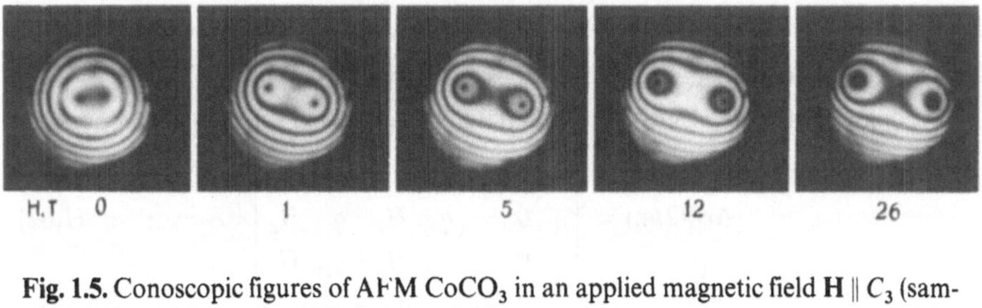

Fig. 1.5. Conoscopic figures of AFM $CoCO_3$ in an applied magnetic field $\mathbf{H} \parallel C_3$ (sample thickness 1.5 mm, $\lambda = 400$ nm, $T = 7$ K).

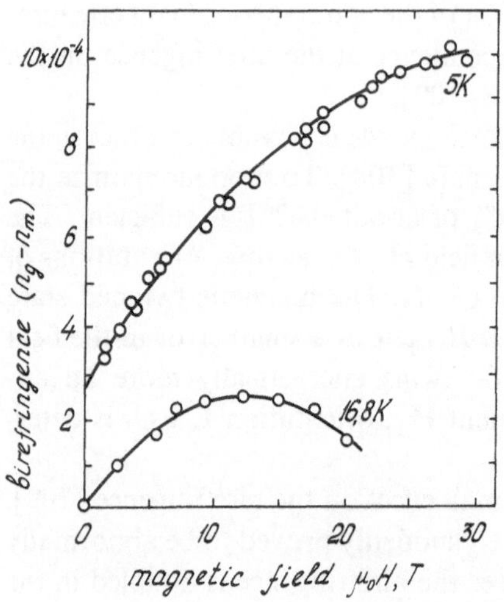

Fig. 1.6. Variation of linear birefringence in AFM $CoCO_3$ in high magnetic fields ($\mathbf{k} \parallel \mathbf{H} \parallel C_3$, $\lambda = 400$ nm).

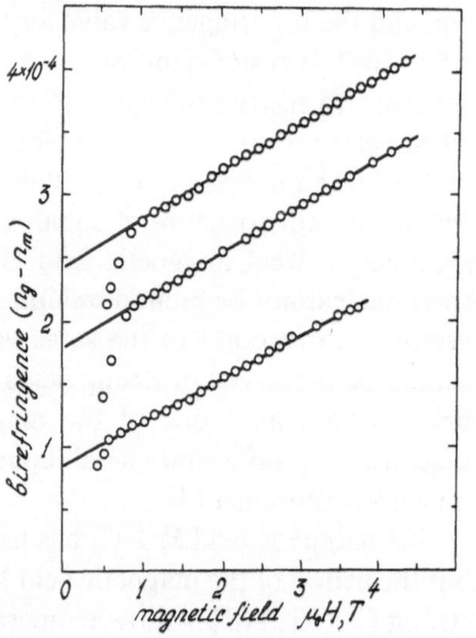

Fig. 1.7. Linear birefringence along the trigonal axis of AFM $CoCO_3$ as a function of the applied magnetic field $\mathbf{H} \parallel C_3$ for $H \ll H_{\mathrm{exch}}$ ($\lambda = 632.8$ nm).

more, projections $M_{1\perp}$ and $M_{2\perp}$ can only decrease. Even if we assume the tilted orientation of the AFM vector $\mathbf{L}$ in $CoCO_3$ [106, 107], then although we can expect in this case an increase of birefringence in the field caused by the rotation of vector $\mathbf{L}$ into the basal plane, variation $\Delta\eta(H)$ should be quadratic and considerably weaker [91]. But if we consider the table of mageto-optic coefficients of the LMOE (see Table 1.2), the behavior of the birefringence in $CoCO_3$ in field $\mathbf{H} \parallel C_3$ becomes qualitatively clear.

In the system of corrdinates $Z \parallel C_3$, $X \parallel C_2$, the matrix of LMOE coefficients $q_{ij\alpha}$ has the following form:

$$q_{ij\alpha}(\underline{2}/\underline{m}) = \begin{vmatrix} 0 & q_{xxy} & q_{xxz} \\ 0 & q_{yyy} & q_{yyz} \\ 0 & q_{zzy} & q_{zzz} \\ 0 & q_{yzy} & q_{yzz} \\ q_{xzy} & 0 & 0 \\ q_{xyx} & 0 & 0 \end{vmatrix} \tag{1.81}$$

The corresponding additions to the dielectric tensor components are equal to

$$\Delta\varepsilon^s_{ij}(\underline{2}/\underline{m}) = \begin{bmatrix} q_{xxz}H_z & 0 & 0 \\ 0 & q_{yyz}H_z & q_{yzz}H_z \\ 0 & q_{yzz}H_z & q_{zzz}H_z \end{bmatrix} \tag{1.82}$$

and the dependence of birefringence on the field can be expressed by the simple formula

$$n_2 - n_1 = n_g - n_m \approx (n_g - n_m)_0 + \frac{1}{2n_0}(q_{xxz} - q_{yyz})H_z. \quad (1.83)$$

Here $(n_g - n_m)_0$ is the spontaneous magnetic birefringence, and $n_0 \simeq \frac{1}{2}(n_g + n_m)$ is the refraction coefficient of an ordinary beam. Small linear rotations in H_z of the indicatrix axes around axis C_2 can be neglected, as relation $q_{yzz}H_z/(\varepsilon_{xx0} - \varepsilon_{zz0}) \lesssim 10^{-3}$ can be expected. Expression (1.82) describes the clearly observed linear dependence of $(n_g - n_m)$ on H_z in fields up to 5 T. The straight line slope at 5 K is equal to $3.8 \times 10^{-5}\ \text{T}^{-1}$ for light with $\lambda = 632.8$ nm. It yields a value of $2.25 \times 10^{-5}\ \text{T}^{-1}$ for the difference of magneto-optic coefficients $(q_{xxz} - q_{yyz})$.

But the given phenomenological explanation of the strong dependence of birefringence on the magnetic field requires changing the sign of the derivative $d(n_g - n_m)/dH_z$ when the field direction is reversed, which was not observed in the first experiments. The constancy of the birefringence variation sign can be connected with the remagnetization of all sublatice moments when the field direction is reversed. The remagnetization or reversal of the sublattice moment directions should take place because of the unambiguous relation between their z- and x-projections, which is caused by the interaction described by the invariant $f \cdot M_z[(L_x + L_y)^3 + (L_x - iL_y)^3]$ in the thermodynamic potential (1.80). The inversion of the directions of all elementary magnetic moments is adequate to the time-inversion operation for the magnetic medium and changes the signs of the magneto-optic coefficients $q_{ij\alpha}$. Also, the simultaneous inversion of signs of $q_{ij\alpha}$ and H_z preserves the effect sign.

The sublattice remagnetization process can be delayed by tilting the magnetic field so that the transverse component $\mathbf{H}_\perp$ is oriented in the required direction. The field transverse component orienting the WFM moment $\mathbf{M}_\perp$ unambiguously orients $\mathbf{L}$ too. The connection between $\mathbf{M}_\perp$ and $\mathbf{L}$ is ensured by the interaction described by the Dzyaloshinskii invariant $d(M_xL_y - M_yL_x)$ in the potential (1.80). In the presence of both field components, the competition of the interactions leads to the dependence of the azimuth angle of the AFM vector on the relation $H_z/H_\perp$. At a sufficiently high value of $\mathbf{H}_\perp$ the AFM vector can be to some extend fixed, and the reduction of birefringence $(n_g - n_m)$ when H_z change sign can be observed [111].

Figure 1.8(a) shows the dependence of the birefringence of the light propagating along C_3 on the strength of the field component $\mathbf{H}_z$, when $\mathbf{H}_\perp$ is almost perpendicular to the C_2 axis. The odd dependence of the birefringence on the sign of $\mathbf{H}_z$ is clearly seen. Also obvious is the dependence of the induced birefringence sign on the specimen AFM state. When the specimen is remagnetized, i.e., when the AFM vector changes its sign which is accomplished by reversing the $\mathbf{H}_\perp$ direction and at a fixed direction $\mathbf{H}_z$, the sign of the induced birefringence is changed.

In the case where the AFM vector is in the symmetry plane and the crystal magnetic symmetry is described by group $2/m$ the matrix of coefficients $q_{ij\alpha}$

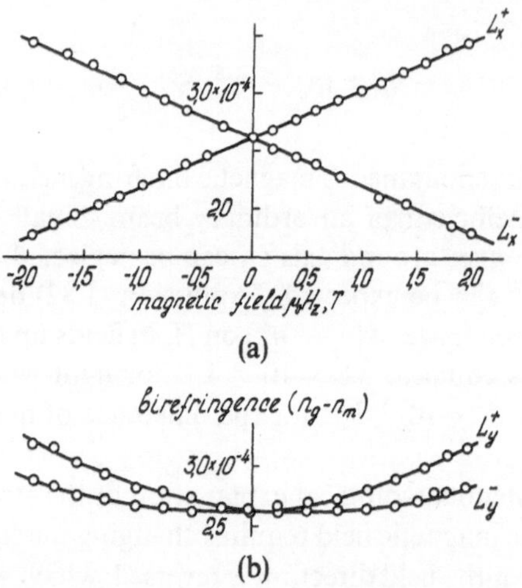

Fig. 1.8. Variation of the dependences of birefringence on the magnetic field strength in AFM $CoCO_3$ ($\mathbf{k} \parallel \mathbf{H} \parallel C_3 \parallel Z$) with the AFM vector $\mathbf{L}$ orientation change: (a) the AFM vector orientation is near the $C_2 \parallel X$-axis; (b) the AFM vector orientation is near the symmetry plane.

has the following form:

$$q_{ij\alpha}(2/m) = \left| \begin{matrix} q_{xxx} & 0 & 0 \\ q_{yyx} & 0 & 0 \\ q_{zzx} & 0 & 0 \\ q_{yzx} & 0 & 0 \\ 0 & q_{xzy} & q_{xzz} \\ 0 & q_{xyy} & q_{xyz} \end{matrix} \right| \tag{1.84}$$

Variations of the components ε_{ij}^s in field H_z are equal to

$$\Delta\varepsilon_{ij}^s(2/m) = \begin{bmatrix} 0 & q_{xyz}H_z & q_{xzz}H_z \\ q_{xyz}H_z & 0 & 0 \\ q_{xzz}H_z & 0 & 0 \end{bmatrix} \tag{1.85}$$

As in the AFM crystal $\varepsilon_{xx0}^s \neq \varepsilon_{yy0}^s$, the birefringence which appears in the field at $\mathbf{k} \parallel z$ nonlinearly depends on H_z. The rotation angle ζ is linear as to H_z of the indicatrix axes x and y around z:

$$\tan 2\zeta = \frac{q_{xyz}}{n_0(n_g - n_m)_0} H_z. \tag{1.86}$$

Here small rotations around directions $C_{\bar{2}}$ can again be neglected.

It is more difficult to fix the position of $\mathbf{L}$ near the symmetry plane in the presence of the field component H_z than in the plane which is perpendicular to it, as the unidirected (as to the AFM vector) anisotropy induced by H_z (member $fM_z[(L_x + iL_y)^3 + (L_x - iL_y)^3]$ in potential (1.80)) causes the (linear in

H_z) rotation of the magnetic vectors, which becomes substantial at weak H_z due to the low value of the anisotropy energy in the basal plane.

It is possible to delay to some extent the process of rotation by means of the field component $\mathbf{H}_\perp$ directed along axis C_2. Figure 1.8(b) shows the dependencies of $(n_g - n_m)$ on H_z when $\mathbf{L}$ is oriented near the symmetry plane. The qualitative difference in the birefringence behavior is evident. Even at the orientation error of $\mathbf{L}$, equal to about 4 degrees, the even character of the $(n_g - n_m)$ variations, when H_z changes sign and is close to a zero value of the derivative $d(n_g - n_m)/dH_z$ at point $H_z = 0$, can be seen. The small difference of the dependencies of $(n_g - n_m)$ on H_z obtained for two AFM states is connected with the error of the AFM vector orientation in the symmetry plane.

Figures 1.9(a), (b) illustrate the variation of the azimuth of the optical axes plane at the initial orientations of the AFM vector near axis C_2 and near the symmetry plane. The rotation of the plane of the optical axes is caused not only by the AFM vector rotation. Expect for the case when the AFM vector is

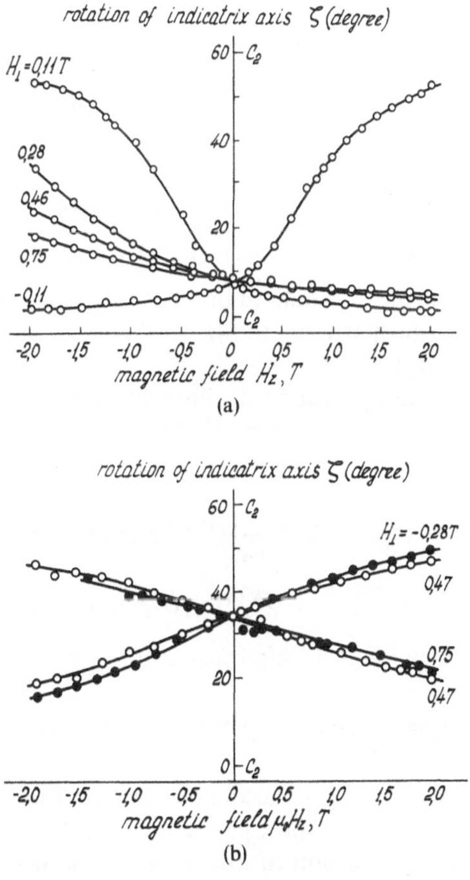

Fig. 1.9. Variation of the plane azimuth of optical axes in cobalt carbonate under an applied magnetic field $\mathbf{H} \parallel C_3 \parallel Z$: (a) initial orientation of the AFM vector near the C_2-axis; (b) initial orientation of the AFM vector near the symmetry plane.

oriented strictly along axis C_2, the plane of optical axes should also rotate at a fixed direction of the AFM vector. Figure 1.9(b) clearly shows the linear dependence of the azimuth of the plane of optical axes on the field H_z strength at orientation **L** near the symmetry plane.

The relation between the azimuth angles of the optical axes plane ξ and the AFM vector ϕ can be found by expressing ε^s_{ij} via the FM and AFM vector projections. According to Tables 1.4 and 1.5 the following expressions can be written for the additions:

$$
\begin{aligned}
\Delta\varepsilon^s_{xx} &= Q^{LL}_{11}L^2_x + Q^{LL}_{12}L^2_y + Q^{LL}_{13}L^2_z + 2Q^{LL}_{14}L_yL_z + Q^{ML}_{15}M_xL_z \\
&\quad + Q^{ML}_{16}M_xL_y + Q^{ML}_{18}M_zL_x + Q^{ML}_{19}M_yL_x, \\
\Delta\varepsilon^s_{yy} &= Q^{LL}_{12}L^2_x + Q^{LL}_{11}L^2_y + Q^{LL}_{13}L^2_z - 2Q^{LL}_{14}L_yL_z - Q^{ML}_{15}M_xL_z \\
&\quad - Q^{ML}_{19}M_xL_y - Q^{ML}_{18}M_zL_x - Q^{ML}_{16}M_yL_x, \\
\Delta\varepsilon^s_{zz} &= Q^{LL}_{13}(L^2_x + L^2_y) + Q^{LL}_{13}L^2_z + Q^{ML}_{36}(M_xL_y - M_yL_x), \qquad (1.87) \\
\Delta\varepsilon^s_{yz} &= Q^{LL}_{41}(L^2_x - L^2_y) + 2Q^{LL}_{44}L_yL_z + Q^{ML}_{45}M_xL_z + Q^{ML}_{46}(M_xL_y + M_yL_x) \\
&\quad + Q^{ML}_{48}M_zL_x, \\
\Delta\varepsilon^s_{xz} &= 2Q^{LL}_{41}L_xL_y + 2Q^{LL}_{44}L_xL_z + Q^{ML}_{46}(M_yL_y - M_xL_x) \\
&\quad - Q^{ML}_{45}M_yL_z - Q^{ML}_{48}M_zL_y, \\
\Delta\varepsilon^s_{xy} &= 2Q^{LL}_{14}L_xL_z + (Q^{LL}_{11} - Q_{12})^{LL}L_xL_y + Q^{ML}_{15}M_yL_z \\
&\quad + \tfrac{1}{2}(Q^{ML}_{16} + Q^{ML}_{19})(M_yL_y - M_xL_x) + Q^{ML}_{18}M_zL_y.
\end{aligned}
$$

Here the small terms proportional to product M_iM_j are not noted. They are similar to ther terms proportional to L_iL_j but smaller than them by an order of magnitude of $(H/H_E)^2$.

If we diagonalize matrix ε_{ij} and substitute the values of projections M_i and L_j, obtained by minimizing potential (1.80), we can write the approximate linear form in the H_i expression for the birefringence value

$$
\begin{aligned}
n_g - n_m = (n_g - n_m)_0 &+ \frac{1}{n_0}[Q^{ML}_{18}L\chi_{zz}H_z\cos 3\phi \\
&+ \tfrac{1}{2}(Q^{ML}_{16} + Q^{ML}_{19})L\chi_\perp(H_y\cos\phi + H_x\sin\phi)], \qquad (1.88)
\end{aligned}
$$

where the spontaneous magnetic birefringence $(n_g - n_m)_0$ is equal to

$$
(n_g - n_m)_0 = \frac{1}{2\eta_0}[(Q^{LL}_{11} - Q^{LL}_{12})L^2 - (Q^{ML}_{16} + Q^{ML}_{19})LM_\perp]. \qquad (1.89)
$$

Equations (1.88) and (1.89) do not account for the contributions caused by the small WFM moment $M_\parallel$ and the z-projection of the AFM vector. The azimuth angle of the plane of optical axes ξ, the angle between the plane of optical axes, and the vector $\mathbf{L}_\perp(\xi - \phi)$ are determined to the same approximation by the expressions

$$\tan 2\xi = \{2n_0(n_g - n_m)_0 \sin 2\phi - [\alpha\chi_{zz}H_z - \beta(\chi_\perp H_y - M_{\perp 0})] \sin\phi$$
$$+ \beta(\chi_\perp H_x + M_{\perp 0}) \cos\phi\}/\{2n_0(n_g - n_m)_0 \cos 2\phi$$
$$+ [\alpha\chi_{zz}H_z + \beta(\chi_\perp H_y - M_{\perp 0})] \cos\phi$$
$$- \beta(\chi_\perp H_x + M_{\perp 0}) \sin\phi\}, \quad (1.90)$$

$$\tan 2(\xi - \phi) = \{\alpha\chi_{zz}H_z \sin 3\phi + \beta[(\chi_\perp H_y - M_{\perp 0}) \sin\phi$$
$$- (\chi_\perp H_x + M_{\perp 0}) \cos\phi\}/\{2n_0(n_g - n_m)_0 + \alpha\chi_{zz}H_z \cos 3\phi$$
$$+ \beta[(\chi_\perp H_y - M_{\perp 0}) \cos\phi - (\chi_\perp H_x + M_{\perp 0}) \sin\phi]\}, \quad (1.91)$$

where

$$\alpha = 2Q_{18}^{ML}L, \qquad \beta = (Q_{16}^{ML} + Q_{19}^{ML})L.$$

Equation (1.91) shows that, due to the noncollinearity of the sublattices, the plane of the optical axes does not pass through the AFM vector even in the absence of a magnetic field. However, in magnetic fields perpendicular to the C_3 axis, the exit of the optical axes from the plane perpendicular to the field has not been observed experimentally, which indicates the small value of the magneto-optic effect $(Q_{16}^{ML} + Q_{19}^{ML})$.

The experimental data on the direction of the optical axes can be used to study the behavior of the cobalt carbonate magnetic subsystem in an arbitrary directed magnetic field, and to determine the values of the weak interactions which define the noncollinearity and orientation of the crystal sublattice magnetic moments. So, the above-mentioned results make it possible to evaluate the ratio of the coefficients in the thermodynamic potential, which characterize the value of the interaction responsible for the WFM moment $M_\parallel = f \cdot L^3\chi_{zz}$ and the hexagonal anisotropy energy in the basal plane $E_6 = \frac{1}{6}eL^6$. By restricting to small angles $(\phi - \pi/2)$ and $(\zeta - \phi)$ from (1.91) and assuming $\beta = 0$, we obtain

$$\zeta = \phi + \frac{Q_{18}^{ML}L\chi_{zz}H_z}{2n_0(n_g - n_m)_0}. \quad (1.92)$$

And according to (1.88)

$$Q_{18}^{ML}L\chi_{zz}H_z/n_0 = [(n_g - n_m) - (n_g - n_m)_0]_{L\parallel C_3} \quad (1.93)$$

we obtain from these two expressions

$$\frac{d\phi}{dH_z} + \frac{d\zeta}{dH_z} + \frac{d(n_g - n_m)}{2(n_g - n_m)_0 dH_z}. \quad (1.94)$$

On the other hand, dependence of ϕ and H_z is determined by the relation of constangs f and e in potential (1.80), and at small H_z is close to

$$\phi = \frac{\pi}{2} + \frac{f}{3e} \cdot \frac{\chi_{zz}H_z}{2M_0}. \quad (1.95)$$

Finally, relation f/e is determined by means of the variations of the azimuth of the plane of optical axes and the variation of the birefringence value

$$\frac{f}{e} = \frac{6M_0}{\chi_{zz}}\left(\frac{d\zeta}{dH_z} - \frac{d(n_g - n_m)}{2(n_g - n_m)_0 dH_z}\right). \tag{1.96}$$

From the relations shown in Figs. 1.8(a) and 1.9(b) we obtain

$$(d\zeta/dH_z)_{H_\perp \to 0, H_z \to 0} = 12\ \text{degree} \cdot \text{T}^{-1} = 0.21\ \text{T}^{-1}$$

and

$$(d(n_g - n_m)/dH_z)_{L \parallel C_2, H_z \to 0} = 3.8 \cdot 10^{-5}\ \text{T}, \qquad (n_g - n_m)_0 = 2.4 \cdot 10^{-4}.$$

Substituting these values and the values $M_0 = 8.37 \times 10^3$ emu/mol and $\chi_{zz} = 3.5 \times 10^{-2}$ emu/mol [102] into (1.96), we obtain $f/e = 43$. Values f and e can be evaluated as we know the WFM moment $M_\parallel = 7$ emu/mol measured in [104, 105] and the hexagonal anizotropy energy $E_6 = 660$ erg/cm^3 [103]. They yield $f = -3M_0 M_\parallel/\chi_{zz} = -1.3 \times 10^5$ erg/cm^3 and $e = 6 \cdot E_6 = 4 \times 10^3$. The value obtained therefrom, $|f/e| = 35$, agrees with the previous one by the same order of magnitude.

Studying the symmetry of the ligand surrounding the ions $Co^{2+}1$ and $Co^{2+}2$, which form sublattices 1 and 2, we can arrive at some conclusions as to the possible mechanism of the LMOE appearance in $CoCO_3$. Figure 1.10 shows the mutual orientation of the deformed oxygen octahedrons surrounding ions $Co^{2+}1$ and $Co^{2+}2$. The figure also shows the directions of the ion magnetic moments for the symmetrical states $2/\underline{m}$ and $2/m$, and also one of the crystallographic symmetry axes C_2. It can be seen from the figures that when the vectors $\mathbf{M}_1$ and $\mathbf{M}_2$ are rotated toward axis C_3 in the backward and forward directions from the figure plane, those states are realized which differ by mutual arrangement of the magnetic moments and ions O^{2-}. These states differ only by the direction of path-tracing in the basal plane if the initial configuration corresponds to symmetry $2/m$. This character of the rotation inequivalency of the moments $\mathbf{M}_1$ and $\mathbf{M}_2$ "up" and "down" from axis C_3 dictates the necessity of the rotation of the axes of the ion polarizability tensor, and of the axes of the dielectric tensor in the opposite directions at $H \uparrow\uparrow C_3$ and $H \uparrow\downarrow C_3$, even if the moments temselves do not rotate in the basal plane. If the initial location of the moments corresponds to symmetry $\underline{2}/\underline{m}$,

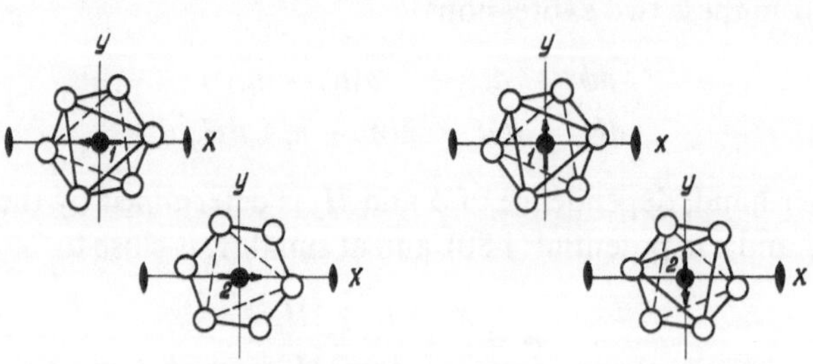

Fig. 1.10. Orientation of magnetic moments of Co_1^{2+} and C_2^{2+} ions with respect to $CoCO_3$ ligand ions. The magnetic symmetry of the crystal states is: (a) $\underline{2}/\underline{m}$; (b) 2/m.

then the states created after the rotation of moments $\mathbf{M}_1$ and $\mathbf{M}_2$ towards axis C_3 in the "up" and "down" directions differ more substantially. At the rotations of the moments "down" (beyond the figure plane) both vectors move towards the ligand ion. At the rotations of the moments "up" (from the figure plane) vectors $\mathbf{M}_1$ and $\mathbf{M}_2$ are directed between the ligand ions. It is obvious that, due to the spin-orbit coupling and the crystallic field action, the polarizability of ions together with their ligand surroundings should oddly depend on the direction of the magnetic field along the crystal trigonal axis. Thus, the LMOE manifestation can be explained in the framework of the single-ion approximation. The microscopic model of the investigated mageto-optic effect is not yet built, although attempts were made [78, 79] to evaluate the role of the quantum mechanical mixing of the electron states of ions Co^{2+} caused by the field $\mathbf{H} \parallel C_3$ in the observed variation of the birefringence.

1.4.2. Reduction of the Optical Class of Tetragonal Antiferromagnetic Cobalt Fluoride in a Longitudinal Magnetic Field

Most, evidently, the LMOE was manifested in a tetragonal AFM crystal of cobalt fluoride. Cobalt fluoride is a well-studied two-sublattice collinear AFM with pronounced piezomagnetic properties [94]. Its magnetic space group is $P\underline{4}_2/mm\underline{m}$ [112] and its point group is $\underline{4}/mm\underline{m}$. The elementary crystallographic cell contains two magnetic ions, the moments of which are directed antiparallel to one another along the crystal tetragonal axis $C_{\bar{4}}$. Below temperature T_N the tetragonal axis is the odd axis $C_{\bar{4}}$. It can be the symmetry axis only in combination with the time-inversion operation, as it requires the sublattice remagnetization. It is obvious that the symmetry operation $C_{\bar{4}}$ is lost when the crystal is placed in the magnetic field $\mathbf{H} \parallel C_{\bar{4}}$. In this case, the crystal should acquire the optical properties of a rhombic crystal. The reduction of the AFM CoF_2 optical class can be clearly seen in the experiment [92, 110] and is illustrated by the conoscopic figures (Fig. 1.11). The appearance of birefringence of the linearly polarized light propagating along axis [001] can be described by the magneto-optic tensor $q_{ij\alpha}$ using the following matrix (see Table 1.2):

$$q_{ij\alpha} = \begin{vmatrix} \cdot & \cdot & \cdot \\ \cdot & \cdot & \cdot \\ \cdot & \cdot & \cdot \\ q_{yzx} & \cdot & \cdot \\ \cdot & q_{yzx} & \cdot \\ \cdot & \cdot & q_{xyz} \end{vmatrix}. \tag{1.97}$$

The linear-in-field additions to the symmetrical part of the dielectric tensor can be written as follows:

$$\Delta\varepsilon_{yz}^s = q_{yzx}H_x, \qquad \Delta\varepsilon_{xz}^s = q_{yzx}H_x, \qquad \Delta\varepsilon_{xy}^s = q_{xyz}H_z. \tag{1.98}$$

Taking into account the fact that the natural crystal birefringence far exceeds magneto-optic birefringence, the variation of the crystal optical indicatrix in fields $\mathbf{H} \parallel X$ and $\mathbf{H} \parallel Y$ can be described by the rotation of their axes around the direction of $\mathbf{H}$ and by a small quadratic in H variations of the length of its semiaxes. In field $\mathbf{H} \parallel Z$ the positions of the indicatrix axes do not depend on the field value. The indicatrix axes in the magnetic field, which does not disturb the collinearity of the moments $\mathbf{M}_1$ and $\mathbf{M}_2$, should be directed along the crystallographic directions $[110] \parallel X'$ and $[1\bar{1}0] \parallel Y'$. The changing of the $\Delta\varepsilon^s_{xy}$ sign and the directions of the indicatrix axes should correspond to the changing of the q_{xyz} sign when the directions of moments $\mathbf{M}_1$ and $\mathbf{M}_2$ are reversed. The indicatrix axes n_1 and n_2 should exchange places and the birefringence for the light propagating along [001] should change its sign

$$(n_{x'} - n_{y'}) = \frac{1}{2n_0} q_{xyz} H_z. \tag{1.99}$$

When the orientation of the indicatrix axes is changed, the plane of optical axes are reoriented by 90 degrees. Similar variations of the crystallographic properties should also take place at fixed directions of the sublattice moments, but only when the direction of the magnetic field is reversed. The reorientation of conoscopic figures, when the field direction is reversed and during the AFM remagnetization, can be seen clearly in the photographs (Figs. 1.11 and 1.12). The energy nonequivalence of the AFM$^+$ and AFM$^-$ states in field $\mathbf{H} \parallel [001]$ was caused by spontaneous stresses in the crystal (which occur during growth) and by the directed elastic strains in direction [110]. The conoscopic figure corresponding to the remagnetization (Fig. 1.12, photograph 3) shows the apparent AFM nonuniformity in the specimen—its division into AFM$^+$ and AFM$^-$ domains. The metastable AFM state remains stable in fields up to 6, 10 T at elastic stresses of about 30 kg/cm^2 and 20 kg/cm^2, respectively.

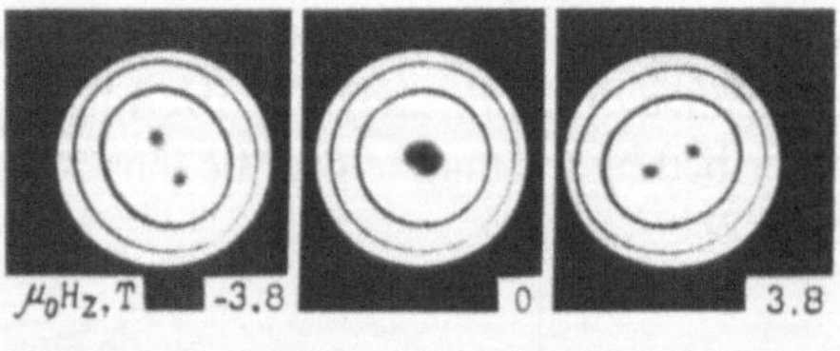

Fig. 1.11. Optical biaxiality in the tetragonal crystal of AFM CoF_2 induced by a longitudinal magnetic field $\mathbf{H} \parallel C_4 \parallel \mathbf{L}$.

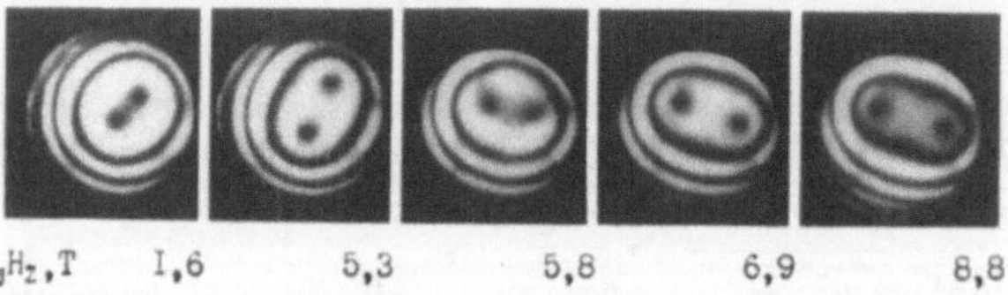

Fig. 1.12. Reorientation of the optical axes arrangement under the AFM remagnetization of cobalt fluoride, $\mathbf{H} \parallel C_4 \parallel \mathbf{L}$.

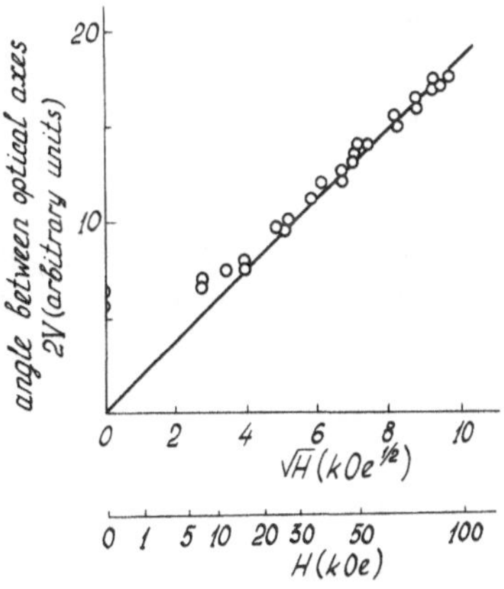

Fig. 1.13. The angle between the optical axes in AFM CoF_2 as a function of the magnetic field strength $\mathbf{H} \parallel C_4$. Elastic stresses in the crystal are about 20 kg/cm^2.

The dependence of the angle between the optical axes $2V$ on H_z approaches the square-root one with constant $2V/H_z^{1/2} = 2.21\ \mathrm{deg} \cdot \mathrm{T}^{-1/2}$ (see Fig. 1.13). The birefringence dependence $\Delta n_{x'y'} = n_{x'} - n_{y'}$ for two AFM states restored from the conoscopic figures is shown in Fig. 1.14. The experimental points fall on the straight lines with tilt $\Delta n_{x'y'}/H_z = 8.2 \times 10^{-6}\ \mathrm{T}^{-1}$. The magneto-optic coefficient q_{xyz} determined for them is close to $2.5 \times 10^{-5}\ \mathrm{T}^{-1}$ for light with a wavelength $\lambda \approx 400$ nm and $T = 11$ K.

The linear-in-H_z dependence of the light birefringence has been observed in AFM CoF_2 at the transverse geometry of the experiment for $\mathbf{k} \parallel [110]$ and $\mathbf{H} \parallel [001]$ [113]. The domain structure in these experiments was not con-

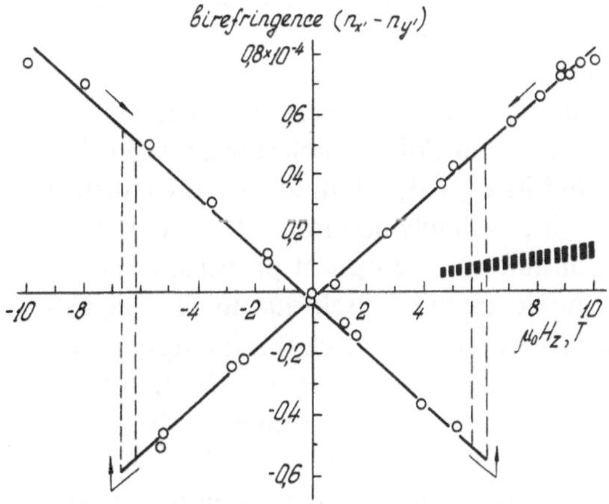

Fig. 1.14. Birefringence of linearly polarized light propagating along the tetragonal axis in AFM CoF_2 induced by the applied magnetic field $\mathbf{H} \parallel C_4$. Two straight lines indicate two AFM states. The shaded section is a contribution to birefringence due to linear magnetostriction and the photoelastic effect.

trolled, but the greatest of the observed tilt values $d(\Delta n_{x'z'})/dH_z$ yields, for q_{xyz}, the value $2.1 \times 10^{-5}\,T^{-1}$, which is close to the above-mentioned value.

It is obvious that when the AFM order vanishes, the LMOE should vanish too. Even more, at the ordering temperature it should vanish in contrast to the spontaneous effect of magnetic birefringence which gradually vanishes at temperature above T_N. The inevitability of vanishing of the effect at T_N can be easily seen if we express the variations of the tensor ε_{ij}^s components through the AFM and FM vector projections L_i and M_i. According to Table 1.2, we can write the following expressions for the linear-in-**H** additions to the dielectric tensor components:

$$\begin{aligned}\Delta\varepsilon_{yz} &= Q^{ML}_{yzxz} M_x L_{z0} + Q^{LL}_{yzyz} L_y L_{z0},\\ \Delta\varepsilon_{xz} &= Q^{ML}_{xzyz} M_y L_{z0} + Q^{LL}_{yzyz} L_x L_{z0},\\ \Delta\varepsilon_{xy} &= Q^{ML}_{xyzz} M_z L_{z0}.\end{aligned} \tag{1.100}$$

Equation (1.100) takes into account that the only nonzero spontaneous component of the AFM vector is projection L_{z0}. Projections L_x and L_y can be induced by the external field directed along axes y and x, respectively, due to the interactions described by the invariant $d(M_xL_y + M_yL_x)$ in the thermodynamic potential of this crystal [102]. The LMOE-induced rotations of the indicatrix proportional to $\Delta\varepsilon_{yz}$ and $\Delta\varepsilon_{xz}$, and the birefringence along axis [001]

$$\Delta n_{x'y'} = \frac{1}{n} Q^{ML}_{xyzz} \chi^{M}_{zz} |L_z \cdot H_z| \operatorname{sign} H_z \cdot \operatorname{sign} L_z, \tag{1.101}$$

are linear in L_z. When the long-range magnetic order vanishes in T_N, not only the AFM vector value fluctuates, but also the sign of its projection. As the fluctuations of L_z of both signs are equally possible, the birefringence $\Delta\eta_{x'y'}$ characterizing the marcroscopic property should vanish when the average value $\langle L_z \rangle$ vanishes. This property of the LMOE is experimentally confirmed and illustrated in Fig. 1.15.

Studying various mechanisms generating the LMOE we, first of all, should note that the AFM crystal CoF_2 is subject to considerable deformation in the field $\mathbf{H} \parallel [001]$ and its crystal cell attains rhombic distortions [114]. Therefore, even if there is absolutely no direct influence of the field on the electron transitions responsible for the optical properties, the birefringence of light along axis C_4 should manifest itself due to the magnetostriction rhombic distortions and the photoelastic effect. It is interesting to extract the so-called primary contribution from the observed LMOE value. This contribution should be observed even in a fixed specimen in the absence of magnetostriction deformations. To do this, the photoelastic coefficient π_{66} has been measured during squeezing of the specimen along the direction [110]. It turned out to be dependent, to a very small extent, on a temperature in the range from room to liquid helium temperatures and is close to the value 0.95×10^{-13} din cm^{-2} for $\lambda = 632.8$ nm. It weakly disperses towards the blue region within the limits of 10 per cent error, caused by the difficulty of obtaining the

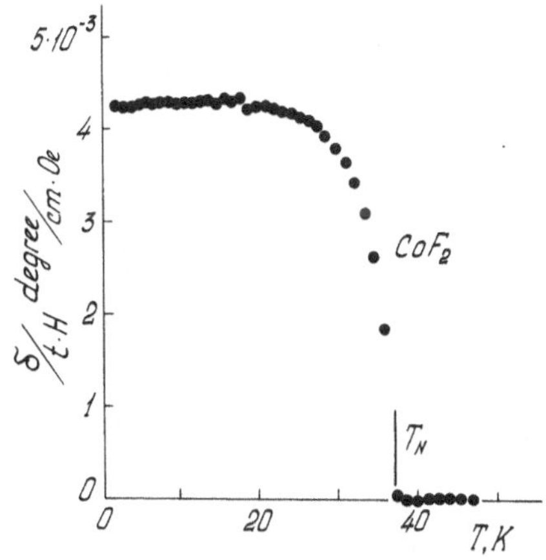

Fig. 1.15. Temperature dependence of field-induced linear birefringence in CoF_2 (**H** ∥ C_4, $\lambda = 632.8$ nm).

uniform deformation. Knowing the magnetostriction deformations along the [110]-direction $(\Delta l/l)_{xy}H^{-1} = 4.9 \times 10^{-6}\ T^{-1}$ [114] and the elastic constant $C_{66} = 8.5 \times 10^{11}$ din cm^{-2} [115], we can find the contribution of the secondary photoelastic effect to the linear birefringence

$$(\Delta n_{x'y'})_{ph}/H = 4.9 \cdot 10^{-6}\, n_0^3 \cdot \pi_{66} \cdot C_{66}.$$

The shaded area in Fig. 1.14 shows the contribution of the photoelastic effect, taking into account possible errors of evaluation. It can be seen that the tetragonal structure distortions are not the decisive ones for the LMOE. The primary effect comprises at least 70 per cent of the observed one.

When studying the mechanisms of the LMOE appearance, it should be underlined that all of them should be connected to the low-symmetry surrounding of ions Co^{2+} by ions F^-.

The deformed ligand octahedrons CoF_2 have rhombic symmetry. Their second-order axes X1, Y1 and X2, Y2 are mutually perpendicular for ions $Co^{2+}1$ and $Co^{2+}2$ of the first and second sublattices. Ions $Co^{2+}1$ and $Co^{2+}2$, together with their ligand surrounding, are low-symmetry subsystems the birefringence of which along the crystal axis [001] exactly compensate each other. But as soon as the equivalence of ions $Co^{2+}1$ and $Co^{2+}2$ is disturbed, the birefringence compensation is disturbed too. Axis [001] ceases to be the optical axis. The magnetic field **H** ∥ [001] is parallel to the ion magnetic moments of one sublattice and antiparallel to the moments of the second one. It disturbs the equivalence of the sublattices. The energy levels of ions $Co^{2+}1$ and $Co^{2+}2$ are shifted in opposite directions. The different values of shifts in the ground and excited states lead to a difference of frequencies of the electron transitions of one and the same type in ions Co^{2+} from different sublattices. Besides, due to the change in the energy distances between the ground and the nearest excited levels in ions $Co^{2+}1$ and $Co^{2+}2$ in a magnetic field, the complex wave functions and spin values of $Co^{2+}1$ and $Co^{2+}2$ will change [116].

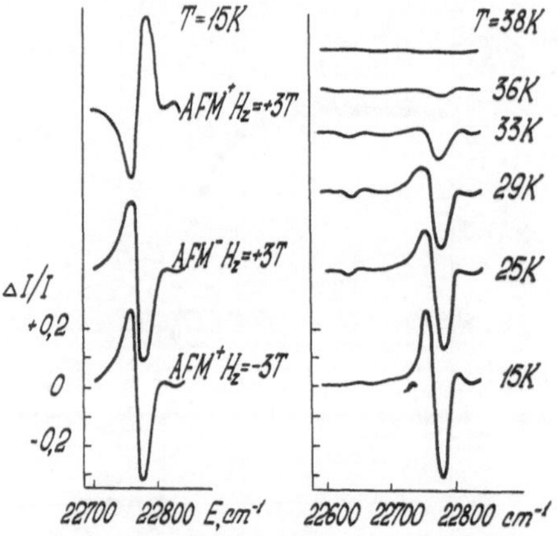

Fig. 1.16. Magnetic linear dichroism in the exciton–magnon absorption.

The changing of states will lead to various changes in the transition oscillator forces in both types of ions. Both mechanisms, causing the sublattice non-equivalence, disturb the compensation of the birefringence properties of the crystal magnetic sublattices. Paer [80, 81, 117] evaluated the possible role of the quantum-mechanical mixing of states in the linear birefringence induced by a magnetic field.

Figure 1.16 shows the dichroism spectrum of the linearly polarized light propagating along the direction [001] in the region of transitions ${}^4\Gamma_4^+ \rightarrow {}^2\Gamma_1^+(2G)$, ${}^2\Gamma_4^+(2P)$ [118]. It also shows the absorption spectrum. All the observed bands are complex ones and are connected to simultaneous excitation, besides the exciton, of other low-energy quasi particles—phonons, magnons, and excitons. The identification of all transitions is difficult, but band 22,769 cm^{-1} is identified as the exciton–magnon absorption [119]. Then the S-shape of all single bands is characteristic of the linear magnetic dichroism spectrum, This shape of dispersion indicates that the dichroism is caused, in the first place, by the band splitting. It is possible that the small asymmetry of the dichroism curves is connected not only with the exciton–magnon band asymmetry, but also with the changing of the electron transition probabilities in ions $Co^{2+}1$ and $Co^{2+}2$, due to the different mixing of quantum-mechanical states in them [81].

The given region of the spectra reflects the behavior inside the 3*d*-transitions which slightly affect the birefringent properties in the region of transparence. It is impossible to prefer one of these two mechanisms for the fundamental absorption dichroism and birefringence in the ion transparence region. But, as it would appear, the greater value of the effect in CoF_2 is connected with the presence of the orbital moment in the ground state of ion Co^{2+}. Let us note that the LMOE has also been observed in FeF_2 where its value is substantially smaller. Attempts to record it in MnF_2 in the transparent region have failed, but the dichroism in the region 3*d*—transitiòns connected with the LMOE can be measured easily.

1.4.3. Orthorhombic Noncollinear (Canted) Antiferromagnetic $DyFeO_3$

All orthorhombic magnetic crystals, whose chemical and magnetic cells coincide, can be divided into three groups as to the possibility of the LMOE manifestation. In the first group, which comprises symmetrical as well as anti-inversion operation crystals with magnetic point groups $\underline{m}\underline{m}\underline{m}$ and $\underline{m}mm$, the LMOE is symmetrically forbidden. In the second group (compensated AFM of classes 222, $mm2$, and mmm) it causes the optical indicatrix rotation around direction **H** if the field is applied in parallel to the symmetry axes. In the third group of orthorhombic magnetic crystals, the LMOE can cause either the indicatrix rotations around the directions perpendicular to **H** at symmetrical orientations of the field, or the indicatrix deformation. This group comprises pyromagnetic cyrstals of center-symmetrical $m\underline{m}\underline{m}$ and non-centrosymmetrical $\underline{m}\underline{m}2$, $\underline{m}m\underline{2}$, and $2\underline{2}\underline{2}$ classes.

The LMOE was studied in orthorhombic dysprosium orthoferrite which was in the compensated AFM state. The magnetic symmetry of this crystal does not remain constant with temperature. At low temperatures, $T < T_{N_2} \approx 2$ K, its point group is not truly known, but there are reasons to consider that the crystal belongs to class $mm\underline{m}$ or class $\underline{m}\underline{m}\underline{m}$ [120], in which the LMOE in forbidden. At higher temperatures, $T_{N_2} < T < T_M \approx 50$ K, it is a compensated AFM with magnetic point group mmm. At temperatures above the Morin point, $T_M < T < T_{N_1}$, its magnetic structure is not completely compensated and is described by group $\underline{m}\underline{m}m$. The transition from the AFM structure to the WFM structure can also take place under the action of a magnetic field. Table 1.2 shows that the LMOE manifestations in the AFM and WFM states of $DyFeO_3$ should be qualitatively different. For the AFM class mmm, coefficients q_{yzx}, q_{xzy}, q_{xyz} differ from zero and the LMOE should lead, first of all, to the indicatrix rotation around the direction of **H** applied along the symmetry axes. For class $\underline{m}\underline{m}m$, q_{xxz}, q_{yyz}, $q_{\underline{z}zx}$, q_{xzx}, q_{yzy} differ from zero and, consequently, in the WFM state at $\mathbf{H} \parallel Z \parallel \bar{2}$ only the lengths of the indicatrix axes vary, and at $\mathbf{H} \parallel X$ or $\mathbf{H} \parallel Y$ the indicatrix rotations will take place, but only around the directions perpendicular to **H**.

Experiments $DyFeO_3$ were also conducted in the longitudinal geometry of experiment $\mathbf{H} \parallel \mathbf{k}$, and the magnetic field was direction along axis $c \parallel z$ [93]. At this orientation of the field the indicatrix axes remain rigidly connected to axes a and b in the WFM state and rotate linearly in **H** in the AFM state. Low transparency of the crystal in the visible part of the spectrum and a great angle between the optical axes did not permit us to use the evident conoscopic method to observe rotations of the indicatrix axes. Besides, small expected values of the indicatrix rotation angles of the order of 10^{-2} radians required detailed analysis of the light polarization variations in a dysprosium orthoferrite plate, as well as a sensitive modulation experimental method.

Figure 1.17 qualitatively shows the light polarization transformation in the AFM $DyFeO_3$ plate in geometry $\mathbf{k} \parallel c$, and $\mathbf{H} \parallel C$ on the Poincaré sphere. Point P corresponds to the linear polarization of the incident light, and points

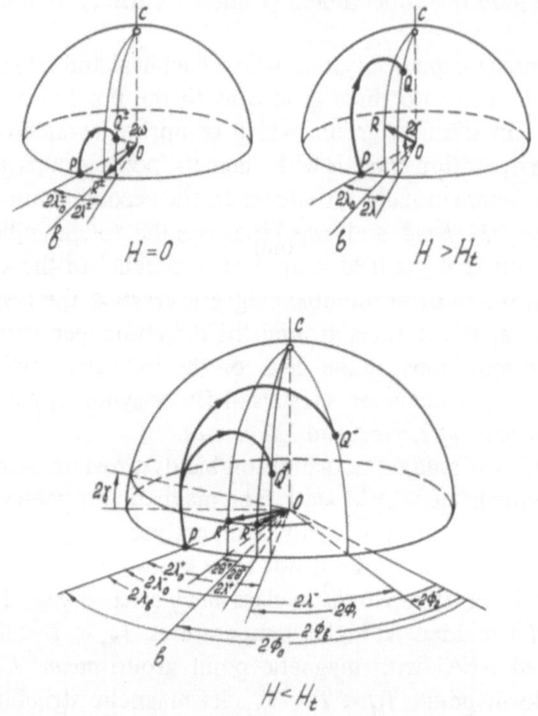

Fig. 1.17. Variations of light polarization in the AFM dysprosium orthoferrite plate for **k** ∥ c ∥ **H** illustrated by a Poincaré sphere.

Q^+ and Q^- to the polarization of the light passing through the AFM$^+$ and AFM$^-$ domains. At $H = 0$ the indicatrix axes X_1 and X_2 are oriented along crystal axes a and b, the Faraday effect is absent, and axis OR is oriented along the fast light wave polarization direction coinciding with axis b. In a magnetic field axis OR leaves the equatorial plane due to the increase of the Faraday rotation, the leaves plane (bc) as well, due to the rotation of the indicatrix axes around axis c. As the indicatrix rotations in opposite directions in the AFM$^+$ and AFM$^-$ states as the result of the LMOE action, the orientations of axis OR for two AFM states are different. In the figure they are denoted by OR^+ and OR^-. Their azimuth angles counted from axis b are equal, by the absolute values,

$$\theta^+ = -\theta^- = \theta = \tfrac{1}{2} \arctan |q_{xyz}H_z|/(\bar{n}_0 \Delta n_{12}^0). \tag{1.102}$$

Here $\bar{n}_0 = \frac{1}{2}(n_{10} + n_{20})$, $\Delta n_{12}^0 = n_{10} - n_{20}$. It is obvious, that OR^+ and OR^- change places when the sign of H_z is changed. Axes OR^+ and OR^- make equal angles with the sphere equatorial plane

$$2\gamma^+ = 2\gamma^- = \arctan(2\rho/\Delta) \tag{1.103}$$

as the Faraday rotation ρ is equal for both AFM states, and the phase shift δ for linearly polarized light in the absence of the Faraday rotation in this geometry of experiment is also independent on the AFM state

$$\delta^+ = \delta^- = \delta = \delta_0 + \frac{\pi(q_{xyz}H_z)^2}{\lambda \bar{n}_0 \Delta n_{120}} \approx \delta_0. \tag{1.104}$$

The polarization of the passed light (points Q^+ and Q^- on the sphere) depends on the AFM state of the plate. The polarization parameters of the light passing through the specimen, which is alternatively in the AFM$^+$ and AFM$^-$ states—azimuth of the polarization ellipse axis λ and the ellipticity $e = a/b = \tan \omega$—can be found from expressions (1.48) and (1.49). The polarization axis azimuths in two AFM states are different because their optical indicatrix axes, from which the azimuth angles are measured, do not coincide. Taking into account that the natural crystal linear birefringence in orthoferrites far exceeds the magnetic circular birefringence [121, 122] and that the expected angle θ is small, the following expressions can be obtained for the azimuth angles of the polarization ellipse axes and also the ellipticity values of polarization of the light passing through the specimen in the AFM$^+$ and AFM$^-$ states from (1.48) and (1.49), valid for small azimuth angles λ_b:

$$\lambda^{\pm} = (\lambda_b \mp \theta)(1 - \cos \delta) + \rho \sin \delta/\delta, \tag{1.105a}$$

$$e^{\pm} = (\lambda_b \mp \theta) \sin \delta + \rho(1 - \cos \delta)/\delta. \tag{1.105b}$$

Here λ_b is the polarization azimuth of the incident light measured from axis b in the crystal. Equations (1.105(a), (b)) show that both the Faraday effect and the LMOE change the azimuth and ellipticity of the light polarization. The Faraday effect contribution can be eliminated by taking the difference $(\lambda^+ - \lambda^-)$ or $(e^+ - e^-)$, i.e., by sequentially measuring the light polarization states after it has passed through the specimen in the AFM$^-$ state, and then in the AFM$^-$ state.

The existence, in the dysprosium orthoferrite plates being investigated, of several sufficiently large AFM domains stable in a wide range of fields made it possible to separate the LMOE and FE contributions into the observed rotation of the light polarization ellipse axis in a magnetic field [93]. Measurements were taken in two specimen regions 1 and 2 with a diameter of about 60 μm located in the AFM$^+$ and AFM$^-$ domains, respectively. The average size of the domains was about 400 μm. By means of an objective, the specimen was projected onto an opaque screen with three holes, two of which limit the specimen regions being investigated (light beams 1 and 2), and the third hole made it possible to pass the light beam (beam 0) outside the specimen. The beam separation scheme is shown in Fig. 1.18. As the domain structure of AFM $DyFeO_3$ did not substantially change during the experiment, the domain position in respect to the selected regions was periodically visually checked by introducing into the optical setup a pivoting mirror directing the light into a visual observation system. Light beams 0, 1, and 2 were sequen-

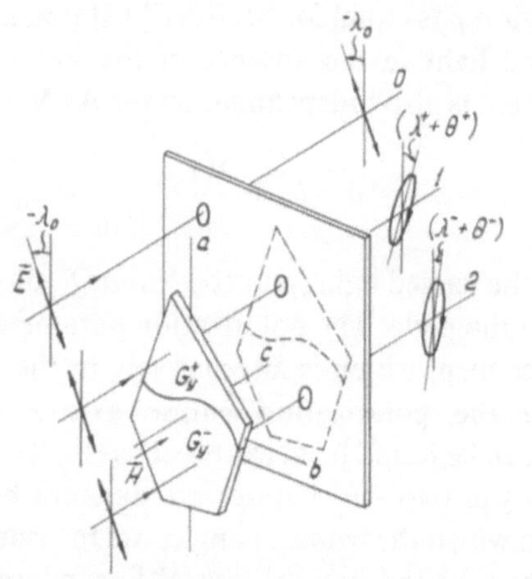

Fig. 1.18. Diagram of magneto-optical measurements in AFM $DyFeO_3$.

tially directed into the measuring scheme comprising a light modulator in the polarization azimuth, analyzer, and PEM with a synchronous detector. The measurement error of the polarization ellipse major axis azimuth did not exceed $\pm 1'$. In order to increase the measurement accuracy, the azimuthes of the polarization axis of the incident and passing light were measured in one system of stationary coordinates. In this frame of reference, the azimuth of the plane of polarization of the incident light is equal to

$$\Phi_0 = \Phi_{10} + \theta^+ + \lambda_0^+ = \Phi_{20} + \theta^- + \lambda_0^- . \tag{1.106}$$

Here Φ_{10} and Φ_{20} are the azimuthes of the principal planes in the specimen regions 1 and 2 in the absence of the field. Naturally, in an ideal case they should coincide. The azimuthes of the polarization axes of the passing light are equal to, for beams 1 and 2, respectively,

$$\begin{aligned} \Phi_1 &= \Phi_{10} + \theta^+ + \lambda^+, \\ \Phi_2 &= \Phi_{20} + \theta^- + \lambda^- . \end{aligned} \tag{1.107}$$

Taking into account (1.105a), we obtain

$$\begin{aligned} \Phi_1 &= (\Phi_0 - \Phi_{10}) \cos \delta^+ + \Phi_{10} + \theta^+(1 - \cos \delta^+) + \gamma^+ \sin \delta^+, \\ \Phi_2 &= (\Phi_0 - \Phi_{20}) \cos \delta^- + \Phi_{20} + \theta^-(1 - \cos \delta^-) + \gamma^- \sin \delta^- . \end{aligned} \tag{1.108}$$

In the ideal case, besides $\Phi_{10} = \Phi_{20}$, $\delta^+ = \delta^-$ also. The expressions for the angles of rotation of the polarization ellipse axes in the AFM^+ and AFM^- domains can be written as follows:

$$\begin{aligned} \Delta\Phi_1(H) &= \Phi_1(H) - \Phi_1(O) = \theta^+(1 - \cos \delta^+) + \gamma^+ \sin \delta^-, \\ \Delta\Phi_2(H) &= \Phi_2(H) - \Phi_2(O) = \theta^-(1 - \cos \delta^-) + \gamma^- \sin \delta^- . \end{aligned} \tag{1.109}$$

Taking into account that $\delta^+ \approx \delta^-$ and assuming that $\theta^+ = -\theta^- = \theta$, and also taking into account relation $\rho/\delta \ll 1$ and substituting $\gamma^+ = \gamma^- = \rho/\delta$, we obtain for the angle of rotation of the indicatrix axes caused by the LMOE and for the Faraday rotation angle

$$\theta(H) = \frac{\theta^+ - \theta^-}{2} = \frac{\Delta\Phi_1(H) - \Delta\Phi_2(H)}{2(1 - \cos\delta)}, \tag{1.110a}$$

$$\rho(H) = \frac{\delta}{2\sin\delta}(\Delta\Phi_1(H) + \Delta\Phi_2(H)). \tag{1.110b}$$

The phase shift was determined from the linear dependence of the azimuth of the polarization axis of the passing light on that of the incident light. The linear dependence of $\Phi_i(\Phi_0)$ had good agreement with the experiment. The slopes of the straight lines for both regions yielded close values independent of the magnetic field: $\delta = 125° + 2\pi k$ and $\delta = 122° + 2\pi k$, were k should be taken equal to 2 to agree with the results of papers [121, 122]. Neglecting the difference between δ^+ and δ^- which can be connected with the difference of the crystal thickness in the investigated regions (the required difference of thickness is 0.14 μm), $\theta(H)$ and $\rho(H)$ were determined with good accuracy in [93].

Figure 1.19 shows the dependences of $\Delta\Phi_1(H)$ and $\Delta\Phi^2(H)$. The difference in the angles of rotation of the polarization axis in two AFM domains can be clearly seen. The azimuth variation in region 1, $\Delta\Phi_1(H)$ (domain AFM$^-$), increases linearly with the rise of the field strength, and the azimuth variation in region 2 (domain AFM$^-$) is practically absent up to the field of the AFM $\rightarrow$ WFM phase transition ($\underline{mm}m \rightarrow m\underline{mm}$), to which the fast increase of the $\Delta\Phi_1$ values corresponds due to the rise of the Faraday rotation contribution. In fields $H > H_t$, where the WFM state $m\underline{mm}$ exists, as expected, $\Delta\Phi_i(H)$ does not depend on H, but the values of $\Delta\Phi_1$ and $\Delta\Phi_2$ in regions 1 and 2 of the already magnetically homogeneous specimen are different. The thickness difference of 0.14 μm is sufficient to explain the difference between δ_1 and δ_2, but it cannot account for the difference between $\Delta\Phi_1$ and $\Delta\Phi_2$, which reaches 8′. It can be supposed that the Faraday rotation in the regions in the same, and the difference between $\Delta\Phi_1$ and $\Delta\Phi_2$ is connected with a small change in the orientation of the plane of the crystal optical axes in regions 1 and 2 after the phase transition. This change should be about 3′. If we take into account that at $H = 0$ the difference between the azimuthes of this plane in regions 1 and 2 reaches 6′, and this explanation seems acceptable. During the phase transition induced by magnetostriction, the internal stresses available in the crystal can be redistributed, which causes a slight variation in the principal optical plane orientation. Near the boundaries of the region of the two-phase magnetic state existence, deviations of the $\Delta\Phi_i(H)$ dependences from the linear dependences, nonmonotonies, and small hysteresises are observed. These peculiarities are partially connected with the existence of the unblanced two-phase state and violation of the AFM monodomainity condition of the investigated

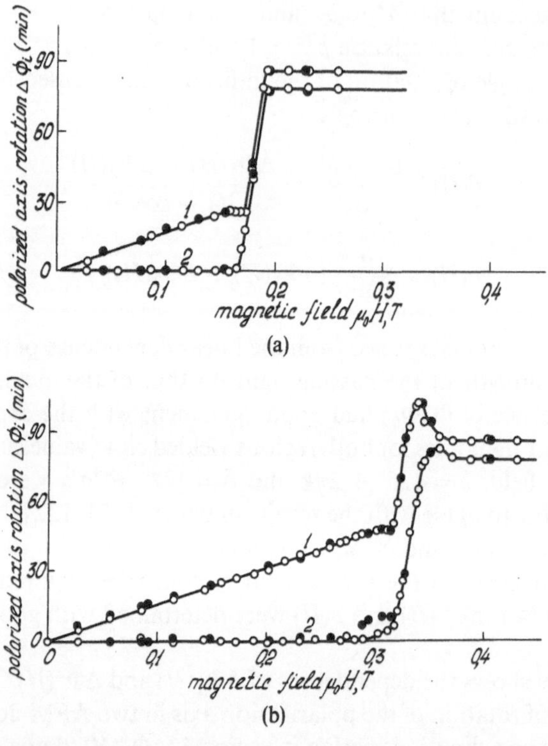

Fig. 1.19. Field dependences of the rotation of the light polarization ellipse axis measured in monodomain sections of AFM $DyFeO_3$ at $\mathbf{k} \parallel \mathbf{H} \parallel c$, $\lambda = 590$ nm: (a) $T = 22$ K; (b) $T = 8$ K. Experimental points ○ and ● correspond to a field increase and decrease, respectively.

regions near the magnetic phase transition. Humps on the curves can be connected with the overlapping of the AFM^+, AFM^-, and WFM^+ domains in the investigated regions. The fact that the value of $\Delta\Phi_2(H)$, corresponding to the AFM^- domain, is equal to zero is explained by the mutual compensation of the Faraday rotation and the indicatrix rotation contributions. According to (1.109) the value $\Delta\Phi_i(H)$ is equal to zero only at definite thicknesses of the crystal $l = (6 + 17k)$ μm (at $k = 2$ we obtain $l = 40$ μm). This requirement for thicknesses of the crystal is determined by the following equality:

$$\frac{\rho}{\theta\delta} = \frac{\varepsilon^a_{xy}}{\varepsilon^s_{xy}} = \frac{1 - \cos\delta}{\sin\delta}. \tag{1.111}$$

The above-mentioned value $\delta = 845°$ yields $\varepsilon^a_{xy}/\varepsilon^s_{xy} = 1.9$. The relationship of $\theta(H)$ determined by means of (1.109) is shown in Fig. 1.20. The points obtained at 8 K and 22 K are on the same straight line with the slope of 50 min/T. Knowing θ, we can calculate the symmetrical nondiagonal component ε^s_{xy} and determine the magneto-optic coefficient q_{xyz}. The values of θ are shown in the figure, and $q_{xyz} = 2.5 \times 10^{-3}$ T^{-1}. The Faraday rotation obtained by means

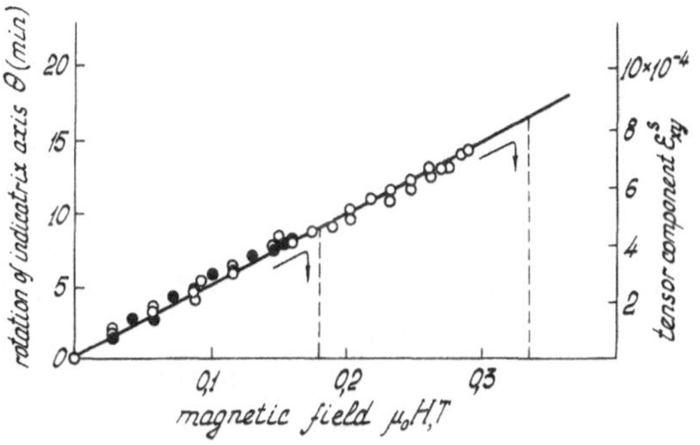

Fig. 1.20. Rotation of optical indicatrix axes about the c-axis in the applied magnetic field $\mathbf{H} \parallel c$, and a nondiagonal symmetry component of the dielectric tensor in AFM $DyFeO_3$. White and black points indicate 8 K and 22 K, respectively. The dashed line indicates an abrupt change in the AFM → WFM phase transition.

of (1.102) is shown in Fig. 1.21. The figure also shows the scale for the antisymmetrical component ε^a_{xy}. The specific Faraday rotation does not depend on temperature in the studied range 8–22 K and is equal to 5.9 deg/cm T. The Faraday magneto-optic constant is equal to $f_{xyz} = 4.9 \times 10^{-3}\,T^{-1}$.

It is interesting to relate the LMOE in $DyFeO_3$ to the behavior of its magnetic subsystem in a magnetic field. Dysprosium orthoferrite has the perovscite rhombically distorted structure with four molecules of $DyFeO_3$ in the elementary cell, described by the spatial group $P_{bnm}(D^{16}_{2h})$. The magnetic elementary cell coincides with the crystal chemical one, contains eight magnetic ions, and requires eight magnetic sublattices for the complete description of the crystal magnetic properties. Taking into account that the main contribu-

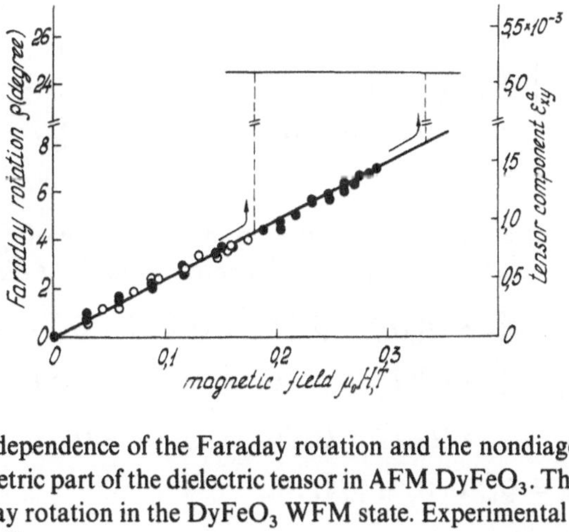

Fig. 1.21. Field dependence of the Faraday rotation and the nondiagonal component of the antisymmetric part of the dielectric tensor in AFM $DyFeO_3$. The horizontal line indicates Faraday rotation in the $DyFeO_3$ WFM state. Experimental points: ●—T = 8 K; ○—T = 22 K.

tion to the magneto-optic effect in orthoferrites in the visual part of the spectrum, away from the electron transition bands in rare-earth ions, is made by the ions of iron [121, 120], the LMOE description in dysprosium orthoferrite can be limited to four iron sublattices. As the location of the iron magnetic moments is close to the collinear one, the two-sublattice model approximation can be used. In this case the FM and AFM vectors are expressed via the magnetic sublattices are follows: $\mathbf{F} = (4\mathbf{M}_0)^{-1}(\mathbf{M}_1 + \mathbf{M}_2 + \mathbf{M}_3 + \mathbf{M}_4)$ and $\mathbf{G} = (4\mathbf{M}_0)^{-1}(\mathbf{M}_1 - \mathbf{M}_2 + \mathbf{M}_3 - \mathbf{M}_4)$. The AFM ordering of the iron ions belongs to type $2_z^- 2_x^- (Z \parallel c, X \parallel a)$ [73, 123]. The thermodynamic potential of dysprosium orthoferrite contains the invariants $d_1 F_z G_x$ and $d_2 F_x G_z$, which describe the appearance of the WFM moment F_z and F_x when the AFM vector is oriented along axes a and c, respectively. In the absence of a magnetic field the AFM state Γ_1 and G_y is realized in which only the y-projection of the AFM vector differs from zero. Above T_M and below T_M in field $\mathbf{H} \parallel c$ at $H > H_t$ state Γ_4 or $F_z G_x$ is realized, in which the FM-vector z-projection and the AFM-vector x-projection differ from zero. At $H > H_t$ projection G_y differs from zero also. In the expansion of ε_{ij}^s into the projections of vectors $\mathbf{F}$ and $\mathbf{G}$, the matrices of coefficients Q_{ijrs}^{FF}, Q_{ijrs}^{GG}, and Q_{ijrs}^{FG} at the adopted canceling of indices have the following form:

$$Q_{ijrs}^{FF}, Q_{ijrs}^{GG} \Rightarrow Q_{\mu\nu} = \begin{pmatrix} Q_{11} & Q_{12} & Q_{13} & 0 & 0 & 0 \\ Q_{21} & Q_{22} & Q_{23} & 0 & 0 & 0 \\ Q_{31} & Q_{32} & Q_{33} & 0 & 0 & 0 \\ 0 & 0 & 0 & Q_{44} & 0 & 0 \\ 0 & 0 & 0 & 0 & Q_{55} & 0 \\ 0 & 0 & 0 & 0 & 0 & Q_{66} \end{pmatrix}, \qquad (1.112)$$

$$Q_{ijrs}^{FG} \Rightarrow Q_{\mu\nu}^{FG} = \begin{pmatrix} 0 & 0 & 0 & 0 & Q_{15} & 0 & 0 & Q_{18} & 0 \\ 0 & 0 & 0 & 0 & Q_{25} & 0 & 0 & Q_{28} & 0 \\ 0 & 0 & 0 & 0 & Q_{35} & 0 & 0 & Q_{38} & 0 \\ 0 & 0 & 0 & 0 & 0 & Q_{46} & 0 & 0 & Q_{49} \\ Q_{51} & Q_{52} & Q_{53} & 0 & 0 & 0 & 0 & 0 & 0 \\ 0 & 0 & 0 & Q_{64} & 0 & 0 & Q_{67} & 0 & 0 \end{pmatrix}. \qquad (1.113)$$

Leaving only projections G_x, G_y, and F_z in the expansion of ε_{ij} we obtain the following expression for the symmetrical components of the dielectric tensor:

$$\begin{aligned} \varepsilon_{xx}^s &= \varepsilon_{xx}^0 + Q_{13}^{FF} F_z^2 + Q_{11}^{GG} G_x^2 + Q_{12}^{GG} G_y^2 + Q_{18}^{FG} F_z G_x, \\ \varepsilon_{yy}^s &= \varepsilon_{yy}^0 + Q_{23}^{FF} F_z^2 + Q_{21}^{GG} G_x^2 + Q_{22}^{GG} G_y^2 + Q_{28}^{FG} F_z G_x, \\ \varepsilon_{zz}^s &= \varepsilon_{zz}^0 + Q_{33}^{FF} F_z^2 + Q_{31}^{GG} G_x^2 + Q_{32}^{GG} G_y^2 + Q_{38}^{FG} F_z G_x, \\ \varepsilon_{xy}^s &= Q_{66}^{GG} G_x G_y + Q_{67}^{FG} F_z G_y, \\ \varepsilon_{xz}^s &= \varepsilon_{yz}^s = 0. \end{aligned} \qquad (1.114)$$

Projections F_z and G_x are transformed in the same way as the magnetic field projections. In state Γ_{14}, realized in field $H < H_t$, projections F_z and G_x are equal to

$$F_z = F_0 \cos\varphi + \chi_{zz}^F H_z, \qquad G_x = G_0 \cos\varphi, \qquad G_y = G_0 \sin\varphi, \tag{1.115}$$

where Γ_0, G_0 are the spontaneous values of the magnetic vectors, and φ is the angle between vector $\mathbf{G}$ and axis a. Angle φ differs slightly from $\pi/2$ and is linearly dependent on the magnetic field strength

$$\cos\varphi = m_0 H_z (2K_2)^{-1}. \tag{1.116}$$

Here m_0 is the WFM moment in state Γ_4, and K_2 is the second-order magnetic anisotropy constant in plane (ab). It follows from expressions (1.114), (1.115), and (1.116) that there is only one symmetrical component of the dielectric tensor of AFM $DyFeO_3$ placed in the magnetic field $\mathbf{H} \parallel c$ linearly dependent on the field strength

$$\varepsilon_{xy}^s = \{(Q_{67}^{FG} F_0 + Q_{66}^{GG} G_0) m_0 (2K_2)^{-1} + Q_{67}^{FG} \chi_{zz}\} G_0 H_z \operatorname{sign} G_y. \tag{1.117}$$

Equation (1.117) makes evident the relation between the magneto-optic coefficients

$$q_{xyz} = \{(Q_{67}^{FG} F_0 + Q_{66}^{GG} G_0) \frac{m_0}{2K_2} + Q_{67}^{FG} \chi_{zz}\} G_0 \operatorname{sign} G_y. \tag{1.118}$$

It can be seen from the expression for ε_{xy}^s that the rotation of the indicatrix axes in $DyFeO_3$, as in $CoCO_3$, is also connected, not only with the rotation of the AFM vector from axis b to axis a (component $Q_{66}^{GG} G_x G_y$ in (1.114)), but also with the bend of sublattices towards axis c (component $Q_{67}^{FG} F_z G_y$). It is impossible to determine the relative contribution of these mechanism.

Let us note that the discovery of the LMOE in dysprosium orthoferrite made it possible to give a simple explanation to the observed hysteresis dependencies on $\mathbf{H}$ of its magneto-optic properties in the region remote from the field of phase transition where the WFM state cannot be realized. The rebuilding of the AFM domain structure in a strong field and the accidental prevailing of one of two types of AFM domains cause the hysteresis of the observed effective rotation of the polarization axis in the magnetic fields, where the existence of the metastable WFM domains cannot be expected.

1.4.4. The Linear Magnetic-Optic Effect in Other Antiferromagnetic Crystals

In this section we will briefly describe other AFM crystals, in which the specific features of the "abnormal" birefringence behavior of linearly polarized light in a magnetic field indicate that the LMOE is obvious in them as well.

When studying magnetic phase transitions in cubic dysprosium–aluminum garnet $Dy_3Al_5O_{12}$ in the transverse geometry of experiment $\mathbf{k} \perp \mathbf{H}$ at

K ∥ [110], **H** ∥ [111], different birefringence in two AFM states of the crystal has been observed [100]. $Dy_3Al_5O_{12}$ is a complex multisublattice noncollinear AFM. It is difficult to express the variations of the dielectric tensor components via the projections of its magnetic components. But, knowing its point magnetic group $m3\underline{m}$ [112], it can be seen from Table 1.2 that the matrix of its magneto-optic coefficients $q_{ij\alpha}$ has identically unequal-to-zero coefficients $q_{yzx} = q_{xzy} = q_{xyz}$ which ensure the linear-in-**H** birefringence of the light propagating along [110]. If the birefringence observed in [100] is actually connected with the LMOE, the linear-in-**H** birefringence should be observed in the longitudinal geometry of experiment too, in particular, at **H** ∥ **k** ∥ [001].

Almost linear regions and hysteresis phenomena, which should also be connected with the LMOE manifestation, can be seen in the dependence of the linear birefringence on the field strength obtained during investigation of the Morin transition in haematite [101, 124]. Subsequent investigations of the α-Fe_2O_3 birefringence in the magnetic field H confirm this assumption. The results given in [125] make it possible to evaluate the magneto-optic constant value $(q_{xxx} - q_{yyx}) = 2.5 \times 10^{-4}\ T^{-1}$ for light with $\lambda = 1.15\ \mu m$ at $T = 254.8$ K.

Somewhat unexpected was the observation of the LMOE, in the longitudinal geometry of experimental at **H** ∥ [001], in the tetragonal manganese–germanium garnet $Ca_3Mn_2Ge_3O_{12}$ [126, 127], as the neutron-diffraction investigations of this AFM have yielded the group of symmetry which forbade the LMOE manifestation in this geometry of experiment. The question of the CaMnGeG magnetic symmetry will be discussed in the next chapter. Here, Fig. 1.22 represents the LMOE-characteristic "butterfly"-type dependence for

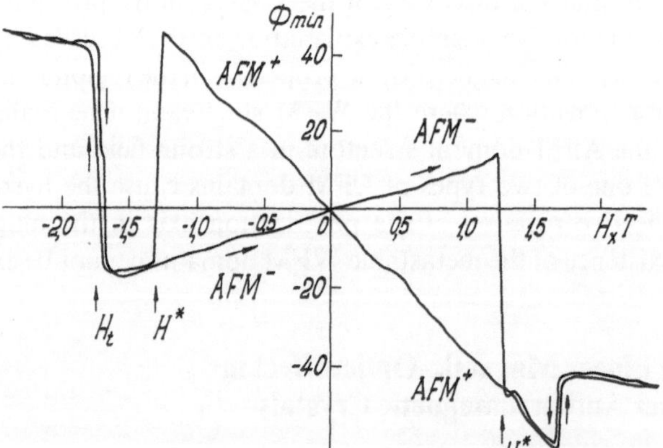

Fig. 1.22. Rotation of the light polarization ellipse axis in the AFM plate of calcium–manganese–germanium garnet in an applied magnetic field **H** ∥ [100] ∥ **k** at $T =$ 9.2 K, $\lambda = 632.8$ nm, $t = 40\ \mu m$. The electric vector of the incident light wave is **E** ∥ [001].

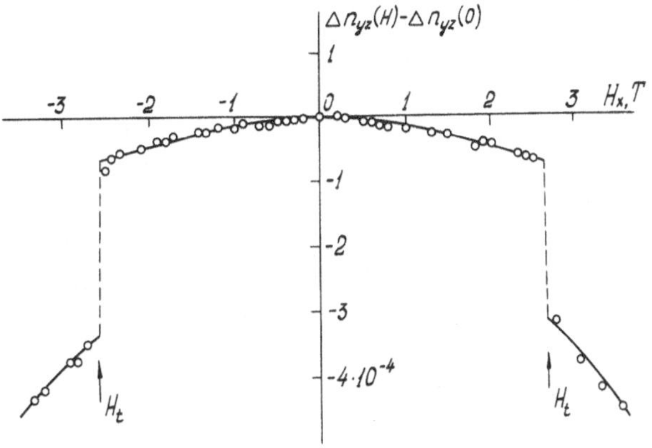

Fig. 1.23. Field-induced linear birefringence in a $Ca_3Mn_2Ge_3O_{12}$ garnet at $\mathbf{k} \parallel \mathbf{H} \parallel [100]$, $T = 6.7$ K, $\lambda = 632.8$ nm.

the angle of rotation of the polarization ellipse axes of the passing light propagating perpendicularly to the crystal tetragonal axis at $\mathbf{k} \parallel [100]$ and $\mathbf{k} \parallel \mathbf{H}$. The butterfly asymmetry is connected with the fact that the rotation of the polarization ellipse axis $\Phi^{\pm}$ is caused not only by the rotation of the optical indicatrix $\theta^{\pm}$ caused by the LMOE, but also by the Faraday rotation ρ which has the same sign for both AFM states

$$\Phi^{\pm}(H) = \theta^{\pm}(H)(1 - \cos\delta) + \frac{\rho(H)}{\delta}\sin\delta. \tag{1.119}$$

In (1.110) δ is the phase difference between the light modes caused by the birefringence $(n_z - n_y) \approx 1.1 \times 10^{-2}$, which slightly varies in the field (see Fig. 1.23). Jumps in field H^* are connected with the AFM remagnetization of the investigated region of the specimen, and jumps in field H_t with the magnetic phase transition. If we independently determine δ and $\Phi^{\pm}(H)$ from (1.119), we find that the specific rotation of the indicatrix around axis [001] is equal to 3.5×10^{-1} deg/T. Its corresponding magneto-optic coefficient

$$q^s_{yzx} = 2n_0(n_z - n_y)(\theta_x/H_x) \tag{1.120}$$

is close to 3×10^{-4} T^{-1} for light with $\lambda = 632.8$ nm and $T = 10.5$ K. The value of the birefringence appearing in geometry $\mathbf{k} \parallel \mathbf{H} \parallel [001]$ is equal to $\pm 5 \times 10^{-5}\, H_z$ where H_z is expressed in tesla, $T = 10$ K, $\lambda = 632.8$ nm.

1.5. Magnetic Gyrotropy Quadratically Dependent on Magnetic Field Strength

As was mentioned above, the Onsager relations for kinetic coefficients generalized for magnetically ordered media remove the prohibition on the even-in-**H** terms in the expansion of the dielectric tensor antisymmetrical components. Tensor $B^{a}_{ij\alpha\beta}$ in (1.11), describing the quadratic dependence of ε^{a}_{ij} on **H**, is identically unequal to zero in the AFM crystals which do not contain the anti-inversion operation $\underline{\bar{1}}$ as the symmetry operation. The quadratic-in-**H** magnetic gyration can also be described by the third-order axial c-tensor $G_{k\alpha\xi}$, dual to the antisymmetrical as to the first two indices c-tensor $B^{a}_{ij\alpha\beta}$: $G_{k\alpha\beta} = e_{kij}B^{a}_{ij\alpha\beta}$. Here e_{kij} is the completely antisymmetrical unit tensor. In contrast to the symmetrical as to two indices axial c-tensor $q^{s}_{ij\alpha}$ describing the LMOE, tensor $G_{k\alpha\beta}$ is symmetrical as to the permutation of all three indices. This property of tensor $G_{k\alpha\beta}$ reduces the number of classes in which the quadratic magnetic rotation is symmetrically allowed. Table 1.6 gives matrices $G_{k\alpha\beta}$ for all Laue magnetic classes in which the quadratic magnetic rotation (QMR) is not forbidden by the general principles of symmetry. Naturally, the QMR is allowed in the same crystal classes, in which the quadratic-in-**H** magnetization is allowed [96, 97], and the symmetry of tensor $G_{k\alpha\beta}$ should be the same, as the symmetry of tensor $G_{k\alpha\beta}$ describing the symmetrical-in-**H** magnetization. The symmetry prohibition on the QMR is removed in the 31 pyromagnetic and 27 AFM classes of crystals. In the crystals without the anti-inversion center it is as in the LMOE, is forbidden in classes $\underline{4}32$, $4\bar{3}m$, $m3m$ and, besides, in classes 422, $4mm$, $\bar{4}2m$, $4/mmm$, 622, $6mm$, $\bar{6}m2$, and $6/mmm$.

The quadratic-in-field rotation of the light polarization plane and circular dichroism were experimentally observed in the AFM fluorides CoF_2 and FeF_2 described by the magnetic point group $\underline{4}/mm\underline{m}$ [62, 63]. Tensor $G_{k\alpha\beta}$ in this class of crystals has nonzero components $G_{xyz} = G_{yxz} = G_{zxy}$. Hence, the magnetic fields with y- and z-, x- and z-, and x- and y-projections can rotate the plane of polarization of the light propagating along axes x, y, and z, respectively. The most easily detected manifestation of the quadratic gyration may be the rotation of the plane of polarization and circular dichroism of the light propagating along tetragonal axis [001] $\parallel z$ in field $H \parallel$ [110]. This geometry was used in the experiments. This is favorable due to the fact that it is free from the linear-in-H magnetic rotation of the plane of polarization of light (the Faraday effect). The LMOE is absent in this geometry too. The latter, although it splits the optical axis due to the appearance of the dielectric tensor components $\varepsilon^{a}_{xz} = \varepsilon^{a}_{yz}$, will not affect the light propagating along [001], as the LMOE-caused rotation of the optical indicatrix makes one of two optical axes coincide with the tetragonal [001]. In the experiments some inconveniences were caused by the magnetic birefringence caused by the Cotton–Mouton effect. But its influence on the rotation of the light polarization axis could be methodically excluded. Besides, the task was simplified by the fact that the values and signs of the linear birefringence caused by the Cotton–Mouton effect are equal for both AFM states, and the quadratic-in-field rotations of the plane of polarization are opposite. All experiments were conducted with monodomain samples. The uniformity of the AFM state was visually checked by means of the LMOE.

Figure 1.24 shows the results of measurement of the rotation of the light polarization plane in the specimen CoF_2 which is alternatively in the AFM^{+}

Table 1.6. Symmetry of the tensor of the quadratic magnetic rotation of the light polarization plane in Shubnikov crystal.

Magnetic Laue group	Tensor matrix[a] $G_{\mu\alpha\beta} \leftrightarrow G_{\mu\nu}$	Magnetic Laue group	Tensor matrix $G_{\mu\alpha\beta} \leftrightarrow G_{\mu\nu}$
1	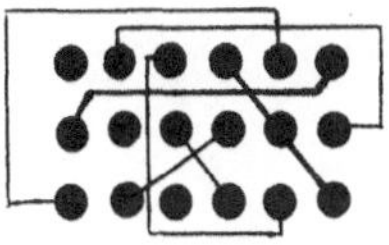	4, 422 6, 6$\underline{22}$	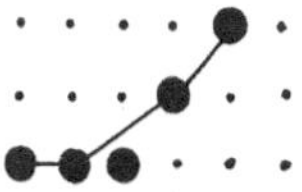
2	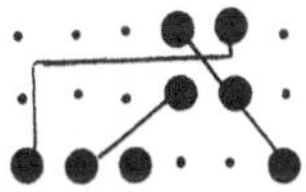	3	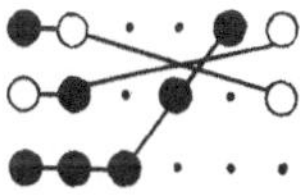
$\underline{2}$	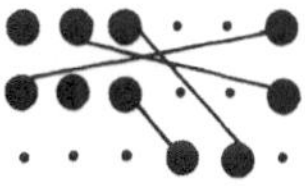	3$\underline{2}$	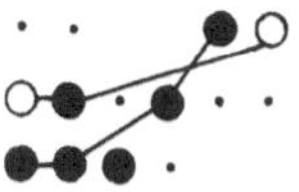
222, 4$\underline{22}$ 23, 4$\underline{32}$	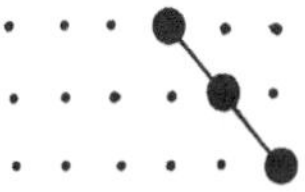	32, 6$\underline{22}$	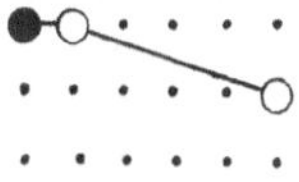
$\underline{222}$	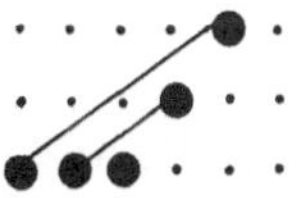	$\underline{6}$	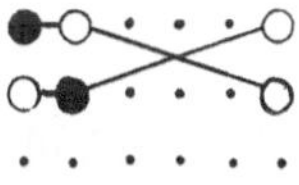
$\underline{4}$	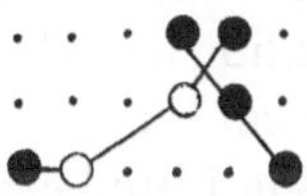	422, 622 432	

[a] The reduction is carried out according to the rule:

$\alpha\beta$	11	22	33	23	13	12
ν	1	2	3	4	5	6.

and AFM$^-$ states. The solid lines (three in the figure) denote the relations

$$\varphi = (\pm BH^2 + CH)l \tag{1.121}$$

which well describe the position of the experimental points for both AFM states for the values of parameters $B = 1.3$ deg/cm T^2 and $C = 0.2$ radn/cm T for light with wavelength 632.8 nm and $T = 15$ K. The linear-in-H addition is caused by a small deviation of the angle between the field and light propagation vector from 90° and is the Faraday rotation. The required deviation of **H**

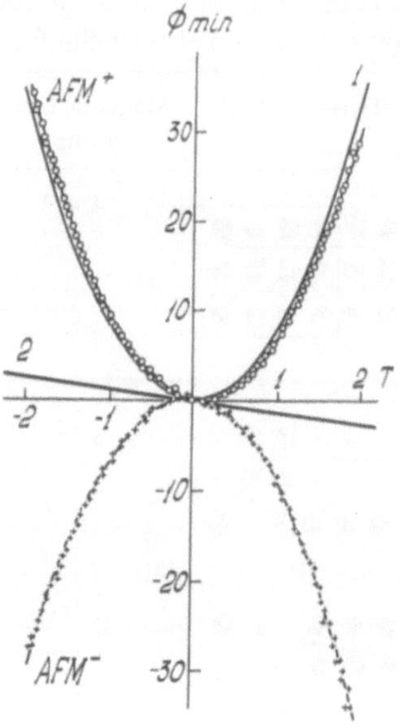

Fig. 1.24. Magnetic rotation of the light polarization plane in AFM CoF_2 at $\mathbf{k} \parallel C_4$, $\mathbf{H} \parallel [001]$. Different experimental points indicate two sample AFM states. Solid lines are quadratic-in-H dependence (line 1), linear-in-H rotations of the light polarization plane due to the Faraday effect induced by small errors in the field orientation (line 2) and the resulting dependence. The sample thickness is 1 mm, $T = 15$ K, $\lambda = 632.8$ nm.

from the perpendicular orientation is about 1°. At $T \to 0$ constant B is equal to 1.4 deg/cm T^2. Similar relations were obtained for FeF_2 for which the value of the quadratic rotation constant under the same conditions is $(7.0 \pm 1.5) \times 10^{-2}$ deg/cm T^2. In MnF_2 quadratic rotation has not been found at the sensitivity of measurement of about 10^{-3} deg/cm T^2.

The homogeneity of the AFM states in MnF_2 specimens was checked by the value of the magnetic linear dichroism in the region of the exciton–magnon absorption connected with the LMOE. The temperature dependence of the rotation shown in Fig. 1.25 demonstrates the disappearance of the effect when crystals CoF_2 and FeF_2 transit into the paramagnetic state. The effects disappear according to the symmetry requirements even when the specimen is in the AFM state, but the field is directed along [100] or [010], when one of the x- or y-projections of the field vanishes.

It is interesting to compare the values of the quadratic- and linear-in-**H** rotations normalized to equal values of the z-projections of the linear and quadratic magnetizations. If we consider the FM and AFM vectors instead of a magnetic field as the action upon the crystal, the nonzero component of the quadratic magnetic gyration vector can be expressed as the sum

$$g_z^{\text{quad.}} = G_{zxy} H_x H_y = G_{zxyz}^{(LL)} L_x L_y L_z + G_{zxyz}^{(ML)} M_x M_y M_z. \qquad (1.122)$$

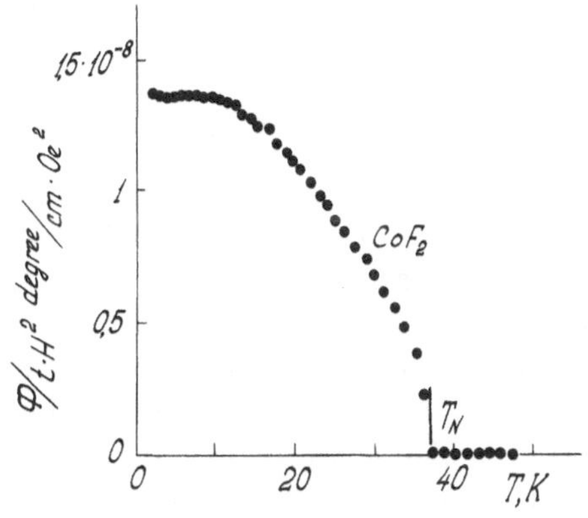

Fig. 1.25. Temperature dependence of the quadratic-in-field magnetic rotation of the light polarization plane in the AFM cobalt fluoride, $\lambda = 632.8$ nm.

A similar expression can be written for the z-projection of the quadratic magnetization

$$M_z^{\text{quad.}} = C_{zxy}H_xH_y = C_{zxyz}^{(LL)}L_xL_yL_z + C_{zxyz}^{(ML)}M_xM_yM_z. \qquad (1.123)$$

Hence, the following relation can exist between the quadratic rotation, quadratic magnetization, and magnetic vectors **M** and **L**

$$\Phi^{\text{quad.}}/l = B^{(M)}M_z^{\text{quad.}} + \alpha L_xL_yL_z + \beta M_xM_yL_z. \qquad (1.124)$$

The last two expressions in (1.124) indicate that quadratic rotation can appear at $M_z^{\text{quad.}} = 0$ too. If we assume $\alpha = 0$ and $\beta = 0$, we obtain for $B_0^{(M)}$ at $T = 4.2$ K the value 2.0 deg/cm emu. To evaluate $M_z^{\text{quad.}}$, the value $C_{zxy} = 0.68$ emu/T^2 was used, which follows from the data of [98] in which it was experimentally found that $M_z = 360$ emu/mol at $H_{[110]} = 5$ T. The proportionality factor $V^{(M)}$, connecting the rotation and magnetization in the Faraday geometry of experiment

$$\Phi/l = V^{(M)}M_z, \qquad (1.125)$$

turned out to be equal to 2.4 deg/cm emu. To determine $V^{(M)}$ the Faraday rotation has been measured, and the longitudinal magnetization was calculated from the susceptibility value $\chi_{zz} = 6.0 \times 10^{-4}$ emu/cm^3 taken from [116]. Although the values of $B^{(M)}$ and $V^{(M)}$ are close and, even more, the possibility of their equality cannot be excluded due to the errors connected with the fact that the results of various experiments have been used, we should underline that the quality of $B^{(M)}$ and $V^{(M)}$ can be only approximate or accidental, as the nature of the magnetizations and those disturbances of the ion energy states which cause the magnetic gyroptropy in the two cases discussed is different. This difference most evidently manifests itself in the magnetic circular dichroism spectra obtained in these two geometries of experiment.

The dichroism has been measured in CoF_2 in the visible region of the absorption spectrum connected with the exciton–magnon and other collective excitations of the ion Co^{2+} 3*d*-shell in the longitudinal ($\mathbf{k} \parallel \mathbf{H} \parallel [001]$) and transverse ($\mathbf{k} \perp \mathbf{H}\, [110]$) geometries of experiment [63, 128]. In the transverse geometry of experiment the circular dichroism was markedly observed. As was expected, the sign of the dichroism does not change when the field reverses its direction, but it changes during the AFM magnetization of the specimen. Figure 1.26 illustrates the change of the dichroism sign when the sublattice magnetic moments in the specimen are reoriented. The figure shows the dichroism spectra obtained sequentially for two AFM states of the specimen: AFM^+ and AFM^-. Figures 1.26 and 1.27 compare the spectrum of the quadratic magnetic circular dichroism with the spectrum of the linear-in-magnetic field circular dichroism obtained in the longitudinal geometry of experiment. The figures also demonstrate the transformation of the circular dichroism spectra in the tilted field oriented in plane (110) as the *z*-component of the field increases. As can be seen from this figure, there are obvious differences between the spectra of the quadratic- and linear-in-**H** dichroisms: the

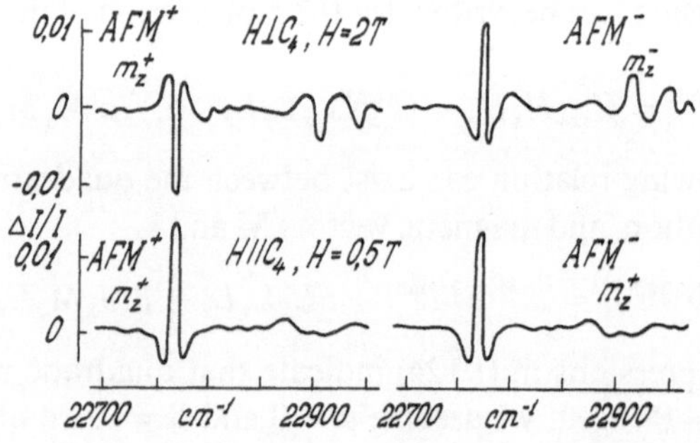

Fig. 1.26. Magnetic circular dichroism in the exciton–magnon absorption of AFM CoF_2.

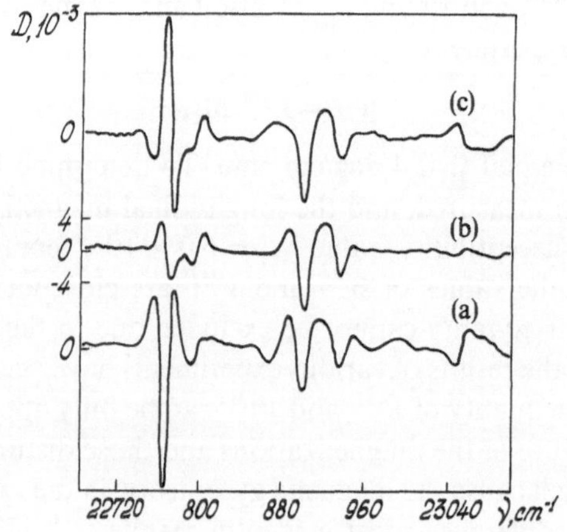

Fig. 1.27. Transformation of magnetic circular dichroism spectra of AFM CoF_2 in a tilted magnetic field. $H_{[110]} = 2.4$ T, $H_{[001]} =$ (a) 0; (b) 0.5 T; (c) 1.0 T.

dichroism better manifests itself in the transverse geometry in the group of bands 22,880–22,950 cm^{-1} than in the longitudinal geometry. The signs of the linear and quadratic dichroisms are different in the region of the exciton–magnon band 22,770 cm^{-1} at one and the same sign of the z-projection of the magnetic moment and at one and the same AFM state of the specimen. It should be noted that the invariability of the AFM state and the constancy of the sign of the magnetic moment z-projection in the experiments, were ensured by the fact that the sign of projection $H_{[110]}$ (during the measurement of the dichroism in the transverse geometry of experiment) and the signs of H_z (during the measurement of the dichroism in the longitudinal geometry) remains the same as they were during the specimen monodomainization. The invariability of the AFM state was visually checked by means of the LMOE. Similar differences in the spectra of the linear and quadratic circular dichroisms were observed in other regions of the CoF_2 spectra as well. The differences between the linear and quadratic dichroisms can be seen more clearly at the simultaneous action of the transverse and longitudinal field components. At $H_{[001]}$ about 0.5 T and $H_{[\bar{1}10]}$ about 2.4 T the dichroism is compensated for in the region of band 23,047 cm^{-1}. The partial compensation revealing the structure of the exciton–magnon absorption is observed in the region of band 22,770 cm^{-1}.

The shape of the dispersion curves of the quadratic magnetic circular dichroism (the isolated bands have an S-shaped dispersion, and the dispersion of the overlapping bands can be represented as the superposition of the S-curves) indicates that the basic mechanism of the dichroism appearance is the removal of the sublattice degeneration of the ion Co^{2+} energy levels and the splitting of the absorption bands.

The degeneration is removed due to the different orientation of the deformed ligand octahedrons surrounding the cobalt ions from the first and second sublattices in respect to direction [110]. Taking into account the orthorhombic symmetry of the local magnetic anisotropy, the shifts of the energy levels of cobalt ions in field $\mathbf{H} \parallel [110]$ can be found. The results of [116], in which the magnetic susceptibility of CoF_2 has been calculated taking into account the rhombic symmetry of the magnetic local anisotropy and the closely spaced Kramers doublet in the spectrum of ion Co^{2+}, make it possible to obtain the expression for the difference of the energies of the ground states of ions $Co^{2+}1$ and $Co^{2+}2$ from different sublattices in field $\mathbf{H} \parallel [110]$:

$$\Delta E_0 = \frac{(f_1 g_{xx}^2 - f_2 g_{yy}^2)\mu_B^2 H^2}{2[1 - f_1 f_2 (z_2 \mathcal{I}_2)^2]}. \tag{1.126}$$

Here $g_{xx} = 2.75$, $g_{yy} = 2.08$ are the g-factors, $z_2 \mathcal{I}_2 = 35\ cm^{-1}$ is the parameter of the exchange interaction of the ion Co^{2+} with its nearest neighbors from the other sublattice, and $f_1 = 0.0480$ and $f_2 = 0.0086$ are determined by the parameters of the axial and rhombic anisotropy and the value of the intersublattice interaction. The evaluation yields $\Delta E \approx 7 \times 10^{-2}\ H^2\ cm^{-1}\ T^{-2}$. In field $H_{[110]} = 2.7$ T we obtain $\Delta E_0 = 0.5\ cm^{-1}$.

The values of the spectroscopic splitting of the exciton and exciton–magnon band are also determined by the splitting of excited states and splitting of the magnon branch in a transverse field at the edge of the Brillouin zone. Although we cannot evaluate the contribution into the splitting of excited states, we can still expect that the splitting of many bands will be the same as the order of magnitude. The evaluation of the dichroism value, caused by this splitting of the bands, indicates that the considered mechanism responsible for the quadratic magnetic circular dichroism, connected with the removal of the energy degeneracy of the levels of ions $Co^{2+}1$ and $Co^{2+}2$, is acceptable.

The other mechanism responsible for the QMR and dichroism can be connected with the variation of the spin length due to the quantum-mechanical mixing of states under the action of field $\mathbf{H} \parallel [110]$, and was investigated in [65]. Due to the variation of the ion states it should lead to a change in the probability of the optical transitions for the left- and right-hand circular polarization. In this case, the dichroism curves should have the Λ-shaped dispersion. In the investigated regions of the spectrum corresponding to the transitions within the $3d$-configuration the dispersion of the bands has not been clearly detected.

From the generalized (for magnetically ordered media) Onsager relations for kinetic coefficients follows the theoretical possibility of the appearance of the Shubnikov crystals without the anti-inversion center, the magneto-optic effects not inherent to para- and diamagnets: the linear (as to magnetic) field strength birefringence of lineary polarized light (LMOE) and quadratic-in-field rotation of the plane of polarization, as well as the field-induced natural gyration, is proportional to the first power of the field strength. The experimental search for the first two effects has shown that both the LMOE and QMR, as well as the linear-in-field dichroism of linearly polarized light and the quadratic-in-field dicroism of circularly polarized light connected with them, have an easily measured value in many antiferromagnets. And more than that, the values of these effects make it possible to visualize the previously invisible collinear or 180-degree antiferromagnetic domains for many types of antiferromagnetics. The high sensitivity of the new magneto-optic effects to the magnetic ordering symmetry enlarges the possibilities of the optical methods for studying the magnetic symmetry of complex magnetically ordered structures and their transformations during phase transitions. The spectra of the linear-in-field dichroism of linearly polarized light and the quadratic magnetic circular dichroism contain new information on the ground and excited energy states of an antiferromagnet and, doubtless, will be useful not only for the identification of electron transitions. Also interesting are the other (not yet investigated) manifestations of magneto-optic effects in nonlinear optics where, in particular, the LMOE allows the doubling of the light frequency in centrosymmetrical crystals, as well as the induction of static magnetization by linearly polarized light.

CHAPTER 2

Magneto-Optical Methods for Investigating the Structure of Antiferromagnetically Ordered Crystals

The spontaneous Faraday rotation and the Cotton–Mouton magneto-optical effect discovered in relatively transparent iron garnets, and then in other ferro and ferrimagnets, have opened a remarkable possibility to observe magnetic domains directly by human eye, thus studying the variety of domain structures and analyzing structural changes induced by a magnetic field. Visual magneto-optical methods have played a highly essential part in the marking and refinement of data recording, data processing, and data transmission techniques using the properties of cylindrical magnetic domains. The new magneto-optical effects, for the first time, provide means to distinguish visually between collinear, 180-degree, or time-reversed antiferromagnetic (AFM) domains. The magneto-optical methods for analyzing AFM domain structures and the AFM domain wall are at their earliest stage of development, with the possibilities far from being exhausted. The special sensitivity of the new effects to the symmetry of the crystal magnetic structure can be used in test experiments aimed at specifying the magnetic symmetry of an antiferromagnet. For example, the linear magneto-optic effect (LMOE) or the quadratic magnetic rotation (QMR) can often provide an unambiguous answer to such questions as to whether the chemical and magnetic crystal unit cells coincide, or whether the crystal possesses an anti-inversion center. The new magneto-optical effects also present new possibilities for investigating multiphase magnetic states at phase transitions.

In this chapter we are going to discuss some experiments in which the new magneto-optical effects were employed to study the structure of magnetic crystals, mainly antiferromagnets.

2.1. Visualization of Collinear or 180-Degree Antiferromagnetic Domains

The possibility of direct or indirect observation of collinear or 180-degree domains in compensated antiferromagnets was discussed more than once.

The presence of AFM domains in nonconducting collinear antiferromagnets and their relative number can be established by the nuclear magnetic resonance (NMR) technique, with the specimen placed in an electric field [129, 130], by observing the diffraction of polarized neutrons [131] through magnetostrictional or piezomagnetic measurements [94, 114, 132 to 134] or through precise magnetometry with the use of SQUIDs [135]. First photographs of energetically equivalent 180-degree AFM domains in a collinear two-sublattice antiferromagnet were obtained in 1978 by the neutron topography technique [136, 137]. The exposure time was 24 hours and the spatial resolution of the image was about 40 μm.

The possibility of visualizing collinear AFM domains through application of optical techniques was discussed in 1963 [3]. In that paper attention was paid to the gyrotropic birefringence that can arise spontaneously in magnetically ordered crystals, and to the externally induced optical effects that should change their sign if the directions off all the elementary magnetic moments in the crystal were reversed. Special emphasis was made on the gyrotropic birefringence, which is allowed by the symmetry of noncentrally symmetric AFM and manifests itself as a spatial dispersion effect. It is characterized by a change of the birefringence sign when the direction of propagation is reversed. However, the analysis of microscopic mechanisms of the effect and estimates performed in [6] have shown it to be too low in magnitude even for be observed. To the authors' knowledge, observation of the effect in magnetically ordered media has not been reported, while a magnetically induced gyrotropic birefringence has reportedly been observed in a diamagnetic crystal [10].

Induced effects have proved better suited for optical visualization of AFM domains. However, the domain structure observed may not be the same as in a free specimen, since the crystal is acted upon by an external agent. Moreover, the AFM domains themselves may become energetically nonequivalent. Time-reversed AFM$^+$ and AFM$^-$ domains were first made visible in the six-sublattice noncollinear antiferromagnet $Dy_3Al_5O_{12}$, under the conditions close to the metamagnetic phase transition in a magnetic field [138, 139]. The domains observed were not energetically equivalent and differed slightly in their magnetic moments. The contrast between the domains was explained by the difference of the Faraday rotation in the domains, owing to the difference of the induced magnetic moments of the AFM$^+$ and AFM$^-$ states [138]. As far as we can judge, the first optical observation of energetically equivalent 180-degree domains in a collinear AFM was carried out in 1979 [140–142] with the aid of the LMOE. Later on, the effect was employed to observe AFM domains in the antiferromagnetic dysprosium orthoferrite [93, 143] and in calcium–manganese–germanium garnet [127, 128]. Visual methods of magneto-optics were employed to test the uniformity of AFM ordering in those crystals and to study the behavior of the AFM domain structure under external influences [95, 99]. The results of these investigations are presented below.

2.1.1. Antiferromagnetic Domains in a Tetragonal Cobalt Fluoride

The symmetry properties of cobalt fluoride (CoF_2) and typical manifestations of the LMOE in the AFM crystal were discussed in Section 1.5.2. In this subsection, we will discuss in more detail the optical wave polarization transformations occurring in AFM domains when the light propagation vector and that of the magnetic field both coincide with the tetragonal axis [001] of the crystal. It is convenient to use the Poincaré sphere representation to describe the polarization variations and to introduce a single immobile reference frame for both domains. The polarization axis azimuth is counted with respect to the [110] direction which coincides with the direction of the longer polarization ellipse axis of the fast light wave in one domain (specifically, AFM^+) and of the slow light wave in the other (AFM^-). In the "longitudinal" geometry (i.e., $\mathbf{k} \parallel \mathbf{H}$) the magnetic field brings about, along with the LMOE, Faraday's rotation too, therefore the OR^+ and OR^- axes, which the Poincaré sphere turns around to represent transformations of the light wave polarization in the AFM^+ and AFM^- domains, are not within the equator plane. They are at angles $2\gamma^+ = \arctan(2\rho/\delta^+)$ and $2\gamma^- = (\pi - 2\gamma^+)$, respectively, to the [110] direction. As before, we have denoted by 2ρ and δ the phase differences between circularly and linearly polarized light waves that would exist if the crystal possessed only circular or linear birefringence, respectively. The intersection points M_1^+, M_2^+ and M_1^-, M_2^- of the OR^+ and OR^- axes with the sphere correspond to polarizations of the normal optical modes capable of propagating through the AFM^+ and AFM^- domains with unchanged polarization. The polarization transformations in the domains are shown in Figs. 2.1(a) and (b). The linear polarization of the incident light wave is represented by point P. In Fig. 2.1(a) the incident polarization is long [110], while in Fig. 2.1(b) it is along [100]. The polarization light leaving the AFM^+ domain is represented by point Q^+ and that at the exit from AFM^- by point Q^-. The polarization parameters in the two cases can be found from know ρ and δ with the aid of (1.50)–(1.57). Since both the Faraday rotation ρ and the linear birefringence δ are of the same magnitudes in both the AFM domains, the phase difference $\Delta = (4\rho^2 + \delta^2)$ between elliptically polarized propagating modes is the same for both domains. However, the sign of the Faraday rotation does not depend on the AFM state, whereas the signs of linear birefringence are opposite for the AFM^+ and AFM^- domains. As a result, the ellipticities and azimuth angles of the polarization ellipses in light beams that have passed the AFM^+ and AFM^- domains, respectively, are, according to (1.50) and (1.54), (1.51) and (1.55), equal, but the directions of tracing the ellipses are opposite. A similar, though simpler situation may be observed in antiferroelectric domains under the conditions of the longitudinal electro-optical Pockels effect. Because of the Faraday rotation, the azimuth angle of the polarization axis, in our case, depends upon the polarization azimuth of the incident plane polarized light in a somewhat more complex way. No matter

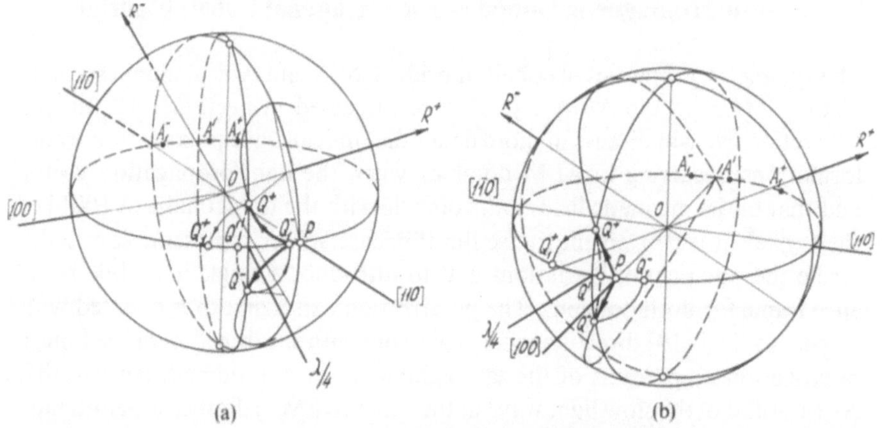

Fig. 2.1. Transformation of the light polarization in AFM domains of CoF_2 at $\mathbf{H} \parallel C_4$, $\mathbf{k} \parallel C_4$. Incident light is linearly polarized (point P) (a) along [110] and (b) along [100]. Polarization of light from the AFM^+ domain is indicated by Q^+, and from the AFM^- domain by Q^-. Q_1^+ and Q_1^- are polarizations of the light transmitted through a quarter-wave plate. The dashed line Q^+Q^- corresponds to the polarization of light transmitted through AFM^+ and AFM^- domains in the overlapping region.

which is the direction of the incident polarization, the azimuth angle changes as the magnetic field vector **H** is changed. Besides, it is different for the light of different wavelengths, since the dispersion dependences $\rho(\omega)$ and $\delta(\omega)$ are not identical.

In order to make domains visible with a linear analyzer, it is necessary to transform the elliptical polarization of light waves into plane polarization. This can be done with a quarter-wavelength linear phase changing plate oriented along the polarization-ellipse axis of the outgoing light wave. The light beams that have passed through the AFM^+ or AFM^- domain (points Q_1^+ and Q_2^- in Figs. 2.1(a) and (b) differ in azimuth angles of the polarizations axis. By rotating the analyzer to cross the polarization plane (points A^+ and A^-, respectively, in the above figures) we can darken one or the other AFM domain.

It may prove convenient to use the experimental scheme in which the specimen is illuminated with an elliptically polarized light wave, such that some of the domains (in Fig. 2.2 it is AFM^-) would transform it into plane polarized. The incident plane polarized light (point P) is transformed by the linear phase changing plate δ_k into elliptically polarized (point P_1). Then the AFM^- domain transforms the polarization back to a plane one (point Q^-), with however an azimuth angle different from the polarization azimuth of the incident light. The AFM^+ domain then leaves the polarization elliptical (point Q^+). Since the Faraday rotation and the birefringence owing to the LMOE are both proportional to **H**, the inclination of the axis OR^- (which the sphere rotates round) remains the same in all magnitudes of the magnetic

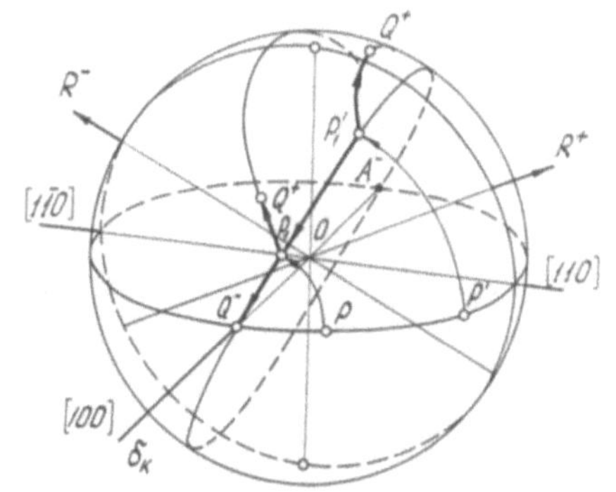

Fig. 2.2. Transformation of light polarization in the AFM domains of CoF_2 at $\mathbf{H} \parallel C_4$, $\mathbf{k} \parallel C_4$, transmitted through a birefringent δ_k-phase plate. Shaded points P, P_1, Q^+, Q^- and $P', P_1', Q^{+\prime}, Q^-$ correspond to different magnetic field strengths.

field. Therefore, light of a given ellipticity can be obtained for any values of **H** with the aid of one and the same phase plate with the phase shift $\delta_k = \pi/2 - 2\rho/\delta$ and a fixed azimuth $\lambda_k = \pi/4$ of its axis. The AFM$^-$ domain is darkened with the aid of a fixed linear analyzer with azimuth angle $\lambda_A = \pi/2 + \lambda_k = 3\pi/4$ (see Fig. 2.2). If the phase shift Δ in the domain is changed, e.g., as a result of a change in the magnetic field strength, then, in order to retain the linear polarization of the light leaving the AFM$^-$ domain, it would be necessary to change the incident polarization in such a way that the difference $(\lambda_k - \lambda)$ would be equal to Δ. The primed points in the figure correspond to a changed value of the magnetic field strength.

To darken the AFM$^+$ domain it is necessary to rotate the δ_k plate by 90°. If the AFM domains happen to be observed at a different temperature, the phase shift introduced by the compensating plate δ_k needs to be changed, since temperature dependences of the Faraday rotation and birefringence may differ essentially.

The structure of 180-degree AFM domains in CoF_2 also can be studied in the longitudinal geometry of the setup without phase-changing plates, using linear polarizers alone. Indeed, the domain projections on the image plane partially overlap and the light that has passed through two domains is polarized differently from the light leaving a single-domain area. The dashed line Q^+Q^- in Fig. 2.1 shows polarization states of the light that has passed through overlapping domains AFM$^+$ and AFM$^-$ with continuously varying thickness. The polarization changes from a right-hand elliptical, passing through the state of linear polarization (point Q') to a left-hand elliptical. The light of polarization Q' can be darkened with the analyzer placed in position A'. This is a simple way to visualize overlapping domains. Note that the azimuth angle λ' of point Q' is different from the azimuth $\lambda^+ = \lambda^-$ of the polarization ellipses Q^+ and Q^-, though only slightly. Depending on the azimuth angle λ of the incident polarization axis with respect to the direction

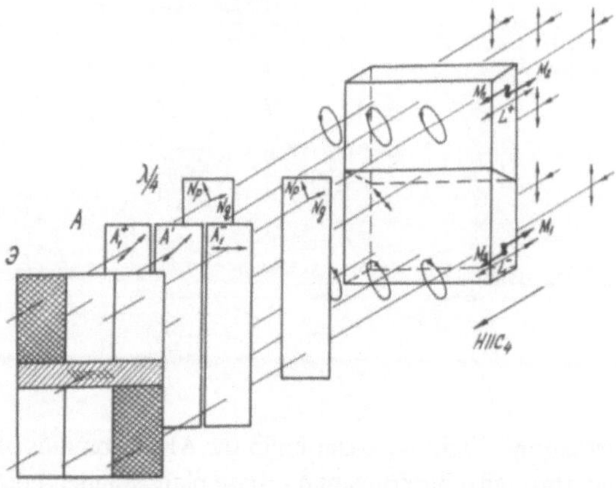

Fig. 2.3. Diagram of the optical observation of 180° AFM domains in AFM CoF_2.

[110], λ' can be either lower (with $\lambda = \pi/4$) or greater than $\lambda^+ = \lambda^-$ (with $\lambda = 0$ or $\pi/2$). It should also be noted that the light, having passed through AFM^+ and AFM^- domains of equal thicknesses, is not plane polarized. The relative thickness of overlapping domains with which the polarization becomes linear depends on the order of the sequence of the AFM^+ and AFM^- domains. All these optical properties of nonuniform specimens arise from the nonorthogonality of normal optical modes pertaining to one domain, relative to the modes of the other.

The contrast between the domains was created in the experiment [140–142] with the aid of a quarter-wavelength plate. The observation scheme is shown in Fig. 2.3. The incident light was polarized either along [110] ($\lambda = 0°$) or along [100] ($\lambda = 45°$). CoF_2 plates of different thicknesses ranging from 0.5 mm to 2 mm were cut perpendicular to the axis C_4. The aperture angle of the light source was 2°. The position of the light cone axis and the aperture angle could be controlled through the observation of conoscope figures. The photographs of Figs. 2.4–2.8 show the AFM domains that were observed in several specimens prepared from independently grown single crystals, and fixed as loosely as possible during the observation. The specimens were slightly pressed against a cold-conductor with a paper strip. The temperature could be varied from 15 K to $T_N = 38$ K. The photograph were taken in blue light ($\lambda \approx 400$ HM) of spectral width about 20 HM.

Observations in polarized light, with the use of a linear analyzer and with duly satisfied geometrical conditions concerning the source aperture and direction of the light beam, generally revealed the presence of a structure in the specimens. The characteristics shown by the structure, such as the dependence of its contrast upon the magnetic field strength, position of the analyzers and the $\lambda/4$-plate, mobility and disappearence at $T = T_N$, and finally, its sensitivity to the prior history of the specimen, have allowed us to interpret it as the structure of 180-degree AFM domains. By analyzing the domain con-

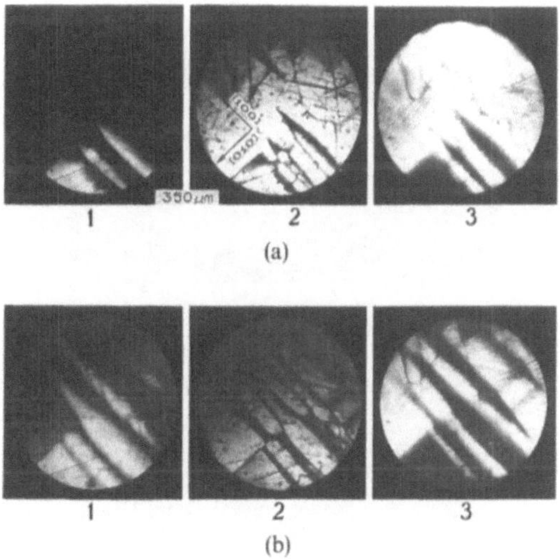

Fig. 2.4. 180° AFM domains in CoF_2. Observation conditions: $\mathbf{k} \parallel C_4 \parallel \mathbf{H}$, $\mathbf{E} \parallel [100]$, $T = 25$ K, $H = 4.9$ T. Photographs 1(a), 3(a), 1(b), and 3(b) are taken using $\lambda/4$-plates for various analyzer positions (A_1^- and A_1^- in Figs. 2.1(a)), photographs 2(a) and 2(b) are obtained without a $\lambda/4$-plate. Photographs 1(b)–3(b) are obtained after the field increase up to 5.0 T and its recovery to the initial value of 49 T.

strast as a function of the $\lambda/4$-plate position, experimentalists could judge whether the domains passed through the entire specimen or not, and then determined the location of oblique walls. While the photographs presented were obtained in rather high fields, domains can be observed quite easily in fields of lower intensity. They can be detected visually in specimens about 1.5 mm thick, with field strengths of 0.05 T.

The domain images shown in Fig. 2.4 were obtained at the $\mathbf{E} \parallel [110]$ polarization, while those of Fig. 2.5 were obtained at $\mathbf{E} \parallel [100]$. Photographs 2(a) and 2(b) of Fig. 2.4 and all the photographs of Fig. 2.5 have been obtained without the $\lambda/4$-plate in the optical setup. With certain positions of the analyzer one can distinguish quite clearly the dark regions separating the areas of equal values of the dimming angle. The analyzer positions providing a maximum contrast between light and dark areas, and the value of contrast, both depend on the magnetic field intensity. Such features of the image are in full agreement with the magneto-optical properties predicted for AFM domain structures. They were all grounds to assume the areas with equal dimming angles for "through" AFM domains. The conclusion has been confirmed by the photographs obtained with a $\lambda/4$-plate. The plate was mounted so that when it was parallel to the analyzer the pass-through AFM^+ and AFM^- domains were of equal optical density. Then the analyzer was rotated clockwise or counterclockwise, according to Fig. 2.1, to dim one or the other AFM domain. The angles of rotation for cases 1 and 3 of Fig. 2.4 were close to

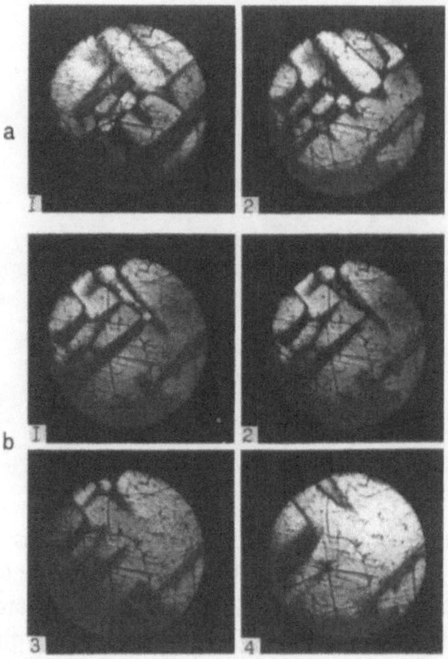

Fig. 2.5. Variation of the AFM domain structure in the CoF_2 sample with the field direction switched on to the opposite one (1(a) and 2(a)). Photographs 1(b), 2(b), 3(b) are obtained after the field is changed stepwise from 2.9 to 4.4 T in 20, 24, and 90 s, and photograph 4(b) is in 90 s after the field increases up to 4.9 T.

$\pm 15°$. Domain images of similar contrast are observed when the incident light is polarized at 45° to the optical axes plane along [100]. Indeed, the contributions of the linear birefringence and the Faraday rotation to the transformed polarization are nearly equal.

Despite the obvious difference of the domain structures observed in different specimens, they all had some features in common. First, the domain structure observations allowed dividing all the specimens into two groups, viz., one with "rigid" domains almost insensitive to external elastic stresses, and the other with "soft" domains whose structure could be easily change by compressing the specimen in the magnetic field. The residual nonuniform mechanical stresses that could be determined by the photoelastic effect were of similar magnitude in both specimens. The distinction between the two specimens could possibly be explained by the presence of uncontrolled impurities or structural microdefects. The specimens subjected to a strong mechanical compression (over 1000 kg/cm^2) acquired "rigidity."

When the specimen was again cooled after heating, the domain structure generally did not duplicate the preceding one (see Fig. 2.6), although individ-

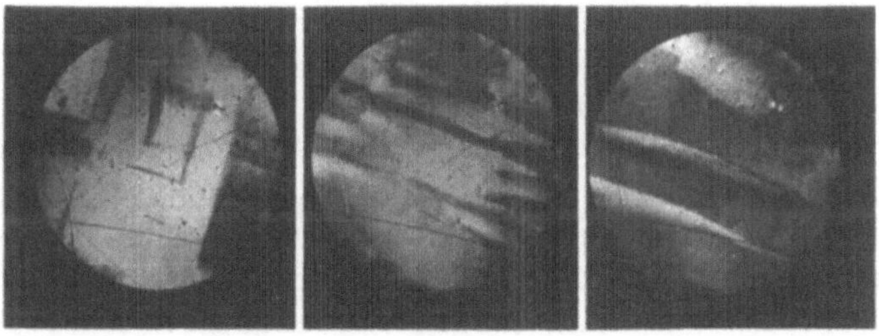

Fig. 2.6. 180° domain structures in AFM CoF_2 observed at temperatures near T_N after repeated heating and cooling, $T_N - T \sim 3$–5 K.

ual domains showed a distinct tendency to appear in the same sections of the specimen. The structure could be varied by producing temperature gradients in the specimen at $T \simeq T_N$ and cooling it in a magnetic field. After multiple transitions through T_N, "stable" domains could change their configuration or disappear, and new "stable" domains arose. A larger number of domains normally existed near T_N than at low temperatures. At $T < T_N$ the size and number of the domains changed as the field intensity was increased beyond a certain magnitude. The domains either increased or decreased in size under the action of the field, depending on the direction of the latter.

The domains produced in the magnetic field often appeared on side faces of the specimen and assumed the form of rectangular blades, lances, or pointed flat lobes (see Figs. 2.4 and 2.5). The lancelike domains advanced slowly as the field was further increased, or retreated if the field was decreased. As soon as the advancing domains reached the opposite face of the specimen the domain structure was immobilized and could not be restored to its prior state in a reduced field (Fig. 2.5). Alterations in the structure were observed upon reversal of the field direction. A change of the domain structure is a slow process lasting for several minutes. The photographs of Fig. 2.5 show the domain motion after a stepwise variation of the field. While the domain structures and their behavior observed in CoF_2 are quite varied, some characteristic features can still be indicated, namely:

(1) preferred orientation of domain walls along the planes (100) and (010) in a magnetic field $\mathbf{H} \parallel [001]$;
(2) appearance of lancelike domains directed along the crystallographic axes of type [100];
(3) a tendency towards production of strip structures crossing at different levels.

The formation and behavior of lancelike AFM domains is quite similar to the behavior of elastic crystal twins in ferroelastic materials. The similarity permits us to suggest that the AFM domain structure formed in CoF_2 in a magnetic field might be closely related to elastic interactions. The reason for the behavior observed may be the larger linear magnetostriction which reaches (in CoF_2) as high a values as $\Delta l/(lH) = 4.9 \cdot 10^{-6}\ T^{-1}$ [114]. The crystal symmetry of the AFM CoF_2 is reduced in field **H** ∥ [001] from 4/*mmm* to *mmm*. Depending on the sign of the AFM vector projection upon the direction of **H**, two orientation states of low symmetry, S_1 and S_2, are possible, which differ in the direction of lattice compression, i.e., along either the [110] or [110] axis. These states may be regarded as elastic states having the paraphase 4/*mmm*.

Among the various types of AFM domain walls allowed by the isotropic exchange interaction, infinitely preferable are those which are crystalline coherent and hence make a minimum contribution to the crystal elastic energy. According to [144], coherent boundaries between states S_1 and S_2 are mirror planes which are present in the paraphase, but are lost at the transition of the orientation states S_1 and S_2. In the case of CoF_2, only two coherent boundaries are possible, namely the (100) and (010) planes. The domain walls appearing in the magnetic field have a pronounced tendency to be oriented along these planes (see Figs. 2.4–2.8). An equally clear tendency is that of forming strip domain systems that intersect on different levels, and this is not accidental. Apparently, such formations are due to the elastic stresses that develop near the incoherent walls close to the (001) plane, and at the lancelike ends of the strip domains.

A detaled analysis of domain structures seems to necessitate an account of the piezomagnetic moment produced at the locations of elastic stresses. The presence of a magnetic moment in the domain wall itself may also prove essential. The moment appears because of the AFM vector rotation in the wall and owing to the Dzialoshinskii–Moriya-type interaction.

Figure 2.7 is an illustration of the interaction between a domain wall and a crystal defect of the edge dislocation type, during advance of the wall. Among other domain wall orientations, those along (120)-type plane were also ob-

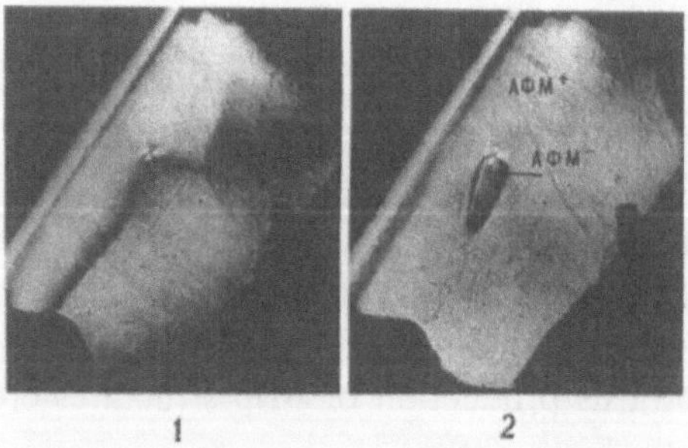

Fig. 2.7. Interaction of the AFM domain with the field of elastic stresses near the crystalline defect of the edge dislocation type: after the sample remagnetization a AFM^- domain is delayed near the defect.

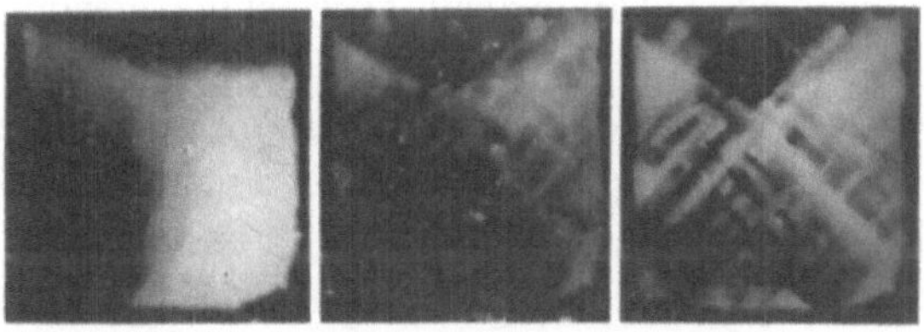

Fig. 2.8. Magnetic field effect on the domain structure formed under AFM ordering: (1) $H_2 = 0$; (2) 1.5 T; (3) 2.2 T; $H_\perp = 0$. Domains were visualized using LMOE in field H_z at about 2.0 T (1, 2, 3).

served. The "softer" specimens often revealed curved boundaries (see Figs. 2.8, 2.9, and 2.11).

Of considerable interest is the behavior of a domain structure near the Neél point. If the specimen is first heated to a higher temperature than T_N and then cooled in a magnetic field $\mathbf{H} \parallel [001]$, one can clearly see an increase in the number of domains as the field strength is raised. The characteristic structure becomes that of a number of domains crossing at different levels and oriented along [100] and [010] (Fig. 2.8). The process of structure formation cannot be monitored, since the contrast between different domains becomes sufficient for observation only 1 or 2 K below the Neél point. With a further decrease in temperature the structural changes become insignificant, the domain structure remaining practically the same from temperaures only 2 or 3 K below T_N. This "freezing" of the structure is most likely due to an increase in lower temperatures in the coupling energy of the domain wall to microinhomogeneities of the crystal. If the crystal temperature approaches the Neél point from below, the process of multiple domain formation in a specimen preliminarily brought to a single-domain state cannot be followed in detail either. Normally, the temperatures at which a similar multitide of domains appears are within 1 K from T_N. Obviously, the subdivision of an AFM specimen into several domains is energetically disadvantageous because the domain wall energy is positive. Therefore, the appearance of a domain structure at the transition through T_N in a magnetic field admittedly is controlled by magnetic ordering kinetics and magnetoelastic interactions. Magnetic ordering fluctuations near T_N are responsible for the formation of a three-dimensional "net" of a more-or-less mobile AFM^+/AFM^- domain structure, just as it should be in second-order phase transitions [145]. The presence of a magnetic field oriented along [001] results in a linear magnetostriction along the [110] and [110] directions, and preferable wall orientation along [100] and [010]. In view of the finite wall thickness, this ultimately results in mechanical strains in and near the wall. The interaction of such strains can produce a periodic structure [146], as is often the case with crystal twinning. The same mechanical strains increase the wall energy and rigidity, thus being the cause of its earlier (at lower $T_N - T$ differences) fixation against crystal imperfections in a growing field.

Another mechanism that could contribute to the domain wall formation is related to the increased entropy contribution of domain walls to the specimen free energy. In the case where the domain wall has no fixed location or its position is fixed only loosely, then its small displacements may result in noticeable changes of the specimen entropy [147]. This contribution increases with temperature and, according to [147, 148], may even render a domain wall thermodynamically advantageous near T_N. This does not seem to occur in the highly anisotropic crystal with CoF_2 is, as the domain structure formation cannot be detected during heating of a single-domain specimen. And yet the entropy contribution of the domain walls may be a favorable factor for the "freezing" of a "live" domain net produced kinetically at temperatures near T_N. The process is not controlled by this mechanism alone. Indeed, the characteristic structure of small-scale domains does not appear if the specimen is cooled without a magnetic field.

The techniques of AFM domain visualization in CoF_2 have allowed us to monitor directly the domain behavior under the action of those external agents which violate the energy equivalence of the AFM domains. For example, this occurs when the specimen is subjected simultaneously to elastic mechanical straining σ_{xy} along [110] and a magnetic field $\mathbf{H} \parallel [001]$. Owing to the piezomagnetic moment $M_z = Q_{zxy}\sigma_{xy}$, which points in opposite directions in the AFM^+ and AFM^- domains, these become energetically nonequivalent, $E^+ - E^- = 2Q_{zxy}\sigma_{xy}H_z$, and hence the wall between the AFM^+ and AFM^- domains is under pressure and starts to move. With sufficiently large σ_{xy} and H_z the specimen may be brought to a single-domain state. This was exactly the way to investigate the piezomagnetic effect in AFM specimens [94]. The magneto-optical methods allow us to follow visually the domain configuration variations in the process [95]. As it turns out, various CoF_2 specimens may demonstrate a different behavior depending on their purity and prior history. The "correct" behavior is shown by the "softer" specimens, in which the remagnetization of the initially uniform AFM material is pronounced quite distinctly. The metastable state is retained up to some threshold value H_{t1} of the magnetic field, after which the walls of small-scale residual domains start to move near to the side faces of the specimen. The residual domains could not be removed even with maximum applied stress $\sigma_{xy} = 600$ kg/cm^2 and magnetic field $H_z = 40$ T. The wall motion practically ceased at $H = H_{t1}$. The difference in the threshold fields was about 0.3 T at $\sigma_{xy} = 600$ kg/cm^2 and up to 1.5 T at $\sigma_{xy} = 200$ kg/cm^2. Both smooth and jumplike motion was observed at the remagnetization which is illustrated in Fig. 2.9. Figure 2.10(a) shows the threshold fields H_{t1} and H_{t2} as functions of the mechanical stress. The values satisfy the condition $\sigma_{xy}H_{ti} = \text{const}$ (see Fig. 2.10(b)). Actually, it means that a domain wall cannot start moving before the pressure against it has reached a threshold value controlled by the energy of coupling to crystal defects. Since the relation between σ_{xy} and H_{ti} is linear, we

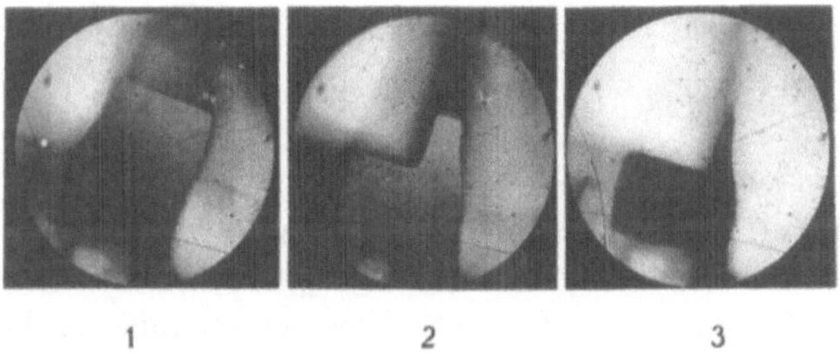

Fig. 2.9. Domain structure change under remagnetization of the AFM CoF_2 sample in a 3.5 T constant magnetic field $\mathbf{H} \parallel C_4$ at pressures varying from 0 to 200 kg/cm^2 along the [110] axis at $T = 15$ T.

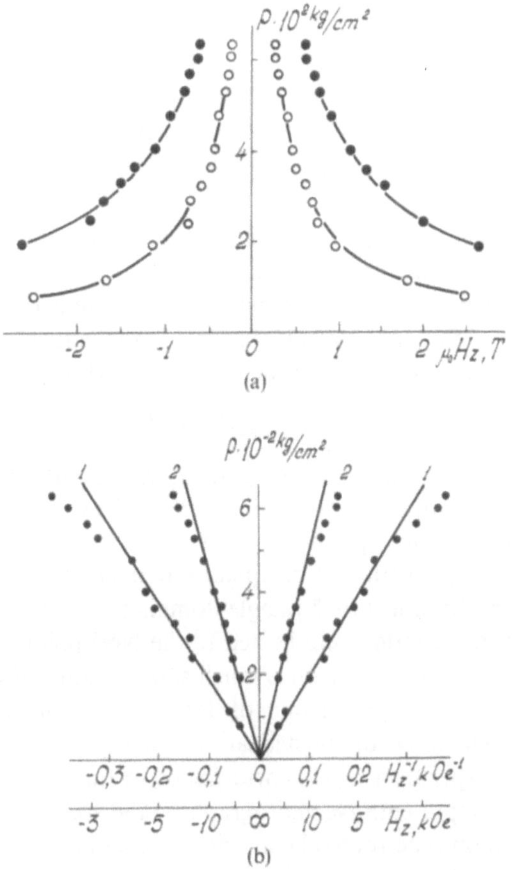

Fig. 2.10. Diagram of the AFM state of CoF_2 in field $\mathbf{H} \parallel C_4$ and pressure P along [110]: in coordinates $a(H, P)$ and (b) in (H^{-1}, P). Lines 1 and 2 correspond to the beginning and end of remagnetization, respectively, with a pressure increase.

are in a position to determine this coercivity energy of the AFM domain wall, i.e.,

$$E_c = Q_{zxy}\sigma_{xy}\overline{H}_{tz}. \tag{2.1}$$

According to [94], the magnitude of the piezomagnetic tensor component Q_{zxy} in CoF_2 is 0.8×10^{-3} Gauss $(kg/cm^2)^{-1}$ and $\overline{H}_t = \frac{1}{2}(H_{t1} + H_{t2})$. For the CoF_2 specimens examined by the authors [95] the coercivity energy was evaluated as 3×10^3 erg/cm^3 at $T \sim 20$ K. However, the range of the threshold field $(H_{t2} - H_{t1})$, within which the remagnetization occurred in the entire specimen volume (see Fig. 2.10), was too wide for the coercivity energy to be determined more accurately. The range width is admittedly associated with the nonuniform distribution of mechanical stresses in the specimen. For the "rigid" specimens no general dependences of the threshold field values upon pressure have been established. In most cases, the applied mechanical stresses had no effect on the magnetization time reversal, which process continued in the magnetic field but without pressure as well.

Another way of making the AFM domains energetically nonequivalent is associated with the **H**-quadratic magnetization effect which is allowed by the symmetry of most piezomagnetic antiferromagnets, viz.

$$M_i = C_{i\alpha\beta}H_\alpha H_\beta. \tag{2.2}$$

In CoF_2 the field-quadratic magnetic moment is perpendicular to the applied field. It assumes its maximum values with the field orientations along [110], [101], or [011]. The domain behavior was studied [99] for the field vector lying in the diagonal plane (110). This is the case in which the additional magnetic energy (different for the AFM$^+$ and AFM$^-$ domains) is

$$E^{\pm} = \pm 3C_{xyz}H_xH_yH_z. \tag{2.3}$$

The difference in these energies, i.e., $6C_{xyz}H_xH_yH_z$, is responsible for the pressure against the domain wall, whose direction can be varied by changing the sign of a field projection. Specifically, it was the z-component whose sign could be changed. Application of the quadratic magnetization effect allowed remagnetizing, and brining it to the single-domain state CoF_2, FeF_2, and even MnF_2 specimens if the latter were heated to the Neél point. The magnetization reversal in CoF_2 occurred over a much shorter range of H_z values (with fixed magnitudes of H_x and H_y) than in the former technique with the application of pressure. The range of magnetization reversal field $H_{t1} - H_{t2}$ is small in the same CoF_2 specimen which corresponds to the diagram of Fig. 2.10. Figure 2.11 shows representative photographs of the domains that are formed under magnetization time-reversal conditions. As can be seen from their appearance, the energetically favored domains in this case develop near distinctly pronounced (001) planes of the specimen. The domain boundaries are of characteristically smooth, rounded shapes and essentially nonplanar. At high temperatures, the domain boundaries varied closely in step with the increasing field, whereas at about 10 K and below the magnetization, reversal

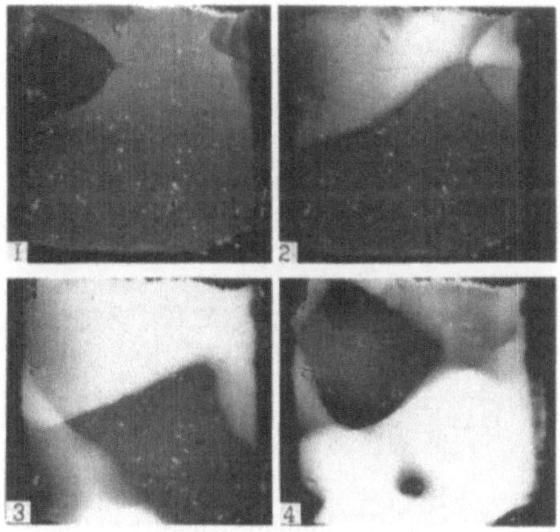

Fig. 2.11. Domain structures formed in CoF_2 after AFM remagnetization in the canted magnetic field $\mathbf{H} \parallel [11h]$, $H_{[1\,1\,0]} = 0.5$ T, H_z changes from 0.88 to 0.90 T.

occurred almost "instantaneously" and the boundary positions could not be fixed visually. Multiple reversals resulted in irreversible changes in the domain configurations. In particular, the domain geometries grew simpler, the walls became nearly planar, and the range of the magnetization reversal fields somewhat increased. These irreversible effects seem to be evidence for a redistribution of crystal defects by a moving domain wall.

Figure 2.12 shows the z-component, at the magnetization reversal point, as a function of the product $(H_x H_y)^{-1}$ of the field components. The measured points lie very neatly on a straight line, which is evidence for the association of the assumed magnetization reversal mechanism in a tilted magnetic field $\mathbf{H} \parallel$

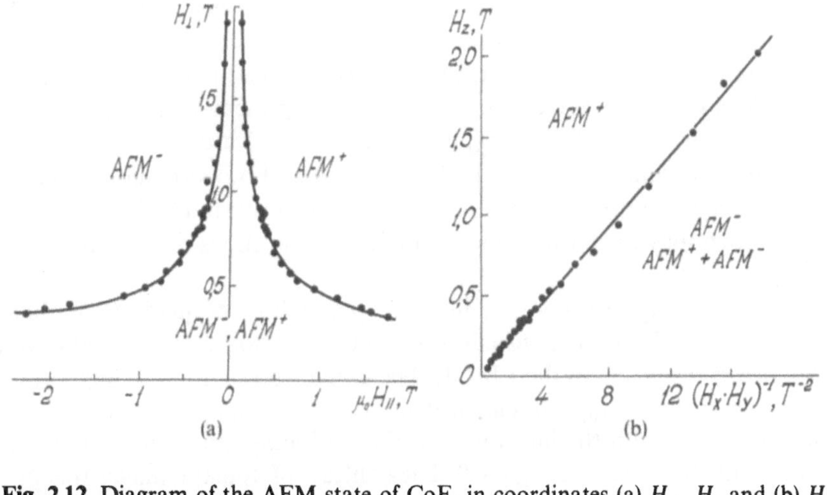

Fig. 2.12. Diagram of the AFM state of CoF_2 in coordinates (a) $H_\perp$, $H_\parallel$ and (b) H_z, $(H_x H_y)^{-1}$. Here $\mathbf{H}_\perp \parallel [110]$, $\mathbf{H}_\parallel \parallel [001]$.

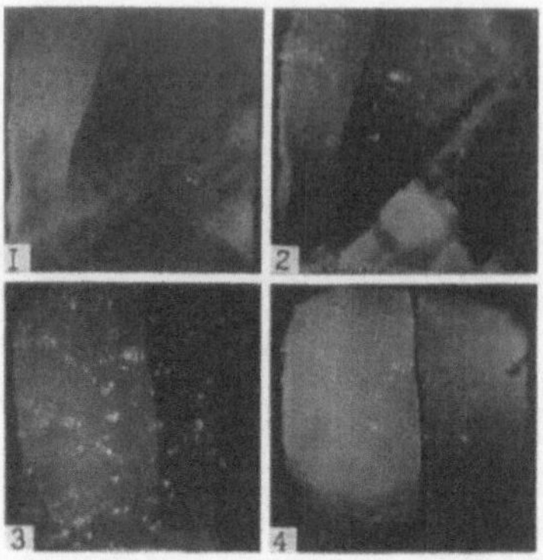

Fig. 2.13. Variation of the AFM CoF_2 domain structure prepared after sample cooling in field $H_z = 2$ T (see Fig. 2.8, photograph 3) under an applied field $H_z = 0$, $H_{[110]} = 2.2$ T (photographs 1, 2) at 10 K and field effect $H_z = 0$, $H_{[110]} = 2.0$ T on formation of the AFM domain structure after sample cooling from $T > T_N$ to $T < T_N$. Domains were visualized by the LMOE in field $H_z = 1.5$ T, $H_\perp = 0$ at 10 K (photographs 2, 4) and using the quadratic magnetic rotation at $H_z = 0$, $H_{[110]} = 2.0$ T (photographs 1, 3).

[11h] with the quadratic magnetization effect. Using the value $C_{xyz} = 0.68$ emu/T^2, which follows from the measurements of [105], we can evaluate the wall coercivity energy,

$$E_c = 6G_{xyz}H_{tx}H_{ty}H_{tz}. \tag{2.4}$$

The magnitude for $T = 4.2$ K is $9 \cdot 10^3$ erg/cm^3 which is in agreement with the value deduced from the magnetization reversal condition under the joint action of pressure and a magnetic field.

The magnetic moment M_z induced in a field $\mathbf{H} \parallel [110]$ may bring about changes in the domain structure, even without the H_z field component. The photographs of Fig. 2.13 illustrate changes in the structure prepared by cooling a specimen from $T > T_N$ in the field $H_z = 2.0$ T (see photograph 3 of Fig. 2.8). The changes occurred when the only field present was $H_{[110]} = 2.2$ T perpendicular to the [001] axis (photographs 1 and 2 are for $T = 10$ K, and photographs 3 and 4 for $T \sim T_N$). The appearance of two domains that divide the specimen into roughly equal parts admittedly is due to the demagnetizing fields which may arise when the magnetic moment M_z is induced by a field $H_{[110]}$.

The photographs of Fig. 2.13 illustrate possibilities of quadratic Faraday rotation for AFM domain visualization in CoF_2. Photographs 2 and 4 were obtained with the aid of the LMOE. The domain structure that had formed with $H_z = 0$ and $H_{[110]} \neq 0$ remained, even after $H_{[110]}$ was switched off, and did not change with the introduction of H_z. Photographs 1 and 3 were obtained with $H_z = 0$ and $H_{[110]} \neq 0$. The contrast of domain images was produced owing to the QMR of the polarization plane.

2.1.2. The Magnetic Phase Transition in Cobalt Fluoride Induced by a Magnetic Field $\mathbf{H} \parallel C_4$

The possibility of controlling the uniformity of the AFM state in cobalt fluoride specimens has helped to understand the reasons for their unusual behavior in high magnetic fields [149].

The experiments aimed at investigating the behavior of the AFM CoF_2 in intense pulsed magnetic fields $\mathbf{H} \parallel C_4 \parallel \mathbf{L}$ have revealed abrupt changes in the crystal state accompanied by destruction of the specimens [150–152]. At first, it was assumed that the crystal experienced a first-order phase transition of the spin-flop type, with their destruction being a result of magnetic phase lamination. However, a more detailed analysis of the as-profound changes brings forth doubts as to the very possibility of a first-order phase transition. CoF_2 is a representative of compensated tetragonal antiferromagnets characterized by a Dzyaloshinski-type invariant $d(M_xL_y + M_yL_x)$ in their thermodynamical potential. The invariant exists owing to a rhombic symmetry of the local magnetic anisotropy. It manifests itself through the effects occurring when sublattice moments are deflected from the tetragonal axis. As follows from the phenomenological analysis of spin-orientational transitions of the spin-flop type, that has been carried out for this class of antiferromagnets in the model of constant length sublattice magnetic moments, a transition to the noncollinear state should occur, with sufficiently large $d \geq d_{cr}$, through a second-order phase transition [153–155]. Comparing predicted values of the critical parameter d_{cr} with the estimates d_{exp} following from measured data we find that in CoF_2 $d_{exp} \geq d_{cr}$, and hence sharp magnetic transformations do not seem possible. The transition to the noncollinear state must be gradual. Even if the transition retained the spin-flop character because of some invariants so far unaccounted for, i.e., remained a first-order phase transition, distinctions between the collinear and the angular state should not be large. Meanwhile, the fact of specimen destruction was ascribed to mechanical strains in the wall separating the domains of two magnetic phases, i.e., the collinear and the spin-flopped. The fact itself served as a criterion for determination of the phase transition order. It can hardly be expected that separation into domains of two close magnetic phases with only slightly different polar angles of their AFM vectors can be responsible for the specimen destruction. A more rigorous analysis of the behavior shown by cobalt ion magnetic moments does not support the concept of a sharp transition. Field-induced variations of the sublattice moment vector lengths can only result in a still more gradual transition, as in ferrimagnets near the compensation point when the different $/T - T_c/$ increases [156].

On the other hand, a possible cause for the destruction of AFM CoF_2 specimens may be the elastic stresses arising in incoherent domain walls that separate the 180-degree AFM domains, owing to linear magnetostriction. In order of magnitude, the elastic stresses in the wall can approach $(\Delta D/t) \cdot E_{shear}$, where E_{shear} is the shear modulus, t is the wall thickness, and ΔD is the variation of the transverse domain size. The latter value may be as high as $5 \cdot 10^{-7}$ cm with the striction $(\Delta l/l) \cdot H^{-1}$ about 10^{-6} T^{-1} [114] and the domain size of 10^{-2} cm in a 10 T field. The domain wall thickness in CoF_2 may reach several tens of Ångström [148]. Assuming the value to be 10^{-6} cm, we can estimate the maximum possible stress in an incoherent wall as 0.5 E_{shear}. Similar stresses may produce macroscopic destructions in the crystal.

To study the magnetic transformation, the authors [149] appealed to magneto-optical methods. The LMOE allowed us to control the domain structure in the specimens. The experiments were performed in pulsed magnetic fields. It was possible to photograph the conoscopic figure of the CoF_2 plate when the field reached its maximum value. The photographed conoscopic figures allowed us to find the position of the optical axes plane and to evaluate the magnitudes of birefringence for the linearly and circularly polarized light, depending on the field strength. Another experimental technique employed was the photoelectric registration of polarization variations in narrow light beams propagating along the optical axes bissectrix. The presence of domains in the specimen could be established from the type of conoscopic figures and the magnitude of the field-induced linear birefringence. To some extent the AFM domain structure could be controlled by varying the mode of specimen fixation. If a thick (about 3 mm) sample was glued onto the cold-conductor with butvar–phenolic adhesive, the result was a multidomain structure. The corresponding conoscopic pattern was nearly uniaxial even in fields as high as 10 T. Owing to compensation for the linear birefringence by AFM domains that follow one another, the Faraday rotation could be measured by conventional techniques. To obtain a single-domain state, the specimen was compressed along [110] in the course of fixation. The crystal reached the state in a magnetic field lower than 10 T and remained single-domain after the field pulse ended.

Multidomain thick specimens were generally destroyed by first pulse of a 15–20 T intensity, whereas single-domain thick specimens endured multiple applications of field pulses up to 25 T. Think multidomain specimens (of about 0.2–0.4 mm) often remained undestroyed as well.

Field dependences of the linear birefringence, $\delta(H)$, and the Faraday rotation, $\rho(H)$, showed a singularity near $H \sim 18.2$ T, which appeared as a kink rather than a jump. A similar singularity could be observed in the intensity of the light that had passed through a multidomain specimen and an analyzer, unless the specimen had been destroyed at a lower field strength. Single-domain specimens showed no jumplike changes whatsoever.

A similar performance suggests that a first-order phase transition does not occur in CoF_2 specimens placed in a longitudinal magnetic field, and hence the specimen destruction cannot be associated with the supposed separation into a collinear and spin-flopped phase. The specimen is destroyed by the elastic stresses that grow due to the magnetic field in the incoherent domain walls separating the 180-degree AFM domains. The fact that thin multidomain specimens often remain undestroyed can be explained by the smaller size of their domains, and hence by lower mechanical stresses in the walls which prove insufficient for destroying the crystal.

The kink point at $H_{cr} = 18.2 \pm 10$ T, shown by the $\delta(H)$ and $\rho(H)$ dependences, is evidence for a smooth, gradual transition to the noncollinearity, i.e., a second-order phase transition. Some conclusions on the noncollinear state symmetry were made from the analysis of the optical indicatrix at $H > 18.2$ T. By analyzing the conoscopic figures for this range of magnetic fields, the authors have established that the optical axes lying within the (110)-type plane in moderate fields remain in the same plane (to within $\pm 3°$) when the field intensity is raised above 18.2 T, up to the maximum achievable. This leads to the conclusion that the point symmetry of the CoF_2 noncollinear state is $\underline{2}/\underline{m}$, with the sublattice magnetic moments turning within the (110)-type plane. Indeed, the magnetic point groups of the various field-induced noncollinear states are those subgroups of the point group $\underline{4}/m\underline{m}m$ characteristic of the AFM CoF_2 which allows projections along **H** of the axial magnetic vector. Considering only the magnetic structures in which the magnetic unit cell is the same as the crystal cell, we can conceive of two types of field-induced noncollinear structures, characterized by the groups $\underline{2}/\underline{m}$ and $\bar{1}$. Group $\underline{2}/\underline{m}$ describes the angular ($\theta \neq \pi/2$) and the spin-flopped ($\theta = \pi/2$) states in which the sublattice moments lie within the diagonal (110)-type crystal plane. Group $\bar{1}$ describes the states in which the sublattice moments may lie in arbitrary planes, including the state where the AFM vector is within the (100) plane, which ceases being a symmetry plane in an applied field. The nonzero components of the LMOE tensor for the state $\underline{2}/\underline{m}$ are $q_{x'x'z'}$ $q_{y'y'z}$, q_{zzz}, and $q_{y'zz}$, while components $q_{x'zz}$ and $q_{x'y'z}$ are identical zeros. The components have been written in the frames $Z \parallel C_4$ and $X' \parallel [110] \parallel \underline{\bar{2}}$. Among the quadratic Cotton–Mouton components $C_{x'y'zz}$ and $C_{x'zzz}$ are zeros, which $C_{x'x'zz}$, $C_{y'y'zz}$, C_{zzzz}, and $C'_{y'zzz}$ are not. In the state $\bar{1}$ none of the tensor components are zeros. Since $q_{x'y',z}$, $q_{x'zz}$, $C_{x'y'zz}$, and $C_{x'zzz}$ in the state $\underline{2}/\underline{m}$ are zeros, the optical axes lie within (110) type-plane. Their bissectrix may not be pointed along [001], but rather make with it the angle

$$\tan 2\xi = 2(q_{y'zz}H_z + C_{y'zzz}H_z^2)(\varepsilon_{y'y'} - \varepsilon_{zz})^{-1} \tag{2.5}$$

in the same plane. The angle ξ is small, because of $q_{y'zz}H_z$, $C_{y'zzz} \ll (\varepsilon_{y'y'} - \varepsilon_{zz})$, and hence is hard to measure.

In the state $\bar{1}$ the plane of optical axes is a general position plane. Its azimuth angle with respect to the (110) plane may well be a sizable value, since the off-diagonal component $\varepsilon_{x'y'}$ and the difference $(\varepsilon_{x'x'} - \varepsilon_{y'y'})$ of diagonal components are magnitudes of the same order. As follows from the estimates of [149], proceeding from measurements of the longitudinal LMOE [92] and birefringence in an $\mathbf{H} \parallel [110]$ field [197], as well as from calculated values of the AFM vector deviation from [001] in a $\mathbf{H} \parallel [100]$ field [158], a deviation of the AFM vector from the (110) plane by a few degrees would result in a measurable, slightly greater deviation of the optical axes plane from (110). Since the optical axes are known to lie within (110) to $\pm 3°$ for magnetic fields up to 25 T, we must conclude that the sublattice moments are within (110) to about the same accuracy.

2.2. Magnetic Structure of the Antiferromagnetic Calcium–Manganese–Germanium (CaMnGe) Garnet

The high sensitivity of the LMOE to the magnetic ordering symmetry, that has been seen distinctly in cobalt fluoride and dysprosium orthoferrite crystals, may prove useful for establishing the point symmetry of complex AFM crystals where other techniques often fail to be sufficiently effective. Thus has been exactly the case with the multisublattice noncollinear tetragonal AFM garnet $Ca_3Mn_2GeO_{12}$.

2.2.1. The Magnetic Point Symmetry of Manganese–Germanium Garnet

The calcium–manganese–germanium garnet (MnGeG) stands out among other AFM garnets because its magnetic ions, Mn^{3+}, are Jahn–Teller ions. They occupy the octahedric sites of the cubic garnet. The Jahn–Teller distortions of the ligand octahedrons are cooperatively ordered at $T_{J-T} = 516$ K and the crystal symmetry reduces from the cubic, $m3m$, to the tetragonal [159–161]. At $T_N = 13.5$ K the crystal magnetic moments are ordered antiferromagnetically [160, 161]. Since tetragonal deformations of the cubic garnet are very small in the Jahn–Teller transition ($c/a \simeq 1.003$ [161]), admittedly becuase of a nonferrodistortional ordering in local deformations of the oxygen octahedrons, several attempts at establishing the space group of the tetragonal crystal have failed.

The change-over to the tetragonal state is a second-order phase transition, which can be seen from the nature of the alteration of the lattice constants [161] and the increase in the linear birefringence when temperatures decrease near T_{J-T} [162]. Assuming the crystal to retain a symmetry center during the transition, we can describe the tetragonal structure of MnGeG below T_{J-T} in terms of either $4/mmm$ or $4/m$ point groups. The magnetic point groups to describe the AFM MnGeG after magnetic ordering may be $4/mmm$, $4/m\underline{mm}$, $\underline{4}/mm\underline{m}$, $4/m$ or $\underline{4}/\underline{m}$ [16$\underline{4}/m\underline{m}m$, 3–165]. All of these allow the LMOE, however, manifestations of the effect are different. Table 2.1 presents matrices of the magneto-optical tensor q_{ij}^3, responsible for the LMOE, for all groups. Also given are the corrections to symmetrical components of the tensorial dielectric constant which may arise owing to the LMOE with three symmetrical orientations of the magnetic field. It can be seen from the table that the point symmetry of MnGeG can be established through analysis of the LMOE data for different experimental geometries. For example, with $H \parallel C_4$ any manifestations of the LMOE for light beams propagating along C_4 should be absent if the crystal is characterized by either of the groups $4/mmm$, $4/m\underline{m}m$, or $4/m$.

Table 2.1

Magnetic point group		$4/mmm$			$4/m\underline{m}\underline{m}$			$\underline{4}/mm\underline{m}$		
Matrix of the LMOE tensor $q_{\mu\alpha} \leftrightarrow q^{s}_{ij\alpha}$		·	·	·	·	·	q_{13}	·	·	·
		·	·	·	·	·	q_{13}	·	·	·
		·	·	·	·	·	q_{33}	·	·	·
		q_{41}	·	·	·	q_{42}	·	q_{41}	·	·
		·	$-q_{41}$	·	q_{42}	·	·	·	q_{41}	·
		·	·	·	·	·	·	·	·	q_{63}
LMOE-induced corrections to the dielectric tensor components for symmetrical orientations of the magnetic field		·	·	·	·	·	$q_{42}H$	·	·	·
	$H \parallel X$	·	·	$q_{41}H$	·	·	·	·	·	$q_{41}H$
		·	$q_{41}H$	·	$q_{42}H$	·	·	·	$q_{41}H$	·
		·	·	$-q_{41}H$	·	·	·	·	·	$q_{41}H$
	$H \parallel Y$	·	·	·	·	·	$q_{42}H$	·	·	·
		$-q_{41}H$	·	·	·	$q_{42}H$	·	$q_{41}H$	·	·
		·	·	·	$q_{43}H$	·	·	·	$q_{63}H$	·
	$H \parallel Z$	·	·	·	·	$q_{13}H$	·	$q_{63}H$	·	·
		·	·	·	·	·	$q_{33}H$	·	·	·

Table 2.1 (*cont.*)

Magnetic point group	$\underline{4}/m\underline{m}m$			$4/m$			$\underline{4}/m$		
Matrix of the LMOE tensor $q_{\mu\alpha} \leftrightarrow q^s_{ij\alpha}$	·	·	q_{13}	·	·	q_{13}	·	·	q_{13}
	·	·	$-q_{13}$	·	·	q_{13}	·	·	$-q_{13}$
	·	·	·	·	·	q_{33}	·	·	·
	·	q_{42}	·	q_{41}	q_{42}	·	q_{41}	q_{42}	·
	$-q_{42}$	·	·	q_{42}	$-q_{41}$	·	$-q_{42}$	q_{41}	·
	·	·	·	·	·	·	·	·	q_{63}
LMOE-induced corrections to the dielectric tensor components for symmetrical orientations of the magnetic field	·	·	$-q_{42}H$	·	·	$q_{42}H$	·	·	$-q_{42}H$
	·	·	·	·	·	$q_{41}H$	·	·	$q_{41}H$
	$-q_{42}H$	·	·	$q_{42}H$	$q_{41}H$	·	$-q_{42}H$	$q_{41}H$	·
	·	·	·	·	·	$-q_{41}H$	·	·	$q_{41}H$
	·	·	$q_{42}H$	·	·	$q_{42}H$	·	·	$q_{42}H$
	·	$q_{42}H$	·	$-q_{41}H$	$q_{42}H$	·	$q_{41}H$	$q_{42}H$	·
	$q_{13}H$	·	·	$q_{13}H$	·	·	$q_{13}H$	$q_{63}H$	·
	·	$-q_{13}H$	·	·	$q_{13}H$	·	$q_{63}H$	$-q_{33}H$	·
	·	·	·	·	·	$q_{33}H$	·	·	·

In case the point group is $\underline{4}/mm\underline{m}$, $\underline{4}/m\underline{m}m$, or $\underline{4}/m$, the crystal becomes biaxial when placed in a magnetic field, and hence the light beam should experience birefringence. The optical axes would be oriented differently for each of the three groups. For $\underline{4}/mm\underline{m}$ and $\underline{4}/m\underline{m}m$ the axes should lie in the planes (110) and (100), respectively. In the case of group $\underline{4}/m$, the azimuth angle of the optical axes plane is determined by the magneto-optic constant ratio q^s_{xyz}/q^s_{xxz} and may not correspond to either (100) or (110). Moreover, the azimuth angle of the optical axes plane may be different for the light of different frequencies. To discriminate between the groups $4/mmm$, $4/m\underline{m}\underline{m}$, and $4/\underline{m}$, the LMOE should be investigated for other geometries of the experiment.

The crystallo-optical properties of MnGeG placed in a magnetic field were studied experimentally for **k** ∥ **H** ∥ C_4 [126, 166] and **k** ∥ **H** ⊥ C_4 [127]. Among the measured values were rotations of the light polarization axis and the ellipticity of the outgoing light. The plate specimens prepared from a bulk single crystal contained many Jahn–Teller twins with different orientations of the tetragonal axis. They could be almost completely removed by thermal treatment of the mechanically strained plate. The specimen state could be monitored by visual microscopic observations in polarized light.

Visual observations at temperatures below the Neél point have shown the LMOE to occur in the AFM MnGeG placed in a magnetic field. Indeed, the observed changes in the light polarization were of opposite sign for two opposite magnetic field orientations. The polarization changes were of sufficient magnitude to allow visual observation of the AFM domains. It proved possible to reverse the specimen magnetization with an applied magnetic field, which made it easier to separate the LMOE and the Faraday rotation from the total polarization changes observed. The fact that the LMOE was observed in the crystal is unambiguous evidence for the coincidence of the chemical and magnetic unit cell in the manganese–germanium garnet; otherwise, the LMOE would be forbidden by the crystal symmetry. This conclusion is in agreement with the results of neutron diffraction investigations [164]. The rotation of the optical indicatrix axis around the field vector **H**, that was observed in the **K** ∥ **H** ∥ [100] geometry (see Fig. 1.23) and proved to be odd in **H**, is allowed in all magnetic symmetries, except $4/m\underline{m}\underline{m}$ and $4/m\underline{m}m$.

The observation of linear birefringence along the tetragonal axis in the longitudinal geometry (i.e., **K** ∥ **H** ∥ [001]) was an unexpected effect, since the LMOE is forbidden for the geometry by the symmetry of class $4/m$, where the AFM $Ca_3Mn_2Ge_3O_{12}$ was believed to belong, according to the results of neutron diffraction measurements [164]. The birefringence was of sufficient magnitude for visual observation of the AFM domains and for monitoring the uniformity of AFM ordering in the specimen areas observed. The experiment clearly revealed the "butterfly"-type field dependence characteristic of the LMOE (see Fig. 2.14). Since the LMOE was observed in the **k** ∥ **H** ∥ [001] geometry, the magnetic groups $4/m$, $4/mmm$, and $4/m\underline{m}\underline{m}$ should be excluded from the list of candidate groups for MnGeG. Indeed, the q^s_{ij} matrices of these classes do not contain the component q^s_{xyz} or unequal components q^s_{xxz} and

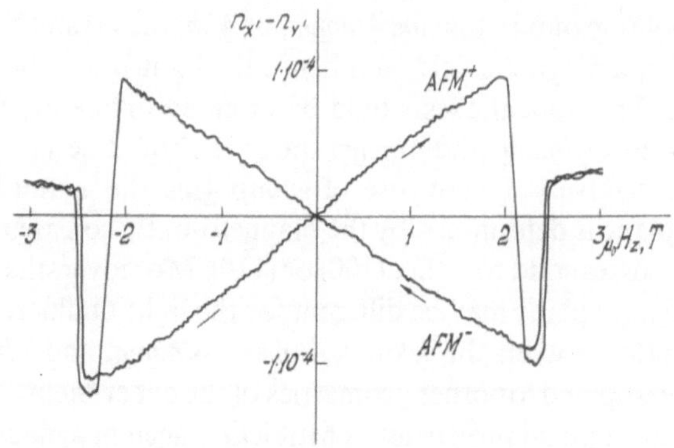

Fig. 2.14. Field-induced $H \parallel [001]$) linear birefringence in the AFM garnet $Ca_3Mn_2Ge_3O_{12}$. $T = 11$ K, $\lambda = 632.8$ nm, $\mathbf{k} \parallel [001]$.

q^s_{yyz} that could be responsible for the birefringence of light propagating along the tetragonal axis.

To select one of the three remaining magnetic classes, i.e., $\underline{4}m$, $\underline{4}/mm\underline{m}$, or $\underline{4}/mm\underline{m}$, it was necessary to establish the position of the optical axes plane. To that end, the ellipticity of the light leaving two (AFM$^+$ and AFM$^-$) domains was measured as a function of the azimuth angle of plane polarized incident light, with two opposite directions of the magnetic field. Such measurements have minimized possible errors associated with elastic strains in the crystal, or with deviations of the light propagation vector from the axis C_4. The angle between the axis of the optical indicatrix X' and the [110] direction was close to 10°.

Another piece of evidence for the nonalignement of the optical indicatrix axes with the directions [110] and [110] was provided by dichroism measurements of the plane polarized light in the same geometry. The linear magnetic dichroism was observed between 560 nm and 680 nm, with the incident light polarized along [110]/[110] and along [100]/[010]. For the polarization direction close [100]/[010] the effect was nearly maximal. It reached the value of $2\cdot10^{-4}$ H_z cm^{-1} (with H_z in Oersted) and could easily be measured. A record of the "dichroism versus H_z" dependence is shown in Fig. 2.15. The effect was an order of magnitude lower than the light polarized along [110]/[110]. Spectral measurements have shown the dichroism to be related, not with the fundamental absorption but rather with a weak absorption band centered around 590 nm.

The characteristics of field-induced birefringence and the dichroism of plane polarized light in the antiferromagnet suggest convincingly that the optical indicatrix axes do not coincide with either the [100]- or [110]-type direction. The angle between an axis and one of the crystallographic directions depends on the optical wavelength. Therefore, the magnetic groups $\underline{4}/m\underline{m}m$ and $\underline{4}/mm\underline{m}$ should be excluded from the list of candidates to represent the magnetic class of MnGeG, as these groups require that the optical

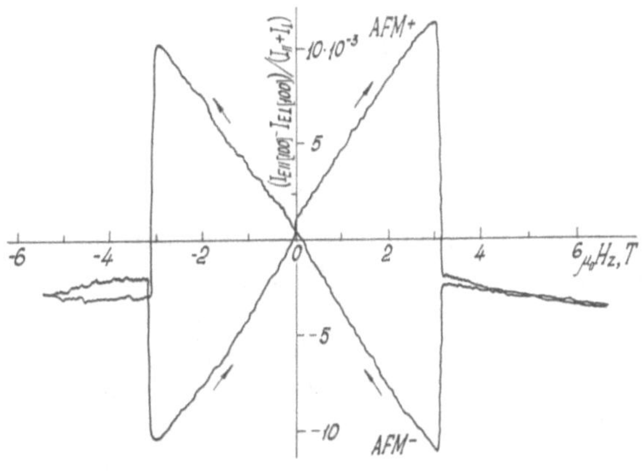

Fig. 2.15. Linear dichroism induced in the AFM garnet $Ca_3Mn_2Ge_3O_{12}$ by a magnetic field **H** ∥ [001]. $T = 7$ K, $\lambda \approx 585$–595 nm, **k** ∥ [001].

axes be oriented along [110] or [100], respectively. Hence, the LMOE characteristics observed can be reconciled with the magnetic group $\underline{4}/m$ alone, permitting nonzero components q^s_{xyz} and q^s_{xxz}/q^s_{yyz} in the LMOE magneto-optic tensor. Schematically shown in Fig. 2.16 is the magnetic structure corresponding to the group $\underline{4}/m$ for the case where the orientation of local magnetic anisotropy axes is controlled by Jahn–Teller distortions. The structure differs from the one of class $4/m$ that was suggested on the basis of neutron

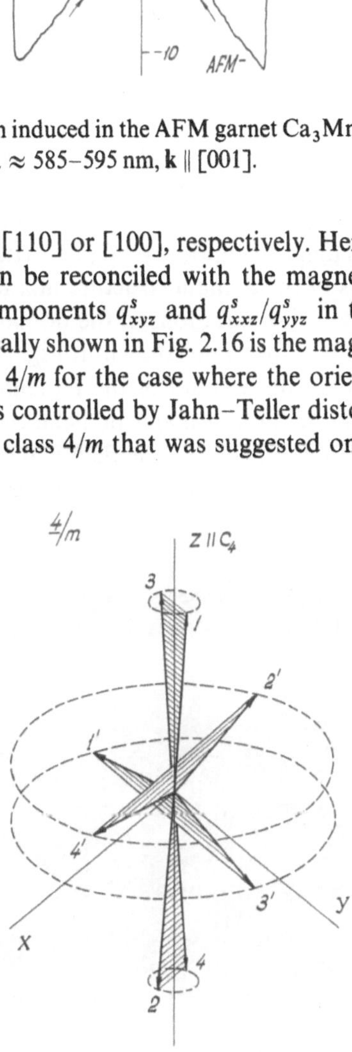

Fig. 2.16. Scheme of location of the sublattice magnetic moments in the AFM garnet $Ca_3Mn_2Ge_3O_{12}$ in an eight-sublattice model, corresponding to the $\underline{4}/m$ group. Figures 1...4 and 1'...4' indicate two groups of equivalent magnetic ions.

diffraction measurements, mainly in the respect that it is a completely compensated AFM structure. It does not permit the weak ferromagnetism allowed in class $4/m$ structures.

2.2.2. Magneto-Optical Observation of Crystalline Twins

Similar to the distinction between piezoelectric characteristics of the Dauphinean and Brazilian quartz twins [167, 168], the magneto-optic properties of twins in magnetically ordered crystals may be different, despite the identical form of their optical indicatrix. The third-order tensors, $q^{s}_{ij\alpha}$ and $G_{\mu\alpha\beta}$, that describe, respectively, the LMOE and the QMR in similar twins, may have components differing in sign. The magneto-optical properties of such twins are very similar, sometimes even identical, to the properties of 180-degree magnetic domains. This may impede interpretation of the nonuniform crystal structure in the antiferromagnet and preclude identification of the AFM^+ and AFM^- domains. On the other hand, this similarity may prove useful in determining the form of the $q_{ij\alpha}$ tensor. It is this situation that has been encountered in the analysis of magneto-optic crystal properties [127, 129].

Visual observations of the domain structure in a manganese–germanium garnet plate showed peculiar variations of the domain configuration during magnetization reversals of the plate. Shown in Fig. 2.17 are photographs of the structure obtained in the vicinity of the remagnetization point for the geometry $\mathbf{H} \parallel [100] \parallel \mathbf{k}$. Without a magnetic field, the dimming angle is the same for all the areas of the plate. As the field is increased, there appears a growing contrast between individual sections, often separated quite distinctly

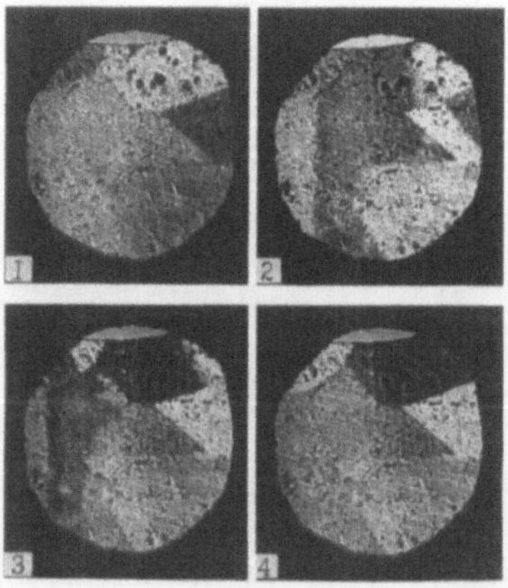

Fig. 2.17. Crystal twins of the Dauphinean type in tetragonal $Ca_2Mn_2Ge_3O_{12}$ crystal visualized by the LMOE. All photographs, except 1 and 4, illustrate AFM remagnetization of a sample. The tetragonal axis lies in the sample plane. $T = 10$ K.

by rectilinear boundaries. Their magneto-optical behavior in the adopted geometry corresponds to the characteristics of 180-degree AFM domains. However, as soon as the field reaches the remagnetization threshold H_t, domains become "recolored," i.e., their magneto-optic characteristics controlled by the LMOE change their sign. After the magnetization time reversal has been completed, the domain configuration remains unchanged. All the domains visible in field $H > H_t$ are equally disadvantageous energetically with respect to the new, "recolored" state. None of the domains grows in size at the expense of a neighbor.

If small areas of a specimen are remagnetized through local heating with a laser beam, propagation of the remagnetization front is almost uninfluenced by boundaries separating the stable domains observed. The walls between the domains did not change their positions after multiple magnetization reversal cycles, nor after multiple heatings to room temperature. This behavior of the domain structure certainly contradicts their interpretation as 180-degree AFM domains. The rigidity shown by the boundaries gives grounds for assuming that the areas observed are nothing other than crystal twins whose optical axes coincide, while the crystallophysical X- and Y-directions are different. The twinning operations for these might be rotations around the second-order axes perpendicular to C_4, which are lost in the phase transition $m3m \cdot \underline{1} \rightarrow 4/m \cdot \underline{1}$, or reflections from the planes passing through the tetragonal axis [170, 171]. The reflection twins and the axial ones are identical, for the phase transition retains the symmetry center. At temperatures above T_N the twins owing to reflections from coordinate and diagonal planes are also identical, since crystal lattices of the second twin components satisfy the symmetry operations $\pm 4_z$ preserved to the AFM ordering temperature. At $T < T_N$ the AFM ordering results in a loss of the operation $\pm 4_z$, while $\pm \underline{4}_z$ remains, and hence twins of the latter two types cease to be identical. Since the retained symmetry operation $\underline{4}_z = 4_z \cdot \underline{1}$ differs from the lost one, 4_z, solely by time inversion, the second components in the twins, owing to reflection from the coordinate and diagonal planes, may obviously differ only in the directions of the ion magnetic moments which are opposite. They are nothing other than AFM^+ and AFM^- domains. The boundary, separating the second components of the crystal twins owing to reflection from the coordinate and the diagonal planes, is a purely AFM boundary, since the crystal lattices of the components are identical. (The latter are related by the symmetry operation $\pm 4_z$ of the magnetically disordered crystal.) Meanwhile, the boundaries between each of the second components and the initial crystal are normal twin boundaries. Magnetically, they are extremely rigid, for they cannot be displaced unless crystal ions are shifted.

Magneto-optical properties of the twin components can be described in terms of the tensors resulting from the initial tensor after the application of twinning operations. Suppose the matrices of the LMOE magneto-optic tensor moduli, for two AFM states of the initial crystal (first twin component), are

$$\begin{array}{cc} \text{AFM}^+ & \text{AFM}^- \\ \left|\begin{matrix} 0 & 0 & q_{13} \\ 0 & 0 & q_{23} \\ 0 & 0 & 0 \\ q_{41} & q_{42} & 0 \\ q_{51} & q_{52} & 0 \\ 0 & 0 & q_{63} \end{matrix}\right|, & \left|\begin{matrix} 0 & 0 & -q_{13} \\ 0 & 0 & -q_{23} \\ 0 & 0 & 0 \\ -q_{41} & -q_{42} & 0 \\ -q_{51} & -q_{52} & 0 \\ 0 & 0 & -q_{63} \end{matrix}\right|, \end{array} \tag{2.6}$$

with $q_{13} = -q_{23}$, $q_{51} = -q_{42}$, and $q_{52} = q_{41}$. Then the matrices for the second reflection twin components take the form

$$\left|\begin{matrix} 0 & 0 & -q_{13} \\ 0 & 0 & -q_{23} \\ 0 & 0 & 0 \\ q_{41} & -q_{42} & 0 \\ -q_{51} & q_{52} & 0 \\ 0 & 0 & q_{63} \end{matrix}\right| \quad \text{and} \quad \left|\begin{matrix} 0 & 0 & q_{13} \\ 0 & 0 & q_{23} \\ 0 & 0 & 0 \\ -q_{41} & q_{42} & 0 \\ q_{51} & -q_{52} & 0 \\ 0 & 0 & -q_{63} \end{matrix}\right| \tag{2.7}$$

AFM$^+$, if reflected by the coordinate plane; AFM$^-$, if reflected by the diagonal plane.	AFM$^-$, if reflected by the coordinate plane; AFM$^+$, if reflected by the diagonal plane.

As can be seen from (2.6) and (2.7), the magneto-optic properties of the twin components are different. In the **k** ∥ **H** ∥ [100] geometry the most important is the magneto-optic modulus q_{41}, responsible for the indicatrix rotation around **k**. It may have different signs for different twin components.

The energy profitability of the AFM$^+$ or the AFM$^-$ state of the crystal twin components in a magnetic field is determined by cubic-in-**H** addition to the crystal magnetic energy, i.e.,

$$\Delta E = -C_{\alpha\beta\gamma} H_\alpha H_\beta H_\gamma. \tag{2.8}$$

The $C_{\alpha\beta\gamma} \leftrightarrow C_{\alpha\mu}$ matrix for the twin components takes the form:

initial crystal, states

$$\begin{array}{cc} \text{AFM}^+ & \text{AFM}^- \\ \begin{bmatrix} \cdot & \cdot & \cdot & C_1 & C_2 & \cdot \\ \cdot & \cdot & \cdot & -C_2 & C_1 & \cdot \\ C_2 & -C_2 & \cdot & \cdot & \cdot & C_1 \end{bmatrix}, & \begin{bmatrix} \cdot & \cdot & \cdot & -C_1 & -C_2 & \cdot \\ \cdot & \cdot & \cdot & C_2 & -C_1 & \cdot \\ -C_2 & C_2 & \cdot & \cdot & \cdot & -C_1 \end{bmatrix} \end{array} \tag{2.9}$$

reflection twin component (reflection from the coordinate plane), states

AFM$^+$ AFM$^-$

or reflection twin component (reflection from the diagonal plane), states

$$\overset{\text{AFM}^-}{\begin{bmatrix} \cdot & \cdot & \cdot & C_1 & -C_2 & \cdot \\ \cdot & \cdot & \cdot & C_2 & C_1 & \cdot \\ -C_2 & C_2 & \cdot & \cdot & \cdot & C_1 \end{bmatrix}}, \overset{\text{AFM}^+}{\begin{bmatrix} \cdot & \cdot & \cdot & -C_1 & C_2 & \cdot \\ \cdot & \cdot & \cdot & -C_2 & -C_1 & \cdot \\ C_2 & -C_2 & \cdot & \cdot & \cdot & -C_1 \end{bmatrix}} \tag{2.10}$$

The energy additions for the AFM^+ and AFM^- states of the initial crystal are

$$-\Delta E_1 = \pm 6C_1 H_x H_y H_z \pm 3C_2 H_z(H_x^2 - H_y^2) \tag{2.11}$$

and the corrections for reflection twins, respectively, for reflection from the coordinate and the diagonal place, can be written as

$$\begin{aligned} -\Delta E_{2\,\text{coord}} &= \pm 6C_1 H_x H_y H_z \mp 3C_2 H_z(H_x^2 - H_y^2), \\ -\Delta E_{2\,\text{diag}} &= \mp 6C_1 H_x H_y H_z \pm 3C_2 H_z(H_x^2 - H_y^2). \end{aligned} \tag{2.12}$$

In (2.11) and (2.12) the upper signs correspond to the AFM^+ state and the lower signs to the AFM^- state.

In the experiment [169], the energy profitability of magnetic states was determined by uncontrollable field deviations from the $\underline{Z} \parallel C_4$ direction. In case the second term in (2.11) and (2.12) is larger than the first one, the preferred state is AFM^+ in the initial crystal and in the twin component owing to reflection from the diagonal plane (which is also the twin component owing to reflection from the coordinate plane but in the AFM^- state). The twin components are characterized by moduli q_{41} of equal magnitude and opposite sign, and equal moduli q_{42} (see the matrices (2.6) and (2.7)). Since the effect of modulus q_{42} on the light polarization is quadratic in H and hence negligibly weak, characteristics of the twin components are nearly identical to characteristics of the AFM^+ and AFM^- domains of a single component. These are exactly the properties that have been observed for neighbor areas separated by rigid boundaries in the experiment with MnGeG. Therefore, the neighbor areas should be regarded as crystal twin components rather than AFM domains. AFM domains separated by mobile boundaries can be observed within the same component of a twin crystal during AFM magnetization reversal. Similar to CoF_2, the AFM domain structure produced in the magnetization reversal process remains almost unchanged when the magnetic field is removed. The photographs of Fig. 2.18 show AFM domain configurations prepared by heating the plate locally with a laser beam, in the presence of the magnetic field in which the initial AFM state is metastable. The increase in temperature reduces the threshold of the remagnetization field, so that the applied field proves sufficient from remagnetizing the specimen in its heated portion. The size of the resultant domain is determined by the amount of energy supplied. After several reversals of the applied field direction, with subsequent local heating runs in which the energy supply was lower for every

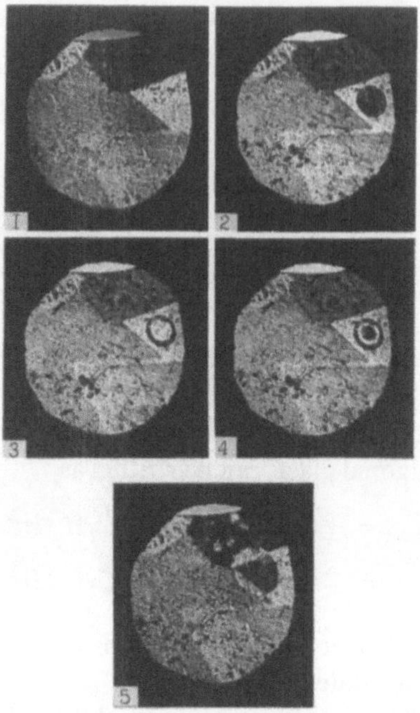

Fig. 2.18. AFM domain formation in a $Ca_3Mn_2Ge_3O_{12}$ plate due to short-duration local heating by a laser beam. Visible sample part diameter is 1 mm.

consecutive time, annular domains can be obtained (see photographs 2 and 3). As can be seen in photograph 3, twin boundaries have every little effect on the formation of AFM domains.

That the twins, owing to reflection from type (100) or type (110) planes in MnGeG, can be observed (with the aid of the LMOE) is a fact in support of the conclusion that the crystal belongs to the class $\underline{4}/m$. Indeed, the twin structure of a class $4/m$ AFM crystal would be undetectable in the experimental geometry employed. A similar analysis of twin magneto-optic characteristics in the class $4/m$ shows the indicatrix rotations around [100], in an **H** ∥ [100] field, to be the same for both components of the crystal twin. The q_{xzx} moduli in the components have opposite signs, as in class $\underline{4}/m$ crystals, however, they have only a negligible effect on the polarization of light propagating along the field. As a result, the twins in class $4/m$ crystals cannot be distinguished with the aid of magneto-optic techniques in the geometry considered.

2.3. The Antiferromagnetic Domains and Two-Phase Magnetic Structures in $DyFeO_3$

The magnetization of condensed media is often accompanied by the formation of a nonuniform magnetic flux through the specimen, with the structure period much greater than the interatomic separation. The nonuniform states arise under the action of demagnetizing fields. Two major types of the nonuniform states are generally distinguished. The first one is a thermodynamically equilibrium two-phase mixture. If spread uniformly over the entire specimen, each of two phases would be metastable in the entire volume of existence of the equilibrium mixture. Such a state can be formed during first-order phase transitions. A similar nonuniform state is known in superconductors as an intermediate state [172, 173], while in magnetic materials, antiferromagnets in particular, it is called a magnetic intermediate state [174]. An equilibrium two-phase state may also arise during the magnetization of nonferromagnetic metals under the conditions favoring the de Haas–van Alfven effect.

The nonuniform state of the other type cannot be considered as a mixture of two coexisting phases, for the boundary separating the "domains" of the two phases is of comparable width with the "domain" size in one of the phases. The magnetic moment of a specimen of this state varies in a magnetic field in a quasi-continuous mode. The least magnitude of variation is determined by the magnetic flux quantum. A nonuniform state of this type can exist in hard superconductors where it is known as the mixed state [172, 173]. The possibility of interpreting the magnetic structure variations observed in certain antiferromagnets in a magnetic field, in terms of a mixed magnetic state, was first noted in [175]. A mixed state can arise, provided the boundary separating the two phases is energetically advantageous. This is the case with Ising antiferromagnets characterized by AFM exchange interactions between the ions of the first and second coordination spheres. In magnetic dielectrics, where the magnetic moments are localized and the interaction scale length is of the order of the lattice parameter, the mixed state may manifest itself in its limiting form, i.e., with the characteristic size of the nonuniformities comparable to the interatomic separation. Following this tradition, such structures may be called commensurable and incommensurable superstructures. The structure period is determined by the relative importance of the interatomic exchange interactions between distant neighbors, and weak interactions of a different nature. A situation, more resembling superconductors, may arise in such magnetic media where the magnetic moment is partially delocalized. The formation of a mixed magnetic state at phase transitions has not been observed directly, however, some of the experimental results available may serve as implicit evidence for the existence of such states in Ising antiferromagnets placed in a magnetic field [175, 176].

In this section, we will present some results concerning the expansion of the intermadiate magnetic state in plate specimens, when the initial state of the plate is magnetically nonuniform ($AFM^+ + AFM^-$) or ($WFM^+ + WFM^-$). The formation of a two-phase magnetic state during a phase transition was first observed in the MnF_2 antiferromagnet under spin-flop phase transition [177]. The appearance of the two-phase structure has been unambiguously detected with the aid of Barkhansen jumps [178]. Some time later, two-phase structures of the "antiferromagnet + paramagnet"-type were photographed during metamagnetic transitions in DyAlG and $FeCl_2$ [179]. A coexistence of different angular and collinear phases was also observed in a Neél ferrimagnet during a field-induced transition to the noncollinear state [180]. Two-phase structures were then observed in $DyFeO_3$ [181–183]. Nouniform magnetic states, close to the thermodynamical equilibrium, were observed in only MnF_2 and $DyFeO_3$. The formation of an equilibrium state in DyAlG or $FeCl_2$ was impeded by the large-scale time for the onset of the domain structure, while in iron garnets the domain structure of coexisting phases was not coherent, in view of the small difference between magnetic moments. The most favorable conditions for studying the formation and development of the intermediate-state domain structure were provided by chemically polished plates of dysprosium orthoferrite crystals. Owing to the characteristically small relaxation times of the domain structure, easily attainable magnetic field strengths and a "convenient" temperature range where the two-phase state exists in $DyFeO_3$, the major properties of two-phase structures can be investigated with the use of the LMOE that is sufficiently high in the crystal. Of interest were the formation of a new magnetic phase from domain boundaries between weakly ferromagnetic (WFM) or AFM domains and the development of bubble domains of a phase in the matrix of the other. A metastable intermediate state of the AFM + WFM was observed at temperature above the Morin point, where uniform AFM states cannot exist.

The rare-earth orthoferrites ($ReFeO_3$) belonging to the crystal class mmm are the object of extensive studies. The interest in these materials is stimulated by the vast variety of magnetic phase transformations that they can undergo in magnetic fields or with temperature variations. The first systematic investigations of different spin-orientation transitions were carried out with the use of rare-earth orthoferrites [123, 184]. The variety of physical characteristics shown by the orthoferrites arises from the presence of eight magnetic ions in the crystal unit cell and the low-symmetry ligand environment of the magnetic ions. The latter fact results in a close resemblance of the characteristic energies of several anisotropic interactions. Indeed, the energies of the antisymmetric exchange interaction, between Fe^{3+} atoms and the symmetric and antisymmetric rare-earth Fe^{3+} exchange interactions, are comparable one another and to the single-ion anisotropy energy of Fe^{3+}. Moreover, the single-ion anisotropy energy of Re^{3+} at low temperatures increases to become comparable to the Re^{3+}–Fe^{3+} exchange interaction energy. Therefore, we will discuss only briefly the AFM$\leftrightarrow$WFM phase transition in $DyFeO_3$. The transition is conditioned by an increased contribution at lower temperatures of Dy^{3+} atoms to the second-order anisotropy constant and by the Dy^{3+}–Fe^{3+} exchange bounding. The transition temperature T_M is close to 50 K. The magnetic state of an orthoferrite is generally described with the aid of eight magnetic vectors including four magnetic moment vectors of the iron sublattices and four of the rare-earth sublattices, all transformed according to irreducible group representations of the orthoferrite crystal symmetry [185, 186].

For our immediate purposes, we may confine ourselves to a simpler two-sublattice model, denoting the states in terms of projections of only two vectors, i.e., the largest-in-magnitude AFM vector **G** and the FM vector **F**. Below T_M, the dysprosium orthoferrite is characterized by a nonzero component G_y, whereas at $T > T_M$ the nonzero components are G_x and F_z. The Morin transition at $T = T_M$ corresponds to jumplike transformations of these projections, viz., $G_y \leftrightarrow G_x F_z$. A magnetic field $\mathbf{H} \parallel c$ first provokes a gradual turning of the AFM vectors in the (ab) plane and then carries it abruptly to the axis a. The turning is accompanied by the appearance and growth of F_z. The transition is often described, in lieu of the complex potential involving a great number of invariants, in terms of the simple potential function [187, 188]

$$\Phi = \Phi_0 + K_2 \cos 2\theta + K_4 \cos 4\theta - M_z \cos\theta \cdot H_z, \tag{2.13}$$

where the anisotropy constants K_2 and K_4 are actually functions of many parameters. K_4 depends mainly on parameters of the iron–iron exchange interaction, while the Fe^{3+}–Re^{3+} interaction prevails in K_2. Since the magnetic moment of the rare-earth sublattices is strongly temperature dependent, because of the weak Fe^{3+}–Re^{3+} exchange bounding, K_2 is a function of temperature too, whereas K_2 is virtually independent of temperature. The first-order phase transition $G_y \leftrightarrow G_x F_z$ is possible with $K_4 < 0$ and occurs spontaneously at $K_2(T) = 0$. A magnetic field $\mathbf{H} \parallel c$ can induce the analogous transition at

$$H_t = \frac{4(K_2 - 4K_4)}{M_z}\left(1 - \sqrt{\frac{4K_4}{4K_4 - K_2}}\right). \tag{2.14}$$

The AFM phase loses its stability at the temperature determined by the equation $K_2(T) = 4K_4$, while the stability of the WFM phase is lost when $K_2(T) = -4K_4$. The transition was investigated experimentally in [184, 189]. As has been found, the changes in magnetic, elastic, and magnetostrictional parameters of the crystal are sufficiently sharp to regard the AFM $\leftrightarrow$ WFM transformation as a first-order phase transition. Characteristic features of the transition were the absence of a sizable temperature hysteresis, and the considerable width (about 2 K) of the temperature range over which the transformation could occur without an external field. To explain the hysteresis-free character of the transition, several models have been suggested.

According to [186], the thermodynamical potential of $DyFeO_3$ at the transition temperature T_M ceases to depend on the azimuth angle of the sublattice magnetic moments in the plane (ab). Within this model, the Morin temperature is in fact a degenerate critical point [190]. This situation is possible in the absence of a fourth-order anisotropy. From the practical point of view, K_4 must be lower the "anisotropy noise" level reflecting the real crystal structure.

According to [191], the transition is hysteresis-free because domain structures exist in both the WFM and AFM states and can change over gradually from one to the other. The 180-degree domain walls serve as nucleation centers in a new (either WFM or AFM) phase. The restructuring process should pass through a thermodynamically equilibrium two-phase state with a coherent domain structure of the type (AFM + WFM$^+$ + AFM + WFM$^-$). According to [123, 191], a similar state should be thermodynamically preferable in an infinite plate over some temperature range which is close to 2 K in the case $DyFeO_3$. This value is close to the width of the temperature range where the Young modulus and the internal friction of dysprosium orthoferrite undergo anomalous variations. Judging by this coincidence and by the type of anomalous performance at the increase of H, the authors concluded that the described domain structure rearrangement was indeed present in the crystal. The model was reportedly in agreement with the results of visual observations of the domain structure in $DyFeO_3$ [192].

In fact, visual observations of an equilibrium domain structure in high quality plates of dysprosium orthoferrite have shown the AFM $\leftrightarrow$ WFM transformation to occur through a normal first-order phase transition in which an incoherent two-phase structure is formed. A coherent two-phase structure can appear only in a magnetic field of sufficient intensity. The WFM phase nucleation without a field is not determined exclusively by the presence or absence of AFM domains. The wide temperature range and the hysteresis-free character of the transition both arise from the high sensitivity of the spin-orientation phase transition to lattice deformations and imperfections of the real crystal. These factors lead to the nucleation of new magnetic phases as a result of a small overheating (or supercooling).

2.3.1. Antiferromagnetic Domains

The multiphase state structure of dysprosium orthoferrite plates was studied visually with the aid of magnet-optical methods. The spontaneous Faraday rotation in the absence of an applied magnetic field is capable of providing a contrast between plate areas with distinct AFM and WFM phases, and allows us to distinguish between the F_z^+ and F_z^- WFM states. As concerns the AFM states G_y^+ and G_y^-, these can be detected visually owing to the significant LMOE in an $\mathbf{H} \parallel c$ magnetic field. The transformation of light wave polarizations by WFM and AFM domains is shown schematically in Fig. 2.19 (in the Poincaré sphere representation). A WFM $DyFeO_3$ belongs to the magnetic class $m\underline{mm}$ in which the q_{xyz} component of the LMOE tensor, responsible for the indicatrix rotation around $z \parallel \mathbf{k}$, is identically equal to zero (see Table 1.2). The nonzero components, q_{xxz} and q_{yyz}, responsible for alteration of the indicatrix axes, do not play a more or less significant part, as the polarization plane of the incident light either lies near the optical axes plane of the crystal or is perpendicular to that plane in the domain observations. It is the geometry in which small variations of the linear birefringence cannot by detected.

The magnetic class of an AFM $DyFeO_3$ is *mmm*. Here the q_{xyz} component is not equal to zero, and hence the optical indicatrix rotates around $\mathbf{H}$ in a field $\mathbf{H} \parallel c$. The rotation directions in AFM$^+$ and AFM$^-$ domains are oppo-

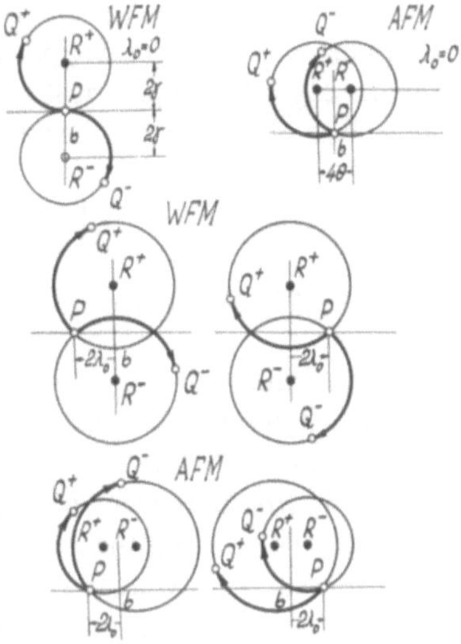

Fig. 2.19. Transformation of light polarization in WFM$^\pm$ and AFM$^\pm$ domains in $DyFeO_3$ (schematic projection of a Poincaré sphere onto a plane). Point P lies on the sphere equator and corresponds to the linear polarization of incident light. Points R^+ and R^- correspond to axes exits on the sphere surface, about which a sphere rotates, the vertical line indicates the crystal ($\mathscr{b}c$)-plane, and its intercept with the equator to the $\mathscr{b}$-axis azimuth from which the λ angles are counted off.

site. While intensity variations of the outgoing light are determined by the Faraday rotation and the angular shift of the indicatrix axes, the domain contrast is determined by the LMOE. The light wave that has passed through a polarizer of azimuth angle λ_0 (with respect to the principal plane of the unperturbed crystal), an elliptically birefringent plate characterized by the angle $2\gamma = \arctan(4\rho/\delta)$ and phase shift $\Delta = (4\rho^2 + \delta^2)$ between the modes, and an analyzer at the angle α to the crossing position with polarizer, has the intensity

$$\begin{aligned}\mathscr{I}/\mathscr{I}_0 = \tfrac{1}{2} - \tfrac{1}{2}\cos 2\lambda_0\{&[\cos^2 2\gamma + \sin^2 2\gamma\cos\Delta]\cos 2(\alpha+\lambda_0)\\ &-[\tan 2\lambda_0\cos(\alpha+\lambda_0) - \sin 2(\alpha+\lambda_0)]\sin 2\gamma\sin\Delta\\ &+\tan 2\lambda_0\cos\Delta\sin 2(\alpha+\lambda_0)\}.\end{aligned} \tag{2.15}$$

The intensity difference between light beams passing through the WFM$^+$ and WFM$^-$ domains, with the polarizer at a small angle to the principal plane, is

$$\left(\frac{\mathscr{I}^+ - \mathscr{I}^-}{\mathscr{I}_0}\right)_{\text{WFM}} = \frac{4\alpha\rho\sin\Delta}{\Delta} = \frac{4\alpha\rho\sin\delta}{\delta}. \tag{2.16}$$

There is no contrast between the domains with $\alpha = 0$, i.e., if the analyzer is crossed with the polarizer.

In the case of AFM^+ and AFM^- domains, the intensity difference vanishes only with the analyzer deviated from the crossing position by the angle $\alpha = -2\lambda_0$. In the only case when $2\lambda_0 = 0$ (the polarization plane of the incident light coincides with the (bc) plane of the crystal), the angle α is zero too:

$$\left(\frac{\mathscr{I}^+ - \mathscr{I}^-}{\mathscr{I}_0}\right)_{\text{AFM}} = 4\theta(2\lambda_0 + \alpha)\left[\frac{\delta^2}{\Delta^2} + \left(1 + \frac{4\rho^2}{\Delta^2}\right)\cos\Delta\right]$$

$$\simeq 4\theta(2\lambda_0 + \alpha)(1 + \cos\delta). \tag{2.17}$$

Equations (2.16) and (2.17) are valid with θ, λ_0, $\alpha \ll 1$, and $(2\rho/\delta)^2 \ll 1$. The dependence of the image contrast upon α is characterized with $\lambda_0 \neq 0$ by a far greater asymmetry for AFM than for WFM domains. Different domains can be identified visually by varying the polarizer and analyzer positions. The identification is made easier by the fact that the domains become contrastly colored, during white light observations, owing to the frequency dispersion of absorption, birefringence, and the Faraday rotation. In a 40 μm-thick plate AFM domains could easily be distinguished in a field of only 0.06 T. In the case where the polarization plane of the incident light coincided with the crystal plane (bc) (i.e., $\lambda_0 = 0$), the visual contrast between the domains reached a maximum with the analyzer deflected by about 1° from the crossing position, and $H = 0.15$ T. The AFM domains could not be distinguished visually with $\alpha > 4°$. The contrast changed to the opposite if the angular shift of the analyzer axis was reversed. The effect could not be observed for $\lambda_0 > 0.5°$. The contrast between AFM domains existed however if the analyzer and the polarizer were crossed ($\alpha = 0$), but the polarization vector of the incident light did not lie in the crystal plane (bc) ($\lambda_0 \neq 0$). With $\lambda_0 \neq 0$, the intensities of the light beams that have passed through the G_y^+ and G_y^- domains, respectively, become equal when the analyzer is rotated by about $2\lambda_0$ from the crossing position.

AFM domains could be observed over the entire temperature range investigated [93], i.e., from helium temperatures to about $T = T_M$. In the vicinity of T_M AFM domains, such as the $G_y \to G_xF_z$ transition, occur at low field intensities. Unlike the domain walls in CoF_2, AFM boundaries in $DyFeO_3$ are not oriented along crystallographic directions. The dominant factors controlling their arrangement are internal imperfections and, especially, surface defects. The number of AFM domains was low in the those plates whose surfaces had no scratches and were chemically etched (see Fig. 2.20). Domains appeared preferably in the same sections, even though the same configurations did not reappear and the shapes changed in the magnetic field. In plates with scratched surfaces, domain walls and individual small domains were often arranged along the boundaries and did not penetrate through the entire specimen. The AFM domain structure could be changed by varying the magnetic field intensity, reversing the field direction, or changing the specimen temperature. The wall motion is a slow process that can continue for tens of seconds after the field has stopped changing. The strongest changes occur in

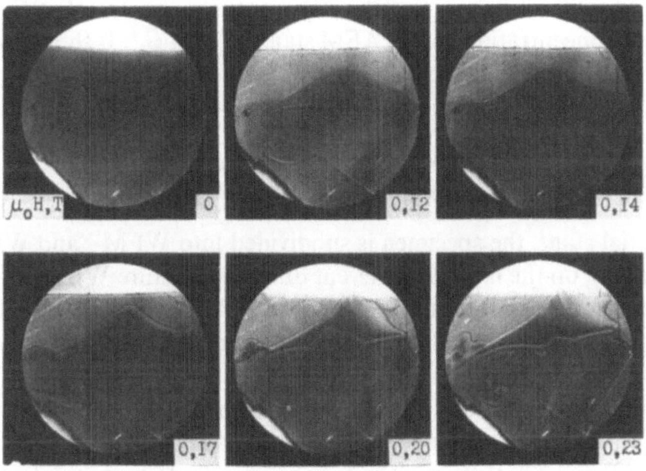

Fig. 2.20. 180° AFM domains in a $DyFeO_3$ plate smoothed chemically.

fields close to that of the AFM ↔ WFM phase transition. If a specimen goes over to the WFM state and then returns to the AFM state, the domain structure is not restored.

It is the existence of a varying nonequilibrium AFM domain structure that brings about a peculiar hysteresis in the field dependence of the polarization ellipse angle of rotation. Figure 2.21 shows such hysteresis curves obtained for two different orientations of the field. The hysteresis could be associated with the presence of a nonequilibrium AFM + WFM two-phase structure produced at the phase transition, and retention of the WFM state, for some reason or another, till $H \to 0$. However, visual observations have clearly

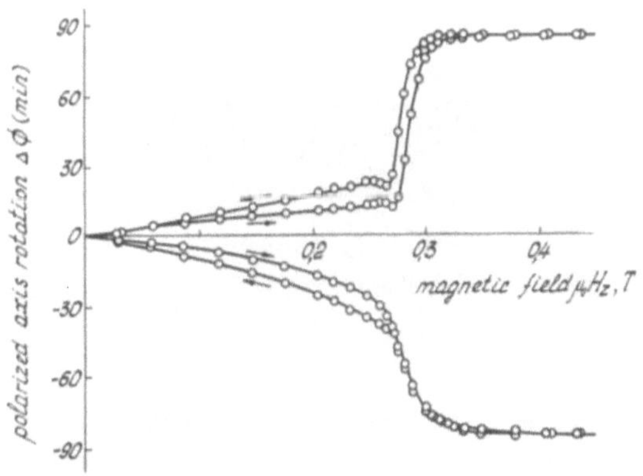

Fig. 2.21. Hysteresis of field dependence of the polarization ellipse rotation angle in multidomain $DyFeO_3$. Stepped variation corresponds to the phase transition AFM → WFM. The bottom curves indicate opposite field direction. $T = 11$ K, $\lambda = 590$ nm.

shown the hysteresis behavior of $\Delta\Phi(H)$ over the range $0 < H < H_t$ to be due to different concentrations of the AFM states G_y^+ *and* G_y^-. It should be emphasized that the magneto-optical hysteresis observed cannot be accompanied by a hysteresis in the crystal magnetic properties.

2.3.2. The Incoherent Two-Phase Domain Structure

At the initial stage, the specimen is subdivided into WFM^- and WFM^+ domains making up the familiar coherent domain structure $\mathrm{WFM}^+ + \mathrm{WFM}^-$. It seems natural to expect that at the WFM → AFM transition, nucleation of the new AFM phase should start at magnetic defects, such as domain walls separating the WFM domains [191, 193]. In the experiment, however, AFM domains were produced and increased in size on several areas of the plate at a time, and obviously were not connected to domain wall expansion (see photograph 1 of Fig. 2.22). If the wall expansion occurred at all, it did so at much smaller separations from the AFM domain than the transverse size of the WFM domain, and hence could not be visually resolved. The formation of individual AFM domains was not associated with crystal nonuniformity. While the motion of the AFM phase front often had a preferred direction, the changes experienced by the front in repeated processes suggest that the gradients in impurity concentration, strain, or temperature never played a decisive role in the formation of AFM boundary details. Demagnetizing fields of the remaining WFM sections exert a considerable influence on the localization of the interface between the AFM and WFM phases. The two major tendencies revealed in the arrangement of the AFM domain boundaries were orientation along or perpendicular to boundaries of the nearest WFM domains, while the latter often tended to be matched with the AFM front.

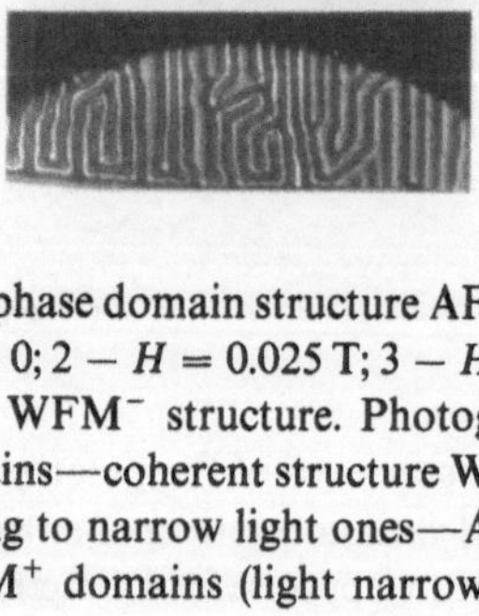

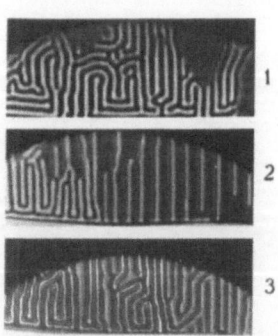

Fig. 2.22. Variation of a two-phase domain structure $\mathrm{AFM} + \mathrm{WFM}^{\pm}$ with a magnetic field increase $H \parallel C$: $1 - H = 0$; $2 - H = 0.025$ T; $3 - H = 0.05$ T. Photograph1: Labyrinth domains—$\mathrm{WFM}^+ + \mathrm{WFM}^-$ structure. Photograph 2: White strip domains separated by wide grey domains—coherent structure $\mathrm{WFM}^+ + \mathrm{AFM}$; Photograph 3: Wide grey strips transforming to narrow light ones—AFM domains appearing from domain walls between WFM^+ domains (light narrow strips) and WFM^- domains (dark wide strips).

The WFM phase appears in an initially AFM specimen in the form of two coupled $WFM^+ + WFM^-$ domains (preferarbly near the edge of the specimen). These develop abruptly to a multidomain WFM region. The presence of the two AFM domains, G_y^+ and G_y^-, whose existence could be detected primarily in a magnetic field, using the LMOE technique, had no effect on the WFM phase formation. Without the field, the AFM boundary is not decorated with WFM domains. The visual observations have allowed us to conclude quite positively that a coherent two-phase state is not formed during the Morin transition in $DyFeO_3$ at $H = 0$.

At first glance, it might seem that the reason for similar behavior is the failure of the domain wall between the WFM domains, $G_x^+ F_z^+$ and $G_x^- F_z^-$, to serve as a nucleation center for the AFM domain G_y. This would have been true if the domain wall magnetic moments had rotated within the place (a, c), as in the majority of orthoferrites. However, in $DyFeO_3$ the (ab) plane becomes an easier plane even at temperatures above T_N [194, 195], and hence the reason for the lack of an intermediate state about T_M should be different.

To establish what it actually is, thermodynamical potentials of a plane-parallel plate in the coherent and incoherent two-phase state were compared (Fig. 2.23). The internal field H_i average size of the domains was considered constant. This is an approximation that proves valid for ellipsoidal specimens of considerable thickness, $z \gg d_{dom}$, with small domain wall thicknesses, $d_w \ll d_{dom}$, and with roughly equal relative concentration ξ_{AFM} and ξ_{WFM} of the two phases (anyway, the values must be essentially different from both zero and unity).

Under the phase transition conditions, the thermodynamical potentials of the plate are equal for the AFM and WFM states, viz.

$$\Phi_{AFM}^{\infty} + \Phi_{AFM}^{H} = \Phi_{WFM}^{\infty} + \Phi_{WFM}^{H} \cdot \tag{2.18}$$

Here Φ_{AFM}^{∞} and Φ_{WFM}^{∞} are potential densities for an infinite crystal in the

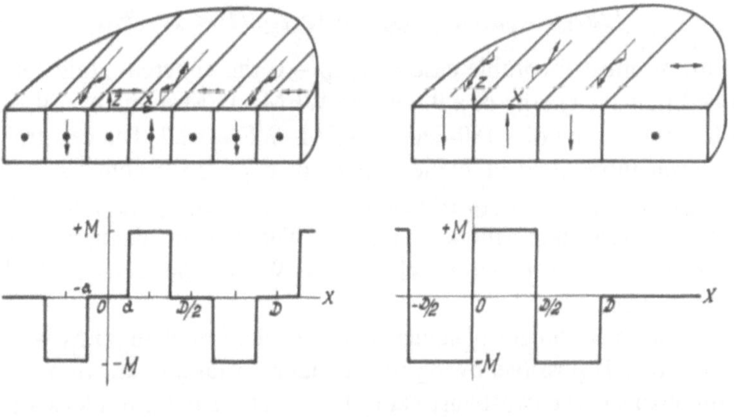

Fig. 2.23. (a) Coherent $WFM^+ - AFM - WFM^-$ and (b) noncoherent $WFM^{\pm} -$ WFM two-phase structures.

uniform AFM and WFM state, respectively, while the potential terms Φ^H_{AFM} and Φ^H_{WFM} that depend on the domain structure and shape of the specimen involve appropriate demagnetizing field energies, $W^{\mathrm{demag}}_{\mathrm{AFM}} = 0$ and $W^{\mathrm{demag}}_{\mathrm{WFM}} \neq 0$, and domain wall energies, $W^{\mathrm{w}}_{\mathrm{AFM}}$ and $W^{\mathrm{w}}_{\mathrm{WFM}}$. The AFM state has been assumed uniform, therefore, $W^{\mathrm{w}}_{\mathrm{AFM}} = 0$. During the phase transition, the nonuniform AFM + WFM state is formed in the plate. The question is: Under what conditions does the term $\Phi^H_{\mathrm{AFM+WFM}}$ of the coherent nonuniform state exceed the nonuniform part $\Phi^{\mathrm{incoh}}_{\mathrm{AFM+WFM}}$ of the potential for the incoherent state? By minimizing the nonuniform state energy, calculated for a model strip domain structure with the aid of the effective charge technique [196], we can obtain, respectively, for the equilibrium period and the energy of the coherent two-phase state

$$D = \frac{\pi\sqrt{\sigma_{\mathrm{AFM-WFM}} z}}{\sqrt{2}\, M \cos(\xi_{\mathrm{AFM}}\pi/2)}, \tag{2.19}$$

$$\Phi^{\mathrm{coh}}_{\mathrm{AFM+AFM}} = \frac{8\sqrt{2}}{\pi} M \sqrt{\sigma_{\mathrm{AFM-WFM}} z}\cos(\xi_{\mathrm{AFM}}\pi/2) + \Phi^{\infty}_{\mathrm{AFM}}\xi_{\mathrm{AFM}} + \Phi^{\infty}_{\mathrm{WFM}}\xi_{\mathrm{WFM}}. \tag{2.20}$$

The energy of the nonuniform incoherent state in which one part of the plate is in the AFM state and the other in the WFM$^+$ + WFM$^-$ state is

$$\Phi^{\mathrm{incoh}}_{\mathrm{AFM}\pm} = \frac{8}{\pi} M \sqrt{\sigma_{\mathrm{WFM-WFM}} z}\, \xi_{\mathrm{WFM}} + \Phi^{\infty}_{\mathrm{AFM}}\xi_{\mathrm{AFM}} + \Phi^{\infty}_{\mathrm{AFM}}\xi_{\mathrm{WFM}}. \tag{2.21}$$

Here $\sigma_{\mathrm{AFM-WFM}}$ and $\sigma_{\mathrm{WFM-WFM}}$ are, respectively, surface energy densities for the walls between AFM and WFM$^\pm$, and WFM$^+$ and WFM$^-$ domains; z is the plate thickness; ξ_{AFM} and ξ_{WFM} are phase concentrations, and M denotes crystal magnetization in the WFM state.

Comparing the potentials of the two two-phase states with equal phase concentrations we find that the coherent two-phase state may be advantageous solely with

$$\sqrt{2\sigma_{\mathrm{AFM-WFM}}/\sigma_{\mathrm{WFM-WFM}}}\cos(\xi_{\mathrm{AFM}}\pi/2) < 1 - \xi_{\mathrm{APM}}. \tag{2.22}$$

Even in the most favorable case of $\xi_{\mathrm{AFM}} \to 0$ the coherent state can arise, provided that the energy of a 90-degree WFM–AFM domain wall is lower than half the energy of a 180-degree WFM–WFM wall. The constraint becomes even more rigid at higher values of ξ_{AFM}, to become $\sigma_{\mathrm{AFM-AFM}} < \frac{1}{4}\sigma_{\mathrm{WFM-WFM}}$ at $\xi_{\mathrm{AFM}} = \xi_{\mathrm{WFM}}$. In fact, $\sigma_{\mathrm{AFM-WFM}} \sim \frac{1}{2}\sigma_{\mathrm{WFM-WFM}}$, hence the coherent two-phase state cannot be realized. The demagnetizing fields do not allow wall expansion between WFM$^+$ and WFM$^-$ domains at the Morin phase transition.

Note that in specimens of sufficient size, formation of an AFM + WFM$^\pm$ superstructure is possible, owing to the magnetostatic interaction of WFM domain blocks in the two-phase state. The effect is similar to block structure formation in ferromagnetic plates [197].

As the volume occupied by the WFM phase reduces in the course of the phase transition, the demagnetizing factor of the WFM domain containing-area reduces too. As a result, the energy density of the nonuniform WFM state is changed and the phase transition itself is "extended" along the temperature scale. The width of this "natural temperature width" of the transition can be estimated from the condition

$$W_{\text{WFM}} - W^*_{\text{WFM}} \approx \frac{dK_2}{dT} z \cdot \Delta T_M, \tag{2.23}$$

where W^*_{WFM} is the nonuniform state energy density of the collapsing group of WFM domains. Assuming the demagnetizing factor of this plate area (of size $D = d^+ + d^- \sim z$) to be equal to π we obtain

$$\Delta T_M \approx \frac{6}{\pi} M \sqrt{\sigma_{\text{WFM-WFM}}/z} \left(\frac{dK_2}{dT}\right)^{-1}, \tag{2.24}$$

with $dK_2/dT = 0.7 \cdot 10^3$ erg/cm^3 K, $\sigma_{\text{WFM-WFM}} = 0.24$ erg/cm^2, $M = 17$ emu, and $z = 4 \cdot 10^{-3}$ cm (see [123, 183]), we have $\Delta T_M \approx 0.3$ K. This magnitude is quite close to the experimentally observed transition width $T \approx 0.4$ K. The importance of the magnetostatic expansion mechanism of the transition temperature range that has been described here is admittedly comparable with other mechanisms, associated with the nonuniformity of real specimens, such as expansion due to mechanical stress gradients, nonuniform impurity concentration, or plate taper.

2.3.3. Coherent Two-Phase Domain Structures

The character of the AFM $\leftrightarrow$ WFM transition in $DyFeO_3$ changes qualitatively in an external magnetic field $\mathbf{H} \parallel c$ above some critical strength value. At greater field intensities a tendency towards formation of a periodic two-phase state appears. The structural alterations are particularly representative in the case of reducing temperature. Even with field values as low as $H \simeq 0.2\, H_{\text{sat}}$ an ordered motion of newly formed AFM domains to the WFM region becomes noticeable. It is accompanied by the annihilation of two WFM$^-$ domains and one WFM$^+$ domain. The result is a distinctly detectable periodic structure WFM$^+$ + AFM (photograph 2 of Fig. 2.20).

At greater field strengths, individual AFM domains are formed in an abrupt manner through the expansion of two domain walls surrounding the energetically disadvantageous WFM$^-$ domain and destruction of its fragment. As the temperature is further decreased, the newly formed AFM domain grows at the expense of the WFM$^-$ domain. The AFM + WFM$^+$ periodic structure that appears, to some extent replicates the former WFM domain structure (see photograph 3 of Fig. 2.20). At a still further decrease in temperature the WFM$^+$ strip domains become unstable, and WFM$^+$ bubble domains are formed instead that later collapse. In plates with few defects, the resultant AFM state is generally either uniform or consists of several AFM domains.

The tendency towards formation of a coherent two-phase structure seems to be initiated by the increased energy of the demagnetizing fields in the $WFM^+ + WFM^-$ structure, owing to changed sizes of the WFM domains. Ultimately, the energy increase makes the $AFM + WFM^+$ structure competitive. Since the two domain structures are close in energy near T_M with relatively weak magnetic fields $H \lesssim H_{sat}$, the characteristic hysteresis-type phase behavior appears, when the phase composition is not repeated at increases and decreases of the magnetic field. An interesting feature of the hysteresis dependences is the lack of a noticeable hysteresis on the effective Faraday rotation versus the field strength curves. The phase hysteresis can be seen quite distinctly owing to the substantial visual contrast between the AFM and WFM states.

An interesting manifestation of hysteresis in the phase composition is observed at temperatures above T_M, where the WFM phase alone is thermodynamically advantageous. If a plate primarily magnetized to saturation is demagnetized, at $H_2^* < H_{sat}$, there appear AFM domains, provided that $T < T_M + \Delta T^*$, with $\Delta T^* \approx 0.3$ K. The resulting periodic structure, $WFM^+ +$ AFM, is preserved in a further decrease of the field strength, down to H_1^*. At $H = H_1^*$ the two-phase structure is replaced abruptly by a labyrinth-type $WFM^+ + WFM^-$ state. The mode of AFM domain appearance and disappearance suggests that the $AFM + WFM^-$ state here is metastable, like the AFM state itself.

It is easy to understand why the metastable AFM phase is formed as a component of the intermediate magnetic state, if we recall that the appearing AFM domains tend to reduce the magnetostatic energy of the plate. Nucleation of AFM domains is possible for higher field intensities than for WFM domains, because of the lower energy of the 90-degree domain wall. The two-phase structure $AFM + WFM^+$ reduces the demagnetizing field energy, thus stabilizing the internal field and precluding formation of the more advantageous domain state $WFM^+ + WFM^-$.

The possibility of producing a visual contrast between 180-degree AFM domains allows us to investigate experimentally the role of the AFM domain wall as a nucleation line of the new phase in the magnetic phase transition. The nucleation mechanism by which the new phase develops from domain walls was suggested in [198–200], analyzing spin flipping in the uniaxial antiferromagnet. Experimental evidence for the reality of this mechanism was provided by the analysis of metamagnetic "AFM-paramagnet" transitions in DyAlG [138] and $FeCl_2$ [179]. Observation of the characteristic domain structures, such as lines or chains of isolated domains that disappeared after repeated magnetization cycles, brought forth the conclusion that AFM domain walls were decorated with paramagnetic domains. In $DyFeO_3$, the role of WFM–AFM domain walls as nucleation surfaces for the AFM phase and, conversely, the WFM phase can be observed directly.

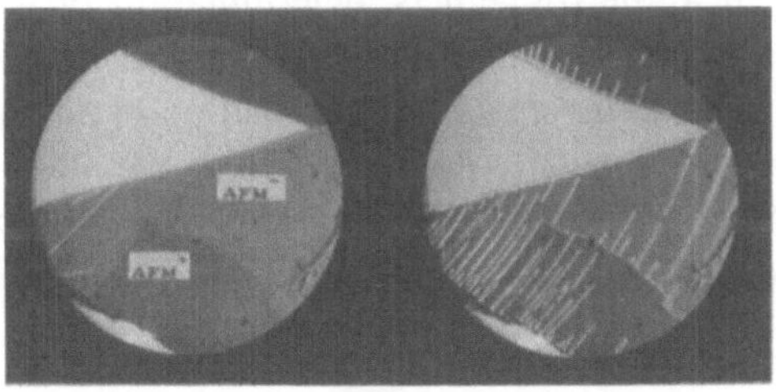

1 2

Fig. 2.24. WFM^+ domains formation from domain wall between AFM^+ and AFM^- domains with a magnetic field $\mathbf{H} \parallel c$ increase.

In the case where the AFM $\leftrightarrow$ WFM transition occurs without a magnetic field, the domain wall has no effect on the AFM domain formation. It does not play a significant role with $H \lesssim H_{sat}$ either. In higher fields, one can clearly see how the WFM domain develops from the AFM domain wall. First, the interface between adjacent domains becomes more and more markedly pronounced, then it gets thicker at certain points located more or less periodically. Finally, the thicker areas develop into strip-type WFM domains and the domain wall configuration may change. The process of WFM domain formation on an AFM domain wall is illustrated in Fig. 2.24. The behavior of the AFM domain wall is in qualitative agreement with the predictions. As follows from approximate estimates, a visually noticeable wall expansion ($\delta > 1$ μm) can occur in the field strength range $(H_t - H) \simeq 10^{-3}$ T. The corresponding temperature range is less than 0.1 K. As the wall is expanded further, its demagnetizing fields should bring about periodic variations in thickness. The demagnetizing field energy of a wall becomes comparable with the nouniform spin rotation energy at wall thicknesses above 1 μm. It is at wall thicknesses of several micrometers, that expansions on the narrow strip AFM domain arising from the AFM wall become observable.

It might seem that a similar wall expansion should accompany formation of the AFM domains from the 180-degree boundary separating the WFM^+ and WFM^- domains. The analyze this process, it would be necessary to take into account the alteration of the magnetizing field energy as a result of the wall expansion and the reduction of the WFM domain size (or a change in the structure period). Along with the finite dimension of the specimen plate, setting limits to the boundary positions, the energy alteration would emphasize the threshold character of AFM domain production from the boundaries between the AFM domains. Admittedly, this is the reason why the formation of AFM domains from WFM walls is observed experimentally in such fields where domain structures $WFM^+ + WFM^-$ and $AFM + WFM^-$ have close periods.

The applications of the LMOE described in this chapter include visual observation of 180-degree AFM domains and the investigation of remagnetization processes in antiferromagnets, determination of magnetic ordering point symmetries in complex multidimensional antiferromagnets, observation of crystal twin structures, and an examination of magnetic phase transformations of the new magneto-optical effect to the study of magnetically ordered materials is far from being exhaustive. We are looking forward to learning of new experiments showing the further potentialities of magneto-optics.

CHAPTER 3

Optical Magnetic Excitations

Discussed in Chapters 3 and 4 are magnetic excitations in antiferromagnetic materials whose stoichiometric composition includes magnetic ions with unfilled 3*d*-shells. Both low-energy magnetic excitations (magnons) and high-energy ones (excitons) are investigated. Davydov magnetic splitting of light absorption lines, which is an experimental proof of the cooperative nature of optical excitations in these crystals, has been found to be particularly pronounced in a cubic antiferromagnetic $RbMnF_3$ in a high magnetic field [201, 202]; this was also observed in more complex spectra without an external field [203, 204]. In this class of crystals optical transitions are mostly spin and parity forbidden. These excitations are imparted from one magnetic ion to another within a sublattice due to resonance exchange interaction. The excitation can be transmitted among ions with opposed orientation of spins due to antisymmetrical exchange interaction between a virtual magnon and an optical exciton.

The main attention is paid to optical magnetic excitations. As for low-energy excitations, they are studied so far as they accompany the processes of high-energy excitation generation or annihilation. In this connection, the generation mechanism discussed here is an interaction of magnetic ions with the electromagnetic wave of optical range, and the object of the experimental study is optical absorption and luminescence spectra of antiferromagnets. Naturally, the subjects touched upon are related to the description of spectra of magnetic excitations, the mechanisms of their generation, migration of magnetic excitions in crystals, and annihilation of excitons.

When describing the problem under investigation, it is possible to emphasize two approaches: a phenomenological one (see, e.g., [205–207]) and a semimicroscopic one [208–210]. Both the approaches are based on the description of magnetic excitations within the framework of the small-radius exciton model (analogous to excitons in molecular crystals); both of them treat magnons as a particular case of excitons. Each approach has a number of simplifying assumptions and limitations. This, however, does not seem to be a big disadvantage as it enables us to obtain concrete results, avoiding unnecessary complications. Both the approaches are consistent and mutually supplement each other. In this connection, the results presented are not described in the framework of a single approach, but in accordance with the formalism used in the original paper. Let us characterize briefly these approaches.

In the semimicroscopic approach [208, 211, 212] the initial Hamiltonian is a single ion Hamiltonian in which the operators of the 3*d*-electron spin and momentum coordinates are used as dynamic variables. A Hamiltonian of a crystal can be reduced to a sum of single-ion Hamiltonians plus interionic interaction. The formation of an effective single-ion spectrum is substantially affected by interactions between 3*d*-electrons of ions in a crystal which is in an antiferromagnetic state; therefore, the scheme used for molecular crystals when obtaining the exciton spectrum cannot to transferred to an antiferromagnetic one. For an antiferromagnet, it is necessary, first, to construct a single-ion state space, and then a crystal state space defined as a product of single-ion spaces. Finally, the initial Hamiltonian is reduced to this space. The reduced Hamiltonian in exciton representation is obtained by means of creation and annihilation operators of crystal excitations. In this formalism, the operator of the ion excitation creation is expressed via Fermi operators of creation and annihilation of 3*d*-electrons in a magnetic ion. The operators of electron (optical) excitations obey the Bose statistics and commute with operators of spin variables. The main advantage of this approach is the possibility of constructing a consistent theory of optical magnetic excitations of an antiferromagnet, in which the parameters of one-particle optical excitation are related with the parameters of a more complex multiparticle spin-optical (exciton–magnon) excitation, both the one-particle and the multiparticle excitations being described by means of three independent parameters, which are experimentally found from data on the purely exciton absorption of light and antiferromagnetic resonance. However, the excited state spectrum is adequately described by Bose operators only at sufficiently low temperatures $T \ll T_N$ (T_N is the Néel temperature). A theory of this kind has been developed for a magnetic Mn^{2+} ion. The number of 3*d*-electrons in manganese coincides with the multiplicity of orbital degeneracy of single-electron *d*-states, due to which it is possible to unambiguously construct the ground state with a maxima projection of full spin. It seems to be possible to extend the main results of the theory to crystals with other magnetic ions.

In the case of phenomenological description, the nature of single-ion wave functions is not studied in detail, and the Hamiltonian of an excited crystal is written as one consisting of two parts—the sum of Hamiltonians of noninteracting ions (zero-order approximation) and the sum of operators of ionic pair interaction energy. The second part of the Hamiltonian describes the transfer of excitations both among ions of one and the same sublattice and among those of opposite sublattices. An actual form of interaction is determined individually for each crystal. An attractive feature of this method is a consistent use of group theory formalism which is done, however, at the expense of neglecting the information on crystal behavior in an external magnetic field, the latter disturbing the collinearity of the crystal magnetic structure. This method enables one to predict the number of Davydov-split exciton bands, and the polarization of excitation light and emission. An advantage of the method is its generality.

The method suggested in [213] has proved to be highly convenient when describing pair exciton–magnon states in antiferromagnetic materials (especially at nonzero temperatures). These authors have introduced a vector exciton operator which in many respects is similar to a spin operator. It should be noted, however, that the latter is used to describe transitions inside ground multiplets, whereas the exciton vector operator combines wave functions of various multiplets. In the simplest case, vector operators are expressed via operators of excitation creation and annihilation, just as spin operators are expressed via magnon Bose operators in Holstein–Primakoff approximation, i.e., in this case, excitons are described with the help of a modified Heisenberg model. The introduction of vector excition operators has considerably simplified the description of exciton–magnon processes at nonzero temperatures, as it made it possible to express all the necessary characteristics via the usual spin correlation functions.

To describe exciton–magnon interaction (between optical and magnetic excitations) in a phenomenological approach, such a form of the Hamiltonian is taken [214] which is used in the spin excitation–impurity interaction theory, the magnetic optically excited ion playing the role of impurity. In this case, there are two parameters—the width of the exciton band and the exciton–magnon interaction—that cannot the unambiguously determined without additional assumptions.

A transition from microscopic exciton formalism to that of vector exciton operators was realized for a particular case of antiferromagnetic oxygen [209]. In this microscopic approach, the Hamiltonian of the electron magnetic excitations of a crystal is a generalization of the Heisenberg Hamiltonian, and differs essentially from the Hamiltonian used in the microscopic theory of magnetic excitons. In this theory, the operators of electron excitations are neither Bose nor Fermi operators; they adhere to complex commutation relationships. These relationships reflect the fact that the same electrons of ions determine both magnetic and optical properties. As in the case of the phenomenological approach, the vector electron operators enable one to calculate the energy spectrum of a crystal magnetic excitation for any temperature. Within the framework of this theory, one of the basic problems of exciton physics was touched upon—the problem of boundary conditions of quasi-particle existence, i.e., the situation when the coherent propagation of magnetic excitons in a crystal is substituted by their random migration (jumps). It is also important to note that both the generalized Heisenberg Hamiltonian [209] and the Hamiltonian in the microscopic theory [211] contain information on the mutual orientation of spins in different sublattices, which makes it possible to investigate noncollinear magnetic structures. In general, this theory has marked a new step in the microscopic approach to the problem of magnetic excitations in an antiferromagnet.

Later, an even more general approach to this problem was developed [210]. When studying exciton–magnon interactions, the above-mentioned microscopic theories did not investigate the relationship between the amplitudes of these interactions and the magnetic system symmetry with respect to spin rotation, and the influence of this correlation on the dependence of the energy spectrum of excitons upon the temperature and magnetic field. This problem was solved by using the Hamiltonian formulation invariant with respect to the spin rotations. The Hamiltonian of electrostatic interactions was obtained on the basis of the microscopic approach; it has a tensor form with respect to spin rotations. It should be noted that the exchange part of the Hamiltonian contains both Heisenberg and non-Heisenberg terms. This theory has not yet found its application in the interpretation of magnetic excitations of some crystal; therefore it is only in the introduction that we mention these developments.

In what follows we pay special attention to the exciton semimicroscopic theory, as it reflects adequately enough the nature of the phenomena under consideration. A schematic description of this theory is the subject of the first section. Discussed in the second section are those conditions under which the interaction of the light wave with a magnetically concentrated crystal gives rise to the creation of magnetic excitations in this crystal. Special attention is paid to the generation of pair exciton–magnon excitations, as it is this mechanism that dominates in the class of crystals under consideration. The third section is devoted to the discussion of the results of an experimental study of exciton–magnon and two-exciton transitions in the case of multiquantum excitations in a strong light wave field. Dynamics and decay of excitons are discussed in the fourth and fifth sections, special emphasis being placed on the impact of the magnetic structure of crystals on the character of the exciton motion, the mobility of excitons, and the probability of exciton radiation decay.

3.1. The Exciton Description of Magnetic Excitations

The general principles of describing the optical spectra of dielectric crystals are being developed on the basis of the conception of an exciton as a cooperative noncurrent excitation of a crystal. Two types of excitons are distinguished acdording to the degree of interaction between ions in a crystal as compared with intraionic interactions: Frenkel excitons and Wannier excitons which are two extreme cases of the electron-hole coupling.

"Small-radius" excitons (Frenkel excitons), excitations of a system of ions or molecules that interact weakly with one another, are formed in molecular crystals and dielectrics with unfilled $3d$- or $4f$-shells of magnetic ions. As the interionic interaction is relatively small, the energy spectrum of individual ions is not essentially reconstructed when the ion-to-ion interaction is taken into account; therefore, the eigenfunctions φ^f corresponding to eigenvalues of ion energy ε^f are selected as a basis to construct the wave functions of a crystal state.

A quasi-continuous energy band corresponds to each undegenerated excited state f of an individual ion in a crystal. In complex crystals, the unit cell which contains σ identical translationally inequivalent ions, there occurs an additional, the so-called Davydov splitting into σ exciton bands with energies $E_\mu(\mathbf{k}, f)$, where the exciton band index μ has the values $1, 2, \ldots, \sigma$, and $\mathbf{k}$ is a wave vector.

Spin excitations, i.e., magnons, are a particular case of Frenkel excitons [206]. A one-ionic process corresponding to a spin wave in a crystal is, as in the case of optical excitons, a transition from the ground state to an excited one, but in this case the transition is to the lowest one. In the case of this transition, there changes only the spin projection of the ion ground state with $M_S = S$ to $M'_S = S - 1$. Therefore, the energy of the magnon can be calculated within the framework of the exciton model. When calculating energies of excitons and magnons, the main difference lies in the fact that in the latter case the Heitler–London approximation turns out to be invalid.

The cooperative properties of interacting excitations of an antiferromagnetic crystal are taken into account by means of the general exciton model of spin and optical excitations in antiferromagnetic dielectrics [211]. The initial one-ion Hamiltonian is written as

$$\mathscr{H} = -\sum_i \left(\frac{h^2}{2m} \Delta_i - U_i \right) + \frac{1}{2} \sum_{i \neq j} \frac{e^2}{|r_i - r_j|} - \mathbf{H} \sum_i \hat{\mathscr{M}}_i + \mathscr{H}^{\mathrm{rel}}, \tag{3.1}$$

where index i numbers electrons; $\mathscr{M}_i$ and U_i are the magnetic moment operator and the potential energy of the ith electron in the field of magnetic ions and ligands, respectively; and $\mathscr{H}$ is the external d.c. magnetic field. The Coulomb interaction of $3d$-electrons is described by the second term of (3.1), and all the relativistic interactions are included in $\mathscr{H}^{\mathrm{rel}}$. The second quantization representation of the Hamiltonian (3.1) has the form

$$\mathscr{H} = \sum_{\mathbf{n}\alpha} \mathscr{H}_{\mathbf{n}\alpha} + \frac{1}{2} \sum_{\mathbf{n}\alpha,\, \mathbf{m}\beta}' V_{\mathbf{n}\alpha,\, \mathbf{m}\beta} \tag{3.2}$$

where radius vectors $\mathbf{n}\alpha$ and $\mathbf{m}\beta$ are coordinates of the ions, and $\mathscr{H}_{\mathbf{n}\alpha}$ is the energy operator of an ion occupying the site $\mathbf{n}\alpha$ which describes its internal energy state. It accounts, in particular, for the presence of the crystal field, and for the removal of the degeneracy with respect to spin projections; it has the ion site symmetry in a crystal. The operator $V_{\mathbf{n}\alpha,\, \mathbf{m}\beta}$ takes into account part of the interaction between ions $\mathbf{n}\alpha$ and $\mathbf{m}\beta$ which cannot be included in the operator $\mathscr{H}_{\mathbf{n}\alpha}$: in the case of antiferromagnetic dielectrics, it is a Coulomb interaction of electrons.

As the degeneracy with respect to spin projections is removed in the exchange internal field the wave function of the magnetic ion ground state has the following form: $|\varphi^0_{\mathbf{n}\alpha}\rangle = \overset{+}{a}_{\mathbf{n}\alpha,\lambda_1\uparrow}\overset{+}{a}_{\mathbf{n}\alpha,\lambda_2\uparrow}\cdots a_{\mathbf{n}\alpha,\lambda_j\downarrow}|0\rangle$, where $\overset{+}{a}_{\mathbf{n}\alpha,\lambda\sigma}$ is the operator of electron creation on the site $\mathbf{n}\alpha$ with orbital λ and spin projection $\sigma = \pm\frac{1}{2}(\uparrow\downarrow)$. The number of filled states is $2S$. The wave function of the excited state is

$$\left|\varphi^f_{\mathbf{n}\alpha}\right\rangle = \sum_{\lambda\lambda'} c^f_{\lambda\lambda'}\overset{+}{a}_{\mathbf{n}\alpha,\lambda\downarrow}a_{\mathbf{n}\alpha,\lambda'\uparrow}\left|\varphi^0_{\mathbf{n}\alpha}\right\rangle.$$

The simplest type of magnetic ion excitation is a spin one ($f = s$). In this case, each magnetic electron remains in its quantum state λ:

$$|\varphi^s_{\mathbf{n}\alpha}\rangle = \frac{1}{\sqrt{2S}}\sum_\lambda \overset{+}{a}_{\mathbf{n}\alpha,\lambda\downarrow}a_{\mathbf{n}\alpha,\lambda\uparrow}|\varphi^0_{\mathbf{n}\alpha}\rangle = \ell^+_{\mathbf{n}\alpha}|\varphi^0_{\mathbf{n}\alpha}\rangle = B^+_{\mathbf{n}\alpha}(S)|\varphi^0_{\mathbf{n}\alpha}\rangle.$$

Omitting the procedure of reducing the Hamiltonian to crystal state space, it is possible to write the terms of the Hamiltonian (3.2) as

$$\begin{aligned}\mathscr{H}_{\mathbf{n}\alpha} &= \sum_f \langle\varphi^f_{\mathbf{n}\alpha}|\mathscr{H}_{\mathbf{n}\alpha}|\varphi^f_{\mathbf{n}\alpha}\rangle\,\ell^+_{\mathbf{n}\alpha}(f)\ell_{\mathbf{n}\alpha}(f),\\ V_{\mathbf{n}\alpha,\mathbf{m}\beta} &= \sum_{f,g,k,l}\langle\varphi^f_{\mathbf{n}\alpha},\varphi^g_{\mathbf{m}\beta}|V_{\mathbf{n}\alpha,\mathbf{m}\beta}|\varphi^k_{\mathbf{m}\beta}\varphi^l_{\mathbf{n}\alpha}\rangle\,\ell^+_{\mathbf{n}\alpha}(f)\ell^+_{\mathbf{m}\beta}(g)\ell_{\mathbf{m}\beta}(k)\ell_{\mathbf{n}\alpha}(l),\end{aligned} \tag{3.3}$$

where $\ell^+_{\mathbf{n}\alpha}(f)$ and $\ell_{\mathbf{n}\alpha}(f)$ are operators of creation and annihilation of the state f in the site $\mathbf{n}\alpha$ (f being the set of quantum numbers which characterizes the state). The summation is taken with respect to all possible ion states which are eigenstates of the one-ion operator $\mathscr{H}_{\mathbf{n}\alpha}$.

In many cases the exciton operators $|\varphi^f_{\mathbf{n}\alpha}\rangle = B^+_{\mathbf{n}\alpha}(f)|\varphi^0_{\mathbf{n}\alpha}\rangle$, $|\varphi^f_{\mathbf{n}\alpha}\rangle = B_{\mathbf{n}\alpha}(f)|\varphi^f_{\mathbf{n}\alpha}\rangle$ can be introduced quite formally, without specifying their internal structure. The meaning of the operators introduced is made clear during consideration at the level of one-electron states. It enables us to describe all types of ion excitations by means of definite combinations of one-electron operators of creation and annihilation and to define the direct and exchange contributions of the Coulomb interaction operator ($V_{\mathbf{n}\alpha,\mathbf{m}\beta}$).

The Hamiltonian of the antiferromagnetic crystal excitations is expanded in terms of the exciton operators of spin $b^+_{\mathbf{n}\alpha}$, $b_{\mathbf{n}\alpha}$ and optical $B^+_{\mathbf{n}\alpha}(f)$, $B_{\mathbf{n}\alpha}(f)$ excitations as follows:

$$\Delta\mathscr{H} = \mathscr{H} - E_0, \tag{3.4}$$

where $\mathscr{H}$ is the Hamiltonian (3.2) and

$$E_0 = \sum_{\mathbf{n}\alpha}\langle\varphi^0_{\mathbf{n}\alpha}|\mathscr{H}_{\mathbf{n}\alpha}|\varphi^0_{\mathbf{n}\alpha}\rangle + \frac{1}{2}\sum_{\mathbf{n}\alpha,\mathbf{m}\beta}{}' \langle\varphi^0_{\mathbf{n}\alpha},\varphi^0_{\mathbf{m}\beta}|V_{\mathbf{m}\beta,\mathbf{n}\alpha}|\varphi^0_{\mathbf{m}\beta},\varphi^0_{\mathbf{n}\alpha}\rangle \tag{3.5}$$

is the energy of the ground state. In the expansion, the terms which do not retain the number of optical excitations are omitted (the Heitler–London approximation: the energy of optical excitations exceeds that of resonance interaction by several orders of magnitude). The formal expansion (3.4) has the following form:

$$\Delta\mathscr{H} = \mathscr{H}^s_0 + \mathscr{H}^f_0 + \mathscr{H}^{fs}_{\text{int}}, \tag{3.6}$$

where $\mathscr{H}_0^s$ is the Hamiltonian of spin excitations, $\mathscr{H}_0^f$ is the Hamiltonian of optical excitations, and $\mathscr{H}_{\text{int}}^{fs}$ is the Hamiltonian of optical and spin excitations interaction.

Using the one-electron approximation and taking into account the transformation properties of the spin wave functions of electrons, it is possible to obtain an expression for an interionic interaction operator $V_{\mathbf{n}\alpha,\,\mathbf{m}\beta}$ (Coulomb interaction of electrons) which shows explicitly its dependence on the angles θ_α and θ_β at which their spins are oriented:

$$V_{\mathbf{n}\alpha,\,\mathbf{m}\beta} = \sum_{\substack{\lambda_1\ldots\lambda_4\\ \sigma_1\sigma_2}} \langle \mathbf{n}\alpha\lambda_1, \mathbf{m}\beta\lambda_2 | V | \mathbf{m}\beta\lambda_3, \mathbf{n}\alpha\lambda_4 \rangle \overset{+}{a}_{\mathbf{n}\alpha\lambda_1\sigma_1} \overset{+}{a}_{\mathbf{m}\beta\lambda_2\sigma_2} a_{\mathbf{m}\beta\lambda_3\sigma_2} a_{\mathbf{n}\alpha\lambda_4\sigma_1}$$

$$+ \sum_{\substack{\lambda_1\ldots\lambda_4\\ \sigma_1\ldots\sigma_4}} \langle \mathbf{n}\alpha\lambda_1, \mathbf{m}\beta\lambda_2 | V | \mathbf{n}\alpha\lambda_3, \mathbf{m}\beta\lambda_4 \rangle \Phi_{\sigma_1\sigma_4}$$

$$\times \left(\frac{\theta_\alpha - \theta_\beta}{2}\right) \Phi_{\sigma_2\sigma_3} \left(\frac{\theta_\beta - \theta_\alpha}{2}\right) \overset{+}{a}_{\mathbf{n}\alpha\lambda_1\sigma_1} \overset{+}{a}_{\mathbf{m}\beta\lambda_2\sigma_2} a_{\mathbf{n}\alpha\lambda_3\sigma_3} a_{\mathbf{m}\beta\lambda_4\sigma_4}, \tag{3.7}$$

where

$$\Phi_{\sigma\sigma'}\left(\frac{\theta_\alpha - \theta_\beta}{2}\right) = \delta_{\sigma\sigma'} \cos\frac{\theta_\alpha - \theta_\beta}{2} + (\delta_{\sigma\uparrow}\delta_{\sigma'\uparrow} - \delta_{\sigma\uparrow}\delta_{\sigma'\downarrow}) \sin\frac{\theta_\alpha - \theta_\beta}{2}.$$

The first term of (3.7) describes the direct Coulomb part of the interaction of $\mathbf{n}\alpha$ and $\mathbf{m}\beta$ ions, and does not depend upon the mutual orientation of spins S_α and S_β (the axis to read angles θ in the spin plane is selected arbitrarily).

In the case of one optical and all possible spin excitations, all the coefficients of the expansion (3.6) are expressed by means of only three parameters which characterize the exchange interaction of ions. The matrix element $\langle \varphi_{\mathbf{n}\alpha}^0, \varphi_{\mathbf{m}\beta}^0 | V_{\mathbf{n}\alpha,\,\mathbf{m}\beta} | \varphi_{\mathbf{m}\beta}^0, \varphi_{\mathbf{n}\alpha}^0 \rangle$ the exchange part of which, when freed from the angular dependence of spin orientation of ions $\mathbf{n}\alpha$ and $\mathbf{m}\beta$, gives the parameter

$$\mathscr{I}_{\mathbf{n}\alpha,\,\mathbf{m}\beta} = \frac{1}{2S^2} \sum_{\lambda_1\ldots\lambda_4} \langle \mathbf{n}\alpha\lambda_1, \mathbf{m}\beta\lambda_2 | V | \mathbf{n}\alpha\lambda_3, \mathbf{m}\beta\lambda_4 \rangle$$

$$\times \langle \varphi_{\mathbf{n}\alpha}^0 | \overset{+}{a}_{\mathbf{n}\alpha\lambda_1\uparrow} a_{\mathbf{n}\alpha\lambda_3\uparrow} | \varphi_{\mathbf{n}\alpha}^0 \rangle \langle \varphi_{\mathbf{m}\beta}^0 | \overset{+}{a}_{\mathbf{m}\beta\lambda_2\uparrow} a_{\mathbf{m}\beta\lambda_4\uparrow} | \varphi_{\mathbf{m}\beta}^0 \rangle. \tag{3.8}$$

By means of this parameter we express all the exchange interactions in ionic pairs which are in unexcited or spin-excited states when the orbital states of ions is not changed and coincides with the ground state of a magnetic ion.

The matrix element $\langle \varphi_{\mathbf{n}\alpha}^f, \varphi_{\mathbf{m}\beta}^0 | V_{\mathbf{n}\alpha,\,\mathbf{m}\beta} | \varphi_{\mathbf{m}\beta}^0, \varphi_{\mathbf{n}\alpha}^f \rangle$ yields the exchange parameter

$$\mathscr{I}_{\mathbf{n}\alpha,\,\mathbf{m}\beta}^f = \sum_{\lambda_1\ldots\lambda_4} \langle \mathbf{n}\alpha\lambda_1, \mathbf{m}\beta\lambda_2 | V | \mathbf{n}\alpha\lambda_3, \mathbf{m}\beta\lambda_4 \rangle \langle \varphi_{\mathbf{n}\alpha}^f | \overset{+}{a}_{\mathbf{n}\alpha\lambda_1\downarrow} a_{\mathbf{n}\alpha\lambda_3\downarrow} | \varphi_{\mathbf{n}\alpha}^f \rangle$$

$$\times \langle \varphi_{\mathbf{m}\beta\lambda_2\uparrow}^0 | \overset{+}{a}_{\mathbf{m}\beta\lambda_2\uparrow} a_{\mathbf{m}\beta\lambda_4\uparrow} | \varphi_{\mathbf{m}\beta}^0 \rangle, \tag{3.9}$$

which characterizes all the exchange interactions of an optically excited ion with another unexcited or spin-excited ion, at which the optical excitation remains on the same ion. And, finally, the matrix element $\langle \varphi^f_{n\alpha}, \varphi^0_{m\beta} | V_{n\alpha, m\beta} | \varphi^f_{m\beta}, \varphi^0_{n\alpha} \rangle$ yields the exchange parameter

$$\begin{aligned}\mathscr{M}^f_{n\alpha, m\beta} = \sum_{\lambda_1 \ldots \lambda_4} & \langle n\alpha\lambda_1, m\beta\lambda_2 | V | n\alpha\lambda_3, m\beta\lambda_4 \rangle \langle \varphi^f_{n\alpha} | a^+_{n\alpha\lambda_1\downarrow} a_{n\alpha\lambda_3\uparrow} | \varphi^0_{n\alpha} \rangle \\ & \times \langle \varphi^0_{m\beta} | a^+_{m\beta\lambda_2\uparrow} a_{m\beta\lambda_4\downarrow} | \varphi^f_{m\beta} \rangle, \end{aligned} \tag{3.10}$$

which characterizes all the dynamic processes in an ionic pair, taking into account their eventual spin excitations which transfer optical excitations to another ion.

The cooperative properties of excitations are taken into consideration by transferring to exciton and magnon representations which diagonalize the quadratic forms of Hamiltonians $\mathscr{H}^f_0$ and $\mathscr{H}^s_0$ with the help of uv transformations

$$\mathscr{H}^f_0 = \sum_{k\mu} E_\mu(\mathbf{k}, f) B^+_\mu(\mathbf{k}, f) B_\mu(\mathbf{k}, f), \tag{3.11}$$

$$\mathscr{H}^s_0 = \sum_{\alpha, \mathbf{k}, \mu} |\theta_{\alpha\mu}(\mathbf{k})|^2 \varepsilon_\mu(\mathbf{k}) + \sum_{\mathbf{k}, \mu} \varepsilon_\mu(\mathbf{k}) b^+_\mu(\mathbf{k}) b_\mu(\mathbf{k}), \tag{3.12}$$

where $E_\mu(\mathbf{k}, f)$ and $\varepsilon_\mu(\mathbf{k})$ are the energies of f-type excitons and magnons with wave vectors $\mathbf{k}$ in the band μ, respectively. For a two-sublattice crystal with equivalent sublattices

$$E_\mu(\mathbf{k}, f) = \Delta\varepsilon^f + \mathscr{D}^f - L^f_1(\mathbf{k}) + (-1)^\mu L^f_2(\mathbf{k}) \cos^2 \frac{\theta_1 - \theta_2}{2}, \tag{3.13}$$

where

$$L^f_1(\mathbf{k}) = \sum_{n-m} \mathscr{M}^f_{n\alpha, m\alpha} \exp[i\mathbf{k}(\mathbf{n}\alpha - \mathbf{m}\alpha)]$$

is a sublattice exciton dispersion and

$$L^f_2(\mathbf{k}) = \sum_{n-m} \mathscr{M}^f_{n1, m2} \exp[i\mathbf{k}(\mathbf{n}1 - \mathbf{m}2)]$$

is an intersublattice one; $\Delta\varepsilon^f_\alpha$ is the optical excitation energy of an individual magnetic ion; and $\mathscr{D}^f_\alpha$ is the change of energy of the interaction of all the magnetic ions of a crystal with an individual ion as it changes to an excited state.

In the nearest-neighbor approximation and in the case of the symmetrical arrangement of spins of both the sublattices with respect to the external magnetic field $\mathbf{H}$, (3.13) is transformed as follows:

$$\begin{aligned} E_\mu(\mathbf{k}, f) = \tilde{\Delta}\varepsilon^f + A^f \cos^2\theta + \tilde{z}\mathscr{M}^f_{n\alpha, n\pm 1\alpha} \tilde{\gamma}(\mathbf{k}) \\ - (-1)^\mu z \mathscr{M}^f_{n1, n2} \gamma(\mathbf{k}) \cos^2\theta, \end{aligned} \tag{3.14}$$

where

$$\tilde{\Delta}\varepsilon^f = \Delta\varepsilon^f + \mathscr{I}_{n1, n2} z + \mathscr{I}^f_{n\alpha, n\pm 1\alpha} \tilde{z},$$

$$A^f = -\left[1 - \frac{g^f(S-1)}{gS}\right] 2S^2 \mathscr{I}_{n1, n2} z + 2\mathscr{I}^f_{n1, n2} z,$$

$$\gamma(\mathbf{k}) = \frac{1}{z} \sum_{\boldsymbol{\delta}} e^{i\mathbf{k}\boldsymbol{\delta}}, \qquad \tilde{\gamma}(\mathbf{k}) = \frac{1}{\tilde{z}} \sum_{\boldsymbol{\rho}} e^{i\mathbf{k}\boldsymbol{\rho}},$$

where θ is the angle between the direction of the sublattice spins and the external magnetic field $\mathbf{H}$; $\boldsymbol{\delta}$ and $\boldsymbol{\rho}$ are vectors connecting the magnetic ion to the nearest and the next-nearest magnetic neighbor; z and $\tilde{z}$ are the number of the nearest and the next-nearest neighbors, respectively: and g and g^f are g-factors of the ground and excited states, respectively.

The corresponding expression for the energy of a magnon in the nearest neighbor approximation has the form

$$\varepsilon_\mu(\mathbf{k}) = -S\mathscr{I}_{\mathbf{n}1,\mathbf{n}2}z[1 - \gamma^2(\mathbf{k}) + 2\cos^2\theta\gamma(\mathbf{k})(\gamma(\mathbf{k}) - (-1)^\mu)]^{1/2}. \quad (3.15)$$

There are two ($\mu = 1, 2$) excited branches of the spectrum $E_\mu(\mathbf{k}, f)$ that correspond to one optical excitation of ion $\Delta\varepsilon^f$. With $\mathbf{k} = 0$, the energies of these branches differ by the magnitude

$$\hbar\Delta v_D = 2|L_2^f(k = 0)| \cos^2 \tfrac{1}{2}(\theta_1, \theta_2),$$

which is the Davydov magnetic splitting. The dependence of the Davydov splitting upon the degree of noncollinearity of the magnetic sublattices, which is experimentally observed in $RbMnF_3$, crystal, is shown in Fig. 3.1.

It should be noted that in further discussions the matrix elements $\mathscr{I}_{\mathbf{n}\alpha,\,\mathbf{m}\beta}$, $\mathscr{I}^f_{\mathbf{n}\alpha,\,\mathbf{m}\beta}$, and $\mathscr{M}^f_{\mathbf{n}\alpha,\,\mathbf{m}\beta}$ ((3.8)–(3.10)) are treated as phenomenological parameters of the theory.

In the case of a noncollinear antiferromagnet, a general expression for a Hamiltonian of the exciton–magnon interaction is given by the following equation:

$$\begin{aligned}\mathscr{H}^{fs}_{\mathrm{int}} = \sum_{\mathbf{n}\alpha,\,\mathbf{m}\beta}{}' \, [&V^{fs}_{1\,\mathbf{n}\alpha,\,\mathbf{m}\beta}b^+_{\mathbf{m}\beta} + V^{fs}_{4\,\mathbf{n}\alpha,\,\mathbf{m}\beta}(b^+_{\mathbf{n}\alpha}b_{\mathbf{m}\beta} + b^+_{\mathbf{m}\beta}b_{\mathbf{n}\alpha}) \\ &+ V_{5\,\mathbf{n}\alpha,\,\mathbf{m}\beta}(b^+_{\mathbf{n}\alpha}b^+_{\mathbf{m}\beta} + b_{\mathbf{n}\alpha}b_{\mathbf{m}\beta}) + V^{fs}_{\mathbf{n}\alpha}b^+_{\mathbf{n}\alpha}b_{\mathbf{n}\alpha}]B^+_{\mathbf{n}\alpha}(f)B_{\mathbf{n}\alpha}(f) \\ &+ \sum_{\mathbf{n}\alpha,\,\mathbf{m}\beta}{}' \, [V^{fs}_{2\,\mathbf{n}\alpha,\,\mathbf{m}\beta}b^+_{\mathbf{m}\beta}b_{\mathbf{n}\alpha} + V^{fs}_{3\,\mathbf{n}\alpha,\,\mathbf{m}\beta}(b^+_{\mathbf{n}\alpha}b_{\mathbf{n}\alpha} + b^+_{\mathbf{m}\beta}b_{\mathbf{m}\beta}) \\ &+ V^{fs}_{6\,\mathbf{n}\alpha,\,\mathbf{m}\beta}(b^+_{\mathbf{m}\beta}b^+_{\mathbf{m}\beta} + b_{\mathbf{n}\alpha}b_{\mathbf{n}\alpha})]B^+_{\mathbf{n}\alpha}(f)B_{\mathbf{m}\beta}(f),\end{aligned} \quad (3.16)$$

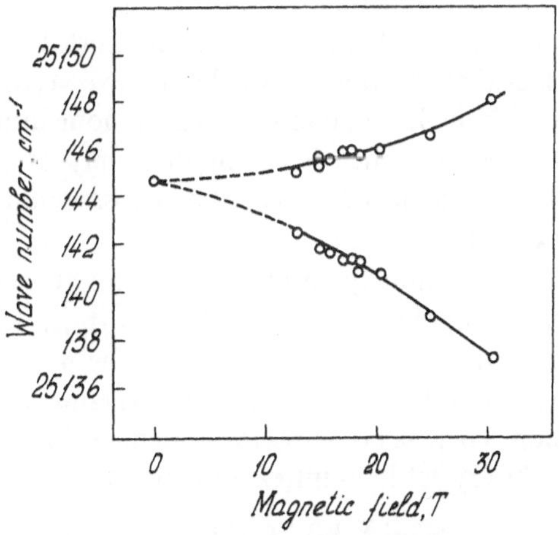

Fig. 3.1. Davydov splitting of the exciton line in an absorption spectrum of antiferromagnetic $RbMnF_3$ in the external magnetic field $H \parallel C_4$. $T = 20.4$ K [202].

where $V^{fs}_{j\mathbf{n}\alpha,\,\mathbf{m}\beta}$ and $V^{fs}_{\mathbf{n}\alpha}$ are matrix elements of the operator derived from (3.7). Their full expression [215] is not given here, it should, however, be noted that they can be expressed by means of parameters $\mathscr{I}_{\mathbf{n}\alpha,\,\mathbf{m}\beta}$, $\mathscr{I}^{f}_{\mathbf{n}\alpha,\,\mathbf{m}\beta}$, and $\mathscr{M}^{fs}_{\mathbf{n}\alpha,\,\mathbf{m}\beta}$. For a collinear antiferromagnet in the nearest neighbor approximation, the values $V^{fs}_{\mathbf{n}\alpha,\,\mathbf{m}\beta}$ and $V^{fs}_{4\mathbf{n}\alpha,\,\mathbf{m}\beta}$ vanish, and expression (3.16) is reduced to

$$\mathscr{H}^{fs}_{\text{int}} = S\mathscr{I}_{\mathbf{n}1\mathbf{n}2} \sum_{\mathbf{n}\alpha,\,\mathbf{m}\beta}' \{[2\Phi(b^{+}_{\mathbf{n}\alpha}b_{\mathbf{m}\beta} + b_{\mathbf{n}\alpha}b_{\mathbf{m}\beta}) + \kappa b^{+}_{\mathbf{m}\beta}b_{\mathbf{m}\beta} + \chi b^{+}_{\mathbf{n}\alpha}b_{\mathbf{n}\alpha}$$
$$+ [\tilde{\Phi}(b^{+}_{\mathbf{m}\beta}b^{+}_{\mathbf{m}\beta} + b_{\mathbf{n}\alpha}b_{\mathbf{n}\alpha}) + \tilde{\kappa} b^{+}_{\mathbf{m}\beta}b_{\mathbf{n}\alpha}]\} B^{+}_{\mathbf{n}\alpha}(f) B_{\mathbf{m}\beta}(f), \tag{3.17}$$

where

$$\Phi = \frac{1}{S\mathscr{I}_{\mathbf{n}1,\,\mathbf{n}2}}\left[\left(\sqrt{\frac{S}{S^f}} - 1\right) S\mathscr{I}_{\mathbf{n}1,\,\mathbf{n}2} - \sqrt{\frac{S}{S^f}\frac{1}{S}}\,\mathscr{I}^{f}_{\mathbf{n}1,\,\mathbf{n}2}\right],$$

$$\kappa = -\frac{1}{S\mathscr{I}_{\mathbf{n}1,\,\mathbf{n}2}}\frac{1}{S}\mathscr{I}^{f}_{\mathbf{n}1,\,\mathbf{n}2}, \qquad \chi = \frac{1}{S\mathscr{I}_{\mathbf{n}1,\,\mathbf{n}2}}\frac{1}{S}\mathscr{M}^{f}_{\mathbf{n}1,\,\mathbf{n}2},$$

$$\tilde{\Phi} = \frac{1}{S\mathscr{I}_{\mathbf{n}1,\,\mathbf{n}2}}\frac{1}{\sqrt{2S(2S-1)}}\mathscr{M}^{f}_{\mathbf{n}1,\,\mathbf{n}2}, \qquad \tilde{\kappa} = \frac{1}{S\mathscr{I}_{\mathbf{n}1,\,\mathbf{n}2}}\frac{1}{S}\mathscr{M}^{f}_{\mathbf{n}1,\,\mathbf{n}2}.$$

Thus, the use of exciton expansion in combination with the microscopic description of the magnetic ion state makes it possible to derive, for the most general noncollinear case, a general Hamiltonian of spin and optical excitations of an antiferromagnetic dielectric with due account taken of all the interactions taking place in it.

3.2. The Mechanism of Magnetic Excitation Generation by Light

3.2.1. One-Particle Processes (Magnetic Dipole Absorption and Emission)

The simplest process of magnetic excitation generation by the electromagnetic field is the following mechanism: an exciton is created by a quantum of light. Due to the fact that the value of the light wave vector is small ($\mathbf{q} \simeq 0$) and in conformity with the law of conservation of momentum, a one-particle absorption in the exciton zones is manifested only at fixed frequencies $(1/\hbar)E_\mu(\mathbf{k} = \mathbf{q} \simeq 0, f)$ in the form of very sharp absorption lines usually referred to as pure-exciton lines.

The crystal-light interaction operator $\mathscr{H}_j$ is composed of one-ion operators $\sum_{\mathbf{n}\alpha}\mathscr{H}_{j\mathbf{n}\alpha}$. For electric dipole transitions $\mathscr{H}_{j\mathbf{n}\alpha} \sim \sum_i \mathbf{r}_{i,\,\mathbf{n}\alpha}$, where $\mathbf{r}_{i,\,\mathbf{n}\alpha}$ is the radius vector of the ith electron in the site $\mathbf{n}\alpha$. The matrix element of the operator $\mathbf{r}_i$ between the states of the same parity is equal to zero in a crystal with center inversion. There is no parity forbiddenness in the magnetic dipole approximation, as the crystal-light interaction operator

$$\mathscr{H}_j = \sum_i \mathbf{h}(\mathbf{L} + 2\mathbf{S})e^{(\mathbf{q}\mathbf{r}_i - \omega t)} \tag{3.18}$$

does not change its state parity. The matrix elements of the orbit and spin operators **L** and **S** between states of the multiplet are not equal to zero. As for the transitions between the ground state with spin S and an excited state with spin $S-1$ (these transitions being the only ones which are discussed in this chapter), in zero-order perturbation theory and the matrix elements of these operators vanish. Taking into consideration the spin-orbit $\lambda\mathbf{LS}$ interaction corrections of the first-order perturbation theory, the wave function has the following form $\varphi^0 = \varphi(^{2S+1}\Gamma^0) + \alpha\varphi(^{2S-1}\Gamma^f)$, where $\varphi(^{2S+1}\Gamma^{0,f})$ are wave functions of the ground and excited states in zero-order approximation with respect to the spin-orbit interaction; and $\Gamma^{0,f}$ are terms of the ground and excited states. The mixing coefficient of wave functions of the ground and excited states is

$$\alpha = \frac{\langle\varphi(^{2S-1}\Gamma^f)|\lambda\mathbf{SL}|\varphi(^{2S+1}\Gamma^0)\rangle}{E(^{2S-1}\Gamma^f) - E(^{2S+1}\Gamma^0)}. \tag{3.19}$$

The matrix element (3.19) of the operator $\lambda\mathbf{LS}$ is not only equal to zero for states with equal $\mathbf{J} = L + S$, the excited states with equal spin $S-1$ are greatly mixed due to the crystalline field. Therefore, the matrix element (3.19) is nonzero for all the states f with $S-1$.

3.2.2. Multiparticle Processes (Electric Dipole Absorption and Emission)

Two-particle absorption is a more complex process: two quasi particles of different, in the general case, types are created in a crystal by a photon. If the quasi-particles do not interact, the absorption takes place at frequencies $1/\hbar\{E_{\mu_1}(\mathbf{k}_1, f_1) + E_{\mu_2}(\mathbf{k}_2, f_2)\}$, where $\mathbf{k}_1 + \mathbf{k}_2 = \mathbf{g} \simeq 0$. Exciton zones being involved lead to broad absorption bands. Interaction of quasi particles usually results in a redistribution of intensities in the absorption bands.

The light absorption of magnetically ordered crystals differs from that of nonmagnetic crystals, first of all by the fact that processes with the participation of spin waves (magnons) become possible. Under these conditions, the influence of an ordered spin system on the crystal optical spectrum turns out to be rather essential, and there arises a possibility of the electric dipole absorption of light in regions of spin-forbidden transitions.

The mechanisms resulting in electric dipole absorption in antiferromagnetic crystals are based on the general theory of pair excitations [216], which is indicative of a possibility for a quantum of light to excite two interacting ions. The probability of such a process turns out to be proportional to the interaction energy squared. It is possible to use, for example, the exchange part of the Coulomb interaction as a type of this interaction in an antiferromagnetic crystal this part being capable of lifting simultaneously the spin and parity of forbiddennesses in case there are ionic pairs in sublattices with opposed spins [217]. A number of other mechanisms were suggested, too. For example, two more mechanisms of the simultaneous excitation of an ionic pair were offered [218, 219]. One of these mechanisms is as follows: in the case of Coulomb interaction between ions, a simultaneous excitation can be caused by the interaction of the electric dipole moment of an ion in one sublattice with an electric quadrupole moment of an ion in another sublattice, the latter moment being induced by the spin-orbit interaction. The second mechanism is based on the interaction of the light wave electric field with spin waves by means of a virtual optical phonon. These mechanisms yield the required result by taking into consideration virtual electric dipole transitions (into odd parity states of ionic configurations) and the interaction of ionic pairs.

A Hamiltonian of crystal–light interaction in the case of an electric dipole transition can be written as

$$\mathscr{H}_\gamma = \mathbf{P}_{\text{eff}}(\mathbf{q} \simeq 0, ff')\mathbf{E}_0 e^{-i\omega t}, \tag{3.20}$$

where $\mathbf{P}_{\text{eff}}$ is the effective electric dipole moment of the crystal, ff' characterizes the number of the excited ion level, and $\mathbf{E}_0$ is the electric vector of the light wave.

The value $\mathbf{P}_{\text{eff}}$ is obtained in the second-order perturbation theory by taking into consideration the exchange interactions between ions of opposite sublattices:

$$\mathbf{P}_{\text{eff}}(\mathbf{q} \simeq 0, ff') = \sum_{\mathbf{n}\alpha,\, \mathbf{m}\beta} \boldsymbol{\pi}_{\mathbf{n}\alpha,\, \mathbf{m}\beta} B^+_{\mathbf{n}\alpha}(f) B^+_{\mathbf{m}\beta}(f'), \tag{3.21}$$

where $\boldsymbol{\pi}_{\mathbf{n}\alpha,\, \mathbf{m}\beta}$ is a dipole moment of transition in a pair of ions $\mathbf{n}\alpha$, $\mathbf{m}\beta$. The absorption of light in the direction $\mathbf{E}_0 \parallel \tau$ is expressed by means of the Green function

$$K^\tau(\nu) \propto \lim_{\delta \to 0} \operatorname{Im} G(P^\tau_{\text{eff}}, P^\tau_{\text{eff}}, \nu + i\delta). \tag{3.22}$$

Expression (3.22) describes the shape of a two-particle light absorption band.

Exciton–Magnon Transitions

If the interaction between quasi particles is not taken into account in (3.22), the dependence of the probability of two-particle exciton–magnon absorption on the frequency will be as follows:

$$K^\tau(\nu) \propto \frac{1}{N} \sum_{\mathbf{k}} |\pi^\xi(\mathbf{k})|^2 u^2(\mathbf{k}) \delta(h\nu - E_{\mathbf{k}} - \varepsilon_{\mathbf{k}}), \tag{3.23}$$

where

$$\boldsymbol{\pi}(\mathbf{k}) = \sum_{\mathbf{n}\alpha,\, \mathbf{m}\beta} \boldsymbol{\pi}_{\mathbf{n}\alpha,\, \mathbf{m}\beta} e^{i\mathbf{k}(\mathbf{R}_{\mathbf{n}\alpha} - \mathbf{R}_{\mathbf{m}\beta})}.$$

The coefficients $\boldsymbol{\pi}_{\mathbf{n}\alpha,\, \mathbf{m}\beta}$ are usually determined by a group-theoretical method [216].

In a free quasi-particle approximation, and not taking into consideration the exciton dispersion, the line shape of the exciton–magnon absorption reflects the density of the magnon state distribution [219, 220]. It goes without saying that a more exact description of the shape and position of exciton–magnon bands requires taking into account the interaction between optically and spin excited ions, and, in some cases, taking into account the exciton dispersion as well [214, 221, 222].

When describing exciton transitions in magnetic crystals with one-ion excitations of $S, M_S \to S-1, M_S'$-type, where $M_S' = M_S, M_S \pm 1$, it is sometimes more convenient to use exciton operators $\boldsymbol{\sigma}_{\mathbf{n}\alpha}$ and $\boldsymbol{\sigma}^+_{\mathbf{n}\alpha}$ [223]. One-ion vector operators $\boldsymbol{\sigma}_{\mathbf{n}\alpha}$ transfer a magnetic ion $\mathbf{n}\alpha$ from the ground state to an excited state, and the operators $\boldsymbol{\sigma}^+_{\mathbf{n}\alpha}$ do vice versa. The matrix elements of the operators $\boldsymbol{\sigma}$ and $\boldsymbol{\sigma}^+$ have the following form:

$$\begin{aligned}
\langle S-1, M_S-1|\sigma_-|S, M_S\rangle &= \langle S, M_S|\sigma_-^+|S-1, M_S-1\rangle \\
&= [(S+M_S)(S+M_S-1)/2S(2S-1)]^{1/2}, \\
\langle S-1, M_S+1|\sigma_+|S, M_S\rangle &= \langle S, M_S|\sigma_+^+|S-1, M_S+1\rangle \\
&= -[(S-M_S)(S-M_S-1)/2S(2S-1)]^{1/2}, \\
\langle S-1, M_S|\sigma_z|S, M_S\rangle &= \langle S, M_S|\sigma_z^+|S-1, M_S\rangle \\
&= -[(S-M_S)(S+M_S)/2S(2S-1)]^{1/2}.
\end{aligned} \tag{3.24}$$

Whereas spin excitations were described hitherto proceeding from the exciton formalism, the Shingagawa–Tanabe method employs a reverse approach: excitons are described with the help of a modernized Heisenberg model. The operator $\sigma^+_{i\mathbf{n}\alpha}\sigma_{k\mathbf{n}\alpha}$ $(i, k = x, y, z)$ relates only terms of the ground multiplet and therefore it can be expressed by means of common spin operators

$$\sigma_i\sigma_k = \frac{\delta_{ik}}{6} + \frac{i\varepsilon_{ikl}}{4} - \frac{(S_iS_k + S_kS_i)}{4S(2S-1)} + \frac{\delta_{ik}(S+1)}{6(2S-1)}, \tag{3.25}$$

where ε_{ikl} is a unit entirely antisymmetrical tensor of the third order. The use of these operators enables one to obtain a concise form of expressing the angle-dependent relationship for all possible types of exciton–magnon processes of light absorption [224]. Using the exciton $\sigma_{\mathbf{n}\alpha}$ and spin $\mathbf{S}_{\mathbf{m}\beta}$ operators for a two-particle excitation, $\mathbf{P}_{\text{eff}}$ can be written as

$$\mathbf{P}_{\text{eff}} = \sum_{\mathbf{n}\alpha,\, \mathbf{m}\beta}' (\boldsymbol{\sigma}_{\mathbf{n}\alpha} \cdot \mathbf{S}_{\mathbf{m}\beta}) \cdot \boldsymbol{\pi}_{\mathbf{n}\alpha,\, \mathbf{m}\beta}. \tag{3.26}$$

For a three-particle excitation, the dipole moment operator is

$$\mathbf{P}_{\text{eff}} = \sum_{\substack{\mathbf{n},\mathbf{m},\mathbf{l} \\ \alpha,\beta,\gamma}} \mathbf{F}_{\mathbf{n}\alpha,\mathbf{m}\beta,\mathbf{l}\gamma}(\mathbf{A}^{(1)}_{\mathbf{n}\alpha} \cdot [\mathbf{A}^{(2)}_{\mathbf{m}\beta} \cdot \mathbf{A}^{(3)}_{\mathbf{l}\gamma}]), \tag{3.27}$$

here the index in brackets indicates the type of excitation: the site operator $\mathbf{A}_{\mathbf{n}\alpha}$ can be both a spin operator $\mathbf{S}$ and an exciton operator $\boldsymbol{\sigma}$.

In the case of a four-particle excitation

$$\mathbf{P}_{\text{eff}} = \sum_{\substack{\mathbf{n},\mathbf{m},\mathbf{l},\mathbf{r}\\ \alpha,\beta,\gamma,k}} \{\mathbf{Q}'_{\mathbf{n}\alpha,\,\mathbf{m}\beta,\,\mathbf{l}\gamma,\,\mathbf{r}k} = ([\mathbf{A}^{(1)}_{\mathbf{n}\alpha}\cdot\mathbf{A}^{(2)}_{\mathbf{m}\beta}][\mathbf{A}^{(3)}_{\mathbf{l}\gamma}\cdot\mathbf{A}^{(4)}_{\mathbf{r}k}])$$

$$+ \mathbf{Q}''_{\mathbf{n}\alpha,\,\mathbf{m}\beta,\,\mathbf{l}\gamma,\,\mathbf{r}k}(\mathbf{A}^{(1)}_{\mathbf{n}\alpha}\cdot\mathbf{A}^{(2)}_{\mathbf{m}\beta})(\mathbf{A}^{(3)}_{\mathbf{l}\gamma}\cdot\mathbf{A}^{(4)}_{\mathbf{r}k})\}. \tag{3.28}$$

The intensity of multiparticle transitions depends on the crystal magnetic structure, as is evident in (3.26) when going from a coordinate system connected with crystallographic axes (x, y, z) to an "eigen" system $(\xi_\alpha;\ \eta_\alpha;\ \zeta_\alpha)$, with the axis ζ_α being oriented along the equilibrium direction of the spin in the sublattice α

$$(\mathbf{A}^{(1)}_{\mathbf{n}\alpha}\cdot\mathbf{A}^{(2)}_{\mathbf{m}\beta}) = \sum_{k_1k_2} G_{k_1k_2}(\theta_\alpha - \theta_\beta)A^{k_1(1)}_{\mathbf{n}\alpha}\mathbf{A}^{k_2(2)}_{\mathbf{m}\beta}, \tag{3.29}$$

where

$$A^i_{\mathbf{n}\alpha} = \sum_k R_{ik}(\theta_\alpha)A^k_{\mathbf{n}\alpha} \qquad (i = \{x, y, z\},\ k = \{+, -, \zeta\},\ A^\pm = A^\xi \pm iA^\eta),$$

$$G_{k_1k_2} = \begin{bmatrix} -\frac{1}{2}\sin^2\theta & \frac{1}{2}\cos^2\theta & \frac{1}{2}\sin 2\theta \\ \frac{1}{2}\cos^2\theta & -\frac{1}{2}\sin^2\theta & \frac{1}{2}\sin 2\theta \\ -\frac{1}{2}\sin 2\theta & -\frac{1}{2}\sin 2\theta & \cos 2\theta \end{bmatrix},$$

where θ is the angle between the direction of the external magnetic field $\mathbf{H}$ and the magnetic moment of the sublattice α. Here and below indices i and k are numbered as follows:

$$x(i = 1),\quad y(i = 2),\quad z(i = 3),$$
$$+(k = 1),\quad -(k = 2),\quad \zeta(k = 3).$$

In a similar way, an angle-dependent relationship can be derived from (3.27) and (3.28) which describes three- and four-particle absorption:

$$(\mathbf{A}^{(1)}_{\mathbf{n}\alpha}\cdot[A^{(2)}_{\mathbf{m}\beta}\cdot\mathbf{A}^{(3)}_{\mathbf{l}\gamma}]) = \sum_{k_1k_2k_3} G_{k_1k_2k_3}(\theta_\alpha, \theta_\beta, \theta_\gamma)A^{k_1(1)}_{\mathbf{n}\alpha}A^{k_2(2)}_{\mathbf{m}\beta}A^{k_3(3)}_{\mathbf{l}\gamma},$$

$$G_{k_1k_2k_3} = \begin{vmatrix} R_{1k_1}(\theta_\alpha) & R_{2k_1}(\theta_\alpha) & R_{3k_1}(\theta_\alpha) \\ R_{1k_2}(\theta_\beta) & R_{2k_2}(\theta_\beta) & R_{3k_2}(\theta_\beta) \\ R_{1k_3}(\theta_\gamma) & R_{2k_3}(\theta_\gamma) & R_{3k_3}(\theta_\gamma) \end{vmatrix}. \tag{3.30}$$

For $k_1 = k_2 = k_3 = 2G_{222}(\theta_\alpha, \theta_\beta, \theta_\gamma) \sim \sin\frac{1}{2}(\theta_\alpha - \theta_\beta)\sin\frac{1}{2}(\theta_\alpha - \theta_\gamma) \times \sin\frac{1}{2}(\theta_\beta - \theta_\gamma)$ and for a two-sublattice antiferromagnetic $G_{222} = 0$ as two of the three angles θ_α, θ_β, and θ_γ should be equal

$$([\mathbf{A}^{(1)}_{\mathbf{n}\alpha}\cdot\mathbf{A}^{(2)}_{\mathbf{m}\beta}][\mathbf{A}^{(3)}_{\mathbf{l}\gamma}\cdot\mathbf{A}^{(4)}_{\mathbf{r}\kappa}]) = \sum_{k_1-k_4} G'_{k_1k_2k_3k_4}(\theta_\alpha, \theta_\beta, \theta_\gamma, \theta_\kappa)A^{k_1(1)}_{\mathbf{n}\alpha}A^{k_2(2)}_{\mathbf{m}\beta}A^{k_3(3)}_{\mathbf{l}\gamma}A^{k_4(4)}_{\mathbf{r}\kappa},$$

$$(\mathbf{A}^{(1)}_{\mathbf{n}\alpha}\cdot\mathbf{A}^{(2)}_{\mathbf{m}\beta})(\mathbf{A}^{(3)}_{\mathbf{l}\gamma}\cdot\mathbf{A}^{(4)}_{\mathbf{r}\kappa}) = \sum_{k_1-k_4} G''_{k_1k_2k_3k_4}(\theta_\alpha, \theta_\beta, \theta_\gamma, \theta_\kappa)A^{k_1(1)}_{\mathbf{n}\alpha}A^{k_2(2)}_{\mathbf{m}\beta}A^{k_3(3)}_{\mathbf{l}\gamma}\dot{A}^{k_4(4)}_{\mathbf{r}\kappa}, \tag{3.31}$$

where $G'_{k_1k_2k_3k_4}$ and $G''_{k_1k_2k_3k_4}$ are expressed by means of matrix elements (3.29) of pair excitations as follows:

$$\begin{aligned} G'_{k_1k_2k_3k_4} &= G_{k_1k_3}(\theta_\alpha, \theta_\gamma)G_{k_2k_4}(\theta_\beta, \theta_\kappa) - G_{k_1k_4}(\theta_\alpha, \theta_\kappa)G_{k_2k_3}(\theta_\beta, \theta_\gamma), \\ G''_{k_1k_2k_3k_4} &= G_{k_1k_2}(\theta_\alpha, \theta_\beta)G_{k_3k_4}(\theta_\gamma, \theta_\kappa). \end{aligned} \tag{3.32}$$

The integral coefficient of absorption is determined by the correlator of the crystal dipole moment $K^\tau = \langle (P^\tau_{\text{eff}})^+ P^\tau_{\text{eff}} \rangle \sim |G|^2(\mathbf{E}_0 \parallel \tau)$. Each matrix element is indicative of the character of dependence of the absorption (or luminescence) intensity upon the angle between the magnetic sublattices of a crystal of one of the eight possible types of pair transitions schematically shown in Fig. 3.2(a)–(i). The transitions in which thermally excited ions participate shall be further referred to as "hot," and the transitions taking place at low temperatures including $T = 0$ K shall be further referred to as "cold." The first line of the matrix (3.29) corresponds to the process of exciton–magnon luminescence (Fig. 3.2(a), (b), (c)), and the second and third lines correspond to absorption (Fig. 3.2(d)–(i)). The matrix element G_{21} reflects the process of a

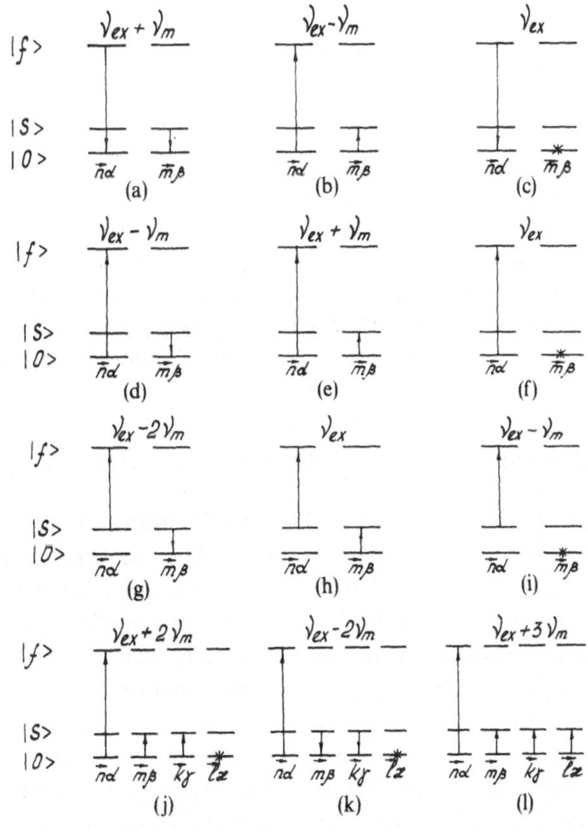

Fig. 3.2. Schematic diagram of electric dipole transitions with participation of one or more magnons [224].

"hot" exciton–magnon absorption, i.e., the creation of an optical excitation in one sublattice with a simultaneous annihilation of a thermally excited magnon in the other (Fig. 3.2(d)). The frequency of this transition is less than that of an exciton line by the value of the magnon energy $\nu_m (\nu = \nu_e - \nu_m)$. The process is described by the operator $\mathbf{S}^+_{n\alpha}\boldsymbol{\sigma}^-_{m\beta}$, the absorption coefficient versus angle dependence being proportional to $\cos^4\theta$.

G_{22} is a "cold" exciton–magnon absorption. The operator is $\mathbf{S}^-_{n\alpha}\boldsymbol{\sigma}^-_{m\beta}$, $K \sim \sin^4\theta$, $\nu = \nu_e + \nu_m$ (Fig. 3.2(e)).

G_{23} is an absorption in crystals without a center inversion in an ion pair (an optical excitation is created in one sublattice, and the ions of the other sublattice take part in the absorption without being excited). The operator is $\mathbf{S}^z_{n\alpha}\boldsymbol{\sigma}^-_{m\beta}$, $K \sim \sin^2 2\theta$, $\nu = \nu_e$ (Fig. 3.2(f)).

G_{31} is an absorption in thermally excited ions of both the sublattices, an optical excitation being created in one sublattice and a spin excitation being annihilated in the other. The operator is $\mathbf{S}^+_{n\alpha}\boldsymbol{\sigma}^z_{m\beta}$, $K \sim \sin^2 2\theta$, $\nu = \nu_e - 2\nu_m$ (Fig. 3.2(g)).

G_{32} is the process of an optical excitation in thermally excited ions of one sublattice and of spin excitation in ions of the other. The operator is $\mathbf{S}^-_{n\alpha}\boldsymbol{\sigma}^z_{m\beta}$, $K \sim \sin^2 2\theta$, $\nu = \nu_e$ (Fig. 3.2(h)).

G_{33} corresponds to an optical transition from a thermally occupied level, ions of the second sublattice take part in the absorption but they are not excited (the transition is possible in crystals without a center inversion in a pair of ions). The operator is $\mathbf{S}^z_{n\alpha}\boldsymbol{\sigma}^z_{m\beta}$, $K \sim \cos^2 2\theta$, $\nu = \nu_e - \nu_m$ (Fig. 3.2(i)).

Some four-particle transitions are schematically shown in Fig. 3.2(i), (k), (l). The operator $\mathbf{S}^z_{n\alpha}\mathbf{S}^-_{m\beta}\mathbf{S}^-_{l\gamma}\sigma^-_{r\kappa}$ describes the process which corresponds to one optical and two spin excitations, a part of the ions taking part in the process without getting excited (Fig. 3.2(j)). The matrix element of such a transition is described by the following expression: $G'_{3,2,2,2}G''_{3,2,2,2} \sim \sin^3\theta\cos\theta$.

The process of optical excitation with annihilation of two magnons (Fig. 3.2(k)) is described by the operator $\mathbf{S}^z_{n\alpha}\mathbf{S}^+_{m\beta}\mathbf{S}^+_{l\gamma}\sigma^-_{r\kappa}$, and the process of exciton and three magnons creation (Fig. 3.2(l)) by the operator $\mathbf{S}^-_{n\alpha}\mathbf{S}^-_{m\beta}\mathbf{S}^-_{l\gamma}\sigma^-_{r\kappa}$. The corresponding matrix elements are $G'_{3,1,1,2}G''_{3,1,1,2} \sim \sin\theta\cos^3\theta$; $G_{2,2,2,2} \sim \sin^4\theta$.

The majority of the processes under consideration were observed experimentally. Consider some concrete examples. The spectrum of an antiferromagnetic crystal MnF_2 (Fig. 3.3) is a classical example of the spectrum in which both the structure of the pure exciton absorption and that of the exciton–magnon absorption are simultaneously well resolved. The purely exciton absorption and luminescence is designated on the figure with E1, 2, and the exciton–magnon absorption with (e–m).

A crystal $CoCO_3$ is very convenient for observing different types of exciton–magnon absorption. Of the nine types of absorption processes shown in Fig. 3.2, all the processes which are possible, for a crystal with inversion center in a pair of interacting ions, have been observed in this crystal [224]. Cobalt carbonate has a crystalline structure belonging to a rhombo-

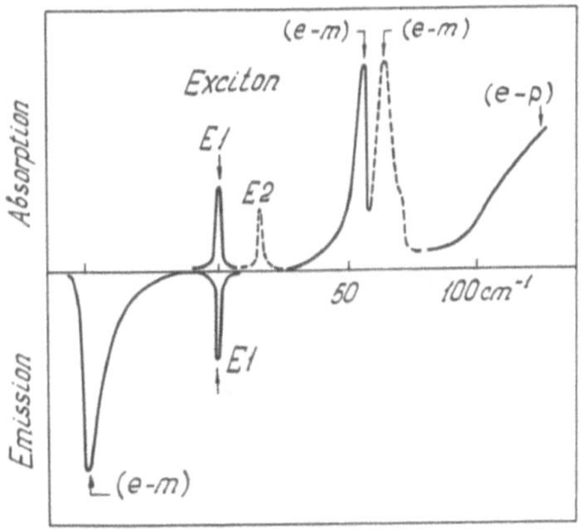

Fig. 3.3. Emission and absorption spectra in regions of frequencies of $^6A_{1g} \rightarrow {}^4T_{1g}$ transitions in MnF_2 crystal at 2 K [268].

hedron crystallographic system $D_{3d}^6(R\bar{3}c)$.[1] When cooled below the Néel temperature ($T_N = 18$ K), the crystal $CoCO_3$ goes over into a weak ferromagnetic state with the ferromagnetism vector **m** and the antiferromagnetism vector **l** oriented in the basic plane. The effective field of anisotropy in the basic plane makes up not more than 0.02 T, the effective field of exchange interaction $H_E = 21$ T, and that of the Moriya–Dzialoshinskii interaction $H_D \approx 3.3$ T. Hence, in an external magnetic field, the transition of $CoCO_3$ into a state close to a saturated paramagnetic one is achieved in fields $2H_E \simeq 42$ T, which are feasible for modern magneto-optical studies. A transition to a saturated paramagnetic state is effected by changing gradually the angle between the magnetic moments of ions in opposite sublattices without phase transformations of any kind. This has made it possible to carry out a comprehensive comparison of experimental results with the theoretically predicted behavior of the exciton–magnon transition intensity depending on the angle between the magnetic moments of ions from the opposite sublattices.

Figure 3.4 shows a part of the $CoCO_3$ absorption spectrum for a $^4\Gamma_4(^4F) - {}^2\Gamma_4(^2P)$ transition at 1.8 K (the bands originating from the magnetic fields at 14 K are shown by a dashed line).

In the spectrum under investigation, there is a magnetic dipole line corresponding to a one-particle pure exciton absorption. All the other lines are of an electric dipole character. It is obvious that the "cold" two-particle exciton–magnon process (operator $\mathbf{S}_{n\alpha}^- \boldsymbol{\sigma}_{m\beta}^-$) has to make the biggest contribution to the absorption. When going over to a normalized magnetization $m = |(\mathbf{M}_1 + \mathbf{M}_2)/2M_0|$, the integral coefficient of absorption of such an exciton–magnon sideband is described by $K \sim \sin^4\theta = (1 - m^2)^2$, where M_0 is the saturation

[1] References to structural and magnetic properties of the crystals which are mentioned will be found in the literature cited.

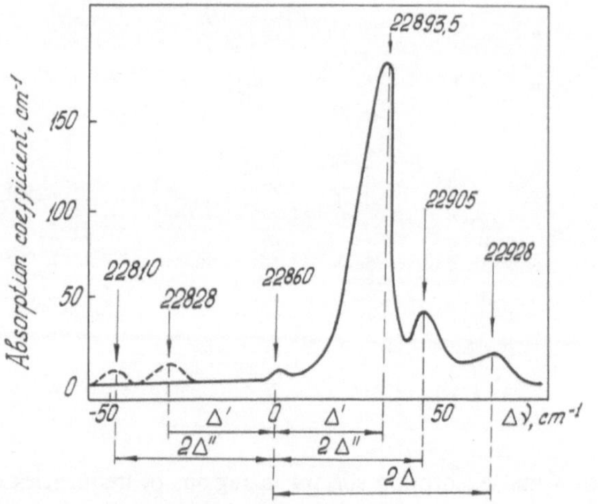

Fig. 3.4. Absorption spectrum of $CoCO_3$ in a region of $^4\Gamma_4 - {}^2\Gamma_4$ transition. $T = 1.8$ K. The dashed line refers to $T = 14$ K [224].

magnetization of sublattices, and $\mathbf{M}_1$ and $\mathbf{M}_2$ are magnetic moments of sublattices. In other words, in a magnetic field which transforms an antiferromagnetic into a paramagnetic state, this absorption has to disappear completely. Figure 3.5 shows experimental and theoretical depenedences of the integral coefficient of such an absorption on the normalized magnetization of $CoCO_3$. The experimental curve does agree well with the theoretical one. This effect of absorption reduction, the so-called "bleaching" of an antiferromagnet, was observed for the first time in an antiferromagnetic ferrous carbonate in the case of a metamagnetic phase transition [225], and somewhat later in MnF_2 [226]. The rest of the exciton–magnon processes are less intensive as they require either the participation of a great number of particles or

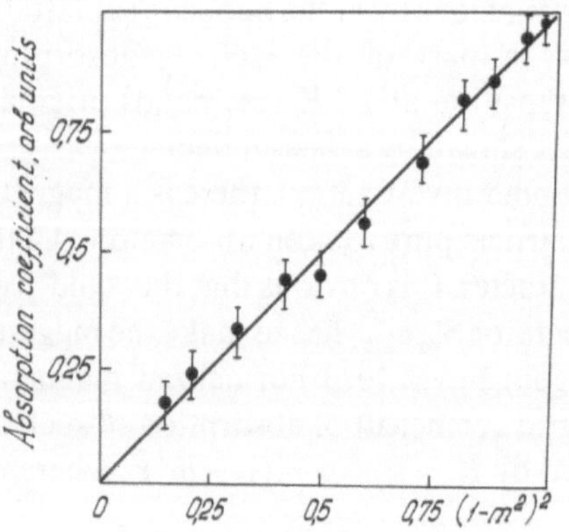

Fig. 3.5. Integral absorption coefficient of exciton band 22,893.5 cm^{-1} versus crystal magnetization ($H \perp C_3$): (—) theory; (●) experiment [224].

Table 3.1

			Calculation			Experiment
$\nu\ (\mathrm{cm}^{-1})$	$\Delta\nu\ (\mathrm{cm}^{-1})$	Identification	$K(\theta)$	θ at K_{max} (rad)	H at K_{max} (T)	H at K_{max} (T)
22,810[a]	2(25)	$\nu_e - 2\nu_m^{(2)}$	$\sin^2\theta\cos^6\theta$	$\pi/6$	~25	≥26
22,828[a]	32	$\nu_e - 2\nu_m^{(1)}$	$\cos^4\theta$	0	~43	>30
22,860[a]	0	ν_e^b	$\sin^2\theta\cos^2\theta$	$\pi/4$	15	20
22,860	0	ν_e	const	—	—	
22,893.5	33.5	$\nu_e + \nu_m^{(1)}$	$\sin^4\theta$	$\pi/2$	0	0
22,905	2(22.5)	$\nu_e + 2\nu_m^{(2)}$ }	$\sin^6\theta\cos\theta$	$\pi/3$	0	0
22,928	2(34)	$\nu_e + 2\nu_m^{(1)}$ }				
22,959	3(33)	$\nu_e + 3\nu_m^{(1)}$	$\sin^8\theta$	$\pi/2$	0	0

[a] At $T \geq 14$ K.
[b] Exciton–magnon band at exciton frequency. $\nu_m^{(1)}$, $\nu_m^{(2)}$ are magnon frequencies at two different points at the boundary of the Brillouin zone.

the availability of a thermally populated excited level. It follows from the above theoretical analysis, that the intervals between the exciton line and its magnon sidebands of all kinds have to be a multiple of the magnon frequency at the boundary of the Brillouin zone (the exciton–magnon interaction being neglected). As follows from Fig. 3.4, this condition is observed in the spectrum under consideration. As for the dependences of the integral intensity of these transitions on the angle θ, they have to differ. Presented in Table 3.1 are values of the angles at which the absorption intensities are maximum for different exciton–magnon processes; also presented is a comparison with an experiment which shows that the exciton–magnon transitions under consideration really take place in an antiferromagnetic $CoCO_3$.

The exciton–magnon transition which corresponds to the operator $\mathbf{S}_{n\alpha}^z \boldsymbol{\sigma}_{m\beta}^-$ (the matrix element is G_{23}, Fig. 3.2(f)) and is realized at the frequency of a pure exciton transition $\nu = \nu_e$ ($K \sim \sin^2 2\theta$) has been observed in $CsMnF_3$ and $RbMnCl_3$ crystals [227]. This obsorption actually increased in the magnetic fields which disturbed the collinearity of the spin system.

And, finally, more complex two-particle processes can take part in antiferromagnets; these are processes of exciton–magnon light absorption in which either doubly spin-excited or spin-optically excited ions participate (Fig. 3.6). The processes shown in Fig. 3.6(a), (b), (c) have been detected in a $RbMnF_3$ crystal [228]. The angle factor for an exciton–two-magnon absorption (Fig. 3.6(a), (b)) is $\sin^6\theta\cos^2\theta$, and that for an exciton–three-magnon absorption (Fig. 3.6(c)) is $\sin^8\theta$, i.e., these factors are similar to angle factors previously obtained for "cold" two- and three-magnon absorption in four-particle processes. It should be noted that a spin-optical excitation, which is characterized by a change of the spin projection by two units and of the spin magnitude by one unit, due to a strong spin forbiddenness, cannot spontaneously migrate in the crystal and is decomposed into optical and spin excitations of two ions which are capable of migrating.

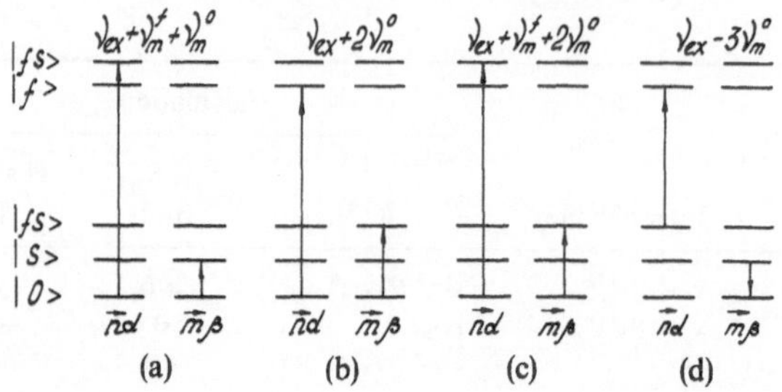

Fig. 3.6. Schematic diagram of exciton–multimagnon processes of light absorption in a pair of magnetic ions.

A "hot" process (Fig. 3.6(d)) of a three-magnon absorption was observed in ferrous fluoride at temperatures close to T_N [229].

Double-Exciton Transitions

The mechanism of a double-exciton light absorption is similar to that of an exciton–magnon absorption. In this case, when expanding the Hamiltonian (3.6), the Hamiltonians of optical excitations of type f and f' are taken into account, as well as the Hamiltonian of optical excitation interaction. As in the case of an exciton–magnon absorption, the light absorption coefficient at optical excitation of two ions into the states f and f' is derived from expression (3.22) taking into consideration (3.21). In the case of a double-exciton transition the states can be both symmetrical and antisymmetrical with respect to $f - f'$ permutation [230]. Then the operator of the dipole moment can be written as

$$\mathbf{P}_{\text{eff}} = \sum_{\mathbf{n}\alpha,\, \mathbf{m}\beta} \pi^{(\nu)}_{\mathbf{n}\alpha,\, \mathbf{m}\beta} C^{+}_{\mathbf{n}\alpha,\, \mathbf{m}\beta}(\nu, ff'), \tag{3.33}$$

where $\pi^{(\nu)}_{\mathbf{n}\alpha,\,\mathbf{m}\beta} = (1/\sqrt{2})(\pi^{ff'}_{\mathbf{n}\alpha,\,\mathbf{m}\beta} - (-1)^{\nu}\pi^{ff'}_{\mathbf{n}\alpha,\,\mathbf{m}\beta})$, $\nu = 1$ in a symmetrical and $\nu = 2$ in an antisymmetrical state, and $C^{+}_{\mathbf{n}\alpha,\,\mathbf{m}\beta}(\nu, ff') = (1/\sqrt{2})(B^{+}_{\mathbf{n}\alpha}(f)B^{+}_{\mathbf{m}\beta}(f') - (-1)^{\nu}B^{+}_{\mathbf{n}\alpha}(f')B^{+}_{\mathbf{m}\beta}(f))$. It follows from these expressions, that in antiferromagnets with the inversion center in an ionic pair, only dipole transitions into antisymmetrical states ($f = f'$) are allowed. However, in case the relations $\pi^{ff'}_{\mathbf{n}\alpha,\,\mathbf{m}\beta} = -\pi^{f'f}_{\mathbf{n}\alpha,\,\mathbf{m}\beta} = -\pi^{ff'}_{\mathbf{m}\beta,\,\mathbf{n}\alpha}$ are not satisfied, double-exciton transitions are possible with both $f \neq f'$ and $f = f'$.

These double-exciton transitions can be exemplified by the MnF_2 absorption spectrum in ultraviolet [231].

3.2.3. Bonded States

The wave function of an antiferromagnet containing two excitons f and f' in different magnetic subsystems can be presented in the following form [232]:

$$\Psi = \sqrt{2/N} \sum_{\mathbf{n},\mathbf{m}} \exp[i\mathscr{K}(\mathbf{n}+\mathbf{m})/2]u_{\mathscr{K}}(\mathbf{n}-\mathbf{m})B^{+}_{\mathbf{n}1}(f;1)B^{+}_{\mathbf{m}2}(f';-1)$$
$$+ \theta_{\mathscr{K}}(\mathbf{n}-\mathbf{m})B^{+}_{\mathbf{n}1}(f';1)B_{\mathbf{m}2}(f;-1)]|gr\rangle, \tag{3.34}$$

where $\mathscr{K}$ is the full momentum of the two-exciton state, the functions $u_{\mathscr{K}}(\mathbf{r})$, $\theta_{\mathscr{K}}(\mathbf{r})$ describe the relative motion of the exciton, N is the number of molecules in the crystal, and $|gr\rangle$ is the wave function of the ground state with respect to which it is assumed that the spins of the first magnetic sublattice are oriented upwards (1) and those of the second one, downwards (-1). Equations for $u_{\mathscr{K}}(\mathbf{r})$ and $v_{\mathscr{K}}(\mathbf{r})$ have the following form:

$$\begin{pmatrix} u_{\mathscr{K}}(\mathbf{r}) \\ \theta_{\mathscr{K}}(\mathbf{r}) \end{pmatrix} = \sum_{\mathbf{r}'} \begin{pmatrix} Y_{\mathscr{K}}(\mathbf{r}-\mathbf{r}')V_{1\mathbf{r}'} & Y_{\mathscr{K}}(\mathbf{r}-\mathbf{r}')V_{2\mathbf{r}'} \\ Y_{\mathscr{K}}(\mathbf{r}'-\mathbf{r})V_{2\mathbf{r}'} & Y_{\mathscr{K}}(\mathbf{r}'-\mathbf{r})V_{1\mathbf{r}'} \end{pmatrix} \begin{pmatrix} u_{\mathscr{K}}(\mathbf{r}') \\ \theta_{\mathscr{K}}(\mathbf{r}') \end{pmatrix}, \tag{3.35}$$

where

$$Y_{\mathscr{K}}(\mathbf{r}) = \frac{2}{N}\sum_{\mathbf{q}} e^{i\mathbf{q}\mathbf{r}}\left[\varepsilon - \sum_{\Delta} \mathbf{M}_{\Delta}^{+} \cos\left(\frac{\mathscr{K}}{2}+\mathbf{q}\right)\Delta + \mathscr{M}_{\Delta}^{f} \cos\left(\frac{\mathscr{K}}{2}-\mathbf{q}\right)\Delta\right]$$

is the Green function of noninteracting excitons, ε is the energy calculated from $E^{f} + E^{f'} + 2\sum_{n\alpha} \mathscr{I}_{01,n\alpha}\langle S_{n\alpha}\rangle$, $\mathscr{M}_{\Delta}^{f} \equiv \mathscr{M}_{\Delta\alpha,0\alpha}^{f}$, and $V_{jn} \equiv V_{jn1,02}^{ff'}$ is the matrix element of static $(j=1)$ and resonance permutation $(j=2)$ interaction. The continuous spectrum of the eigenvalues of (3.34) corresponds to dissociated excitons, and the isolated roots of the equations to bonded excitons, i.e., "biexcitons." Whereas, for states of the first type, the role of exciton–exciton interactions is reduced to the scattering of excitons on one another, the exciton-to-exciton interaction in the bonded state results in such a strong correlation of the motion of the excitations that both excitons act as a single quasi particle. The effective mass of a biexciton created by two different excitons turns out to be dependent on the resonance permutation interaction V_2. This interaction results in the following interesting peculiarity: such a biexciton can move in the crystal even in the case where the mass of one of its constituent excitons is infinitely large.

In accordance with the above-mentioned considerations, the absorption spectrum in the zone of two-exciton states consists of two parts: sharp absorption lines caused by excitation of biexcitons and broad bands of continuous absorption corresponding to dissipative two-exciton states. The stronger the interaction between excitons, the weaker the integral intensity of the continuous absorption. And vice versa, since the interaction decreases and approaches the critical value at which the biexciton state with $\mathscr{K}=0$ splits off from the continuous spectrum states, the integral intensity of the biexciton absorption falls off. A most striking example of such biexciton excitations is the absorption of the low-temperature antiferromagnetic α-phase of solid oxygen [209].

There evidently may appear a new quasi particle—a bonded exciton–magnon state which is similar to a biexciton. This state is quite possible in the antiferromagnets under consideration, it can be realized at appropriate values of exciton–magnon interactions, of the density of states in magnon and exciton bands, and of the width of these bands [233]. Such a bonded exciton–magnon state appears in the MnF_2 absorption spectrum in which there is a narrow peak at the boundary of the broad band of the conventional exciton–magnon absorption in the region of ${}^6A_{1g}({}^6S) \rightarrow {}^4E_g, {}^4A_g({}^4G)$ transition [234].

3.2.4. Parametric Excitation by Optical Pumping

The application of powerful optical radiation to excite high-frequency magnons in magnetics was often discussed in theoretical papers of the last decade. For example, a method of magnon generation based on mixing the radiations of two lasers in a magnetic crystal was recently suggested [235, 236]. The possibility of creating considerable concentrations of high-frequency magnons during double-magnon absorption was also discussed [237]. The most promising, at any rate for antiferromagnetic dielectrics, is the method of magnetic excitation generation based on the powerful laser pumping of two-particle exciton–magnon transitions [238]. In this case, absorption of light creates excitons and magnons with wave vectors of equal values and opposite orientation.

Consider the conditions at which there can energe a parametric generation of magnons in the case of an intensive optical pumping at the frequency of an exciton–magnon transition. The balance equation for the numbers of magnon $n_{\mathrm{m}}(\mathbf{k})$ and exciton $n_{\mathrm{e}}(\mathbf{k})$ populations has the following form [238]:

$$n_{\mathrm{m}}(\mathbf{k})\gamma_{\mathrm{m}} = \lambda[n_{\mathrm{m}}(\mathbf{k}) + n_{\mathrm{e}}(\mathbf{k}) + 1]\varphi(\nu_{12}(\mathbf{k})), \tag{3.36}$$

where $\lambda = (2/\pi)(\mathscr{I}K/n\nu_{\mathrm{em}}\sigma_{12}\Delta\nu)$, K is the light absorption coefficient at the exciton–magnon transition frequency ν_{em}, $\mathscr{I}$ is the light intensity, $\varphi(\nu)$ is the spectral line shape which is assumed to be of Lorentz-type with half-width $\Delta\nu$, and σ_{12} is the two-particle (exciton–magnon) density of states. It is also assumed that the laser-pumping spectrum with $\Delta\nu$ is greater than the quasi-particle relaxation frequencies (γ_{m}, γ_{e}) are the relaxation frequencies of magnons and of excitons, respectively). This assumption is true in the case when pulsed solid-state lasers are used as a pumping source, the width of the generation spectrum of these lasers being, as a rule, greater than 10^9 Hz. Populations of magnons are expressed as follows:

$$n_{\mathrm{m}} = \frac{\lambda\varphi}{1 - \lambda\varphi[(\gamma_{\mathrm{e}} + \gamma_{\mathrm{m}})/(\gamma_{\mathrm{e}} - \gamma_{\mathrm{m}})]}. \tag{3.37}$$

When the denominator in (3.37) vanishes, the number of magnons grows unrestrictedly, this point corresponding to the parametric instability threshold in the system of quasi particles. The threshold is controlled by the smallest of the relaxation frequencies γ_{e}, γ_{m}. Assuming $\gamma_{\mathrm{m}} < \gamma_{\mathrm{e}}$, the following expression was derived for the threshold intensity of radiation:

$$\mathscr{I}_n \sim h\nu_{\mathrm{em}}\gamma_{\mathrm{m}}(\mathbf{k}_{\mathrm{max}})\Delta\nu/Ka_0^3\nu_{\mathrm{m}}(\mathbf{k}_{\mathrm{max}}), \tag{3.38}$$

where a_0 is the lattice parameter, and $\mathbf{k}_{\mathrm{max}}$ are quasi momenta at the boundary of the Brillouin zone. As the light intensity at the exciton–magnon transition frequency increases, the absorption coefficient essentially grows when approaching the threshold power. If the nonlinear decay of quasi particles is not taken into consideration, the absorption coefficient tends to infinity at the threshold. If the nonlinear decay of quasi particles is taken into account, we obtain finite values of the exciton and magnon population numbers, as well as a finite value of $\Delta\nu_m$—the width of the generated quasi-particle distribution line. The full number of quasiparticles and the width of their distribution are governed by the constant α in nonlinear damping, this constant, in its turn, being controlled by the magnon merging processes. The total number of magnons is proportional to α^{-2}, and $\Delta\nu_m \sim \alpha$.

The following ratio between the relaxation frequencies of excitons and magnons is typical for the antiferromagnetic compounds of manganese: $\nu_m > \gamma_e$ [239]. However, as the threshold intensity is governed by the lowest of the relaxation frequencies, the value of $\mathscr{I}_n \simeq 0.5 \times 10^6$ W/cm^2 [238] can be considered to be the upper limit of the quasi-particle parametric excitation threshold.

At attempt was made to detect the change of the light absorption coefficient as a function of the light intensity, this change being caused by the generation of quasi particles [240]. The radiation of a copper vapor pulsed laser ($\lambda = 510.6$ nm) and of the second harmonic of a neodymium laser ($\lambda = 530$ nm) was used as a pumping source. The generation frequencies of the lasers corresponded to the region of the exciton–magnon–phonon absorption of the crystals MnF_2 and $CsMnCl_3 \cdot 2H_2O$ under investigation. The possibility of generating magnons with a boundary quasi momentum, during optical pumping in the spectral region of exciton–magnon–phonon absorption, was considered by Genkin and Golubeva [241]. It was shown that optical pumping in this region of the crystal absorption spectrum results in the following lowering of the quasi-particle generation threshold:

$$\mathscr{I}_n' = \mathscr{I}_n(K_0/K'), \tag{3.39}$$

where K' is the absorption coefficient at the pumping frequency, and K_0 is the absorption coefficient at the exciton–magnon transition frequency. Measurements have shown that up to the exciting light intensity 100 MW/cm^2 the absorption coefficient does not depend upon the intensity, i.e., there is no generation in an exciton–magnon system. Thus, the value of the threshold intensity, at which there can emerge an instability in an exciton–magnon system in the case of pumping into an exciton–magnon–phonon absorption band, is much higher than the theoretically predicted one -0.5 MW/cm^2 [238] which is evidently explained by the following factors:

(i) As the light absorption process in the case of excitation into the exciton–phonon band is a multiparticle one, the pumping of excitons and magnons according to the quasi-momentum conservation law takes places in a much larger energy interval than in the case of a pair absorption process. It must lead to a substantial increase in threshold values of the intensity as compared with a two-particle process.
(ii) In theoretical estimates of the threshold intensity the relaxation frequencies of quasi particles are assumed to be known from experiment and independent of the excitation intensity. In the case of a powerful optical

pumping at a considerable concentration of quasi particles, this assumption is not true. For example, the lifetime of excitons in MnF_2 depends on their concentration [242]. The lifetime of magnons must also depend on the pumping intensity. If the relaxation frequencies of quasi particles increase according to a complex law with the rise in their concentration, it means that the calculations of the threshold intensity for the generation of magnons yielded strongly underrated estimates.

3.2.5. An Estimate of Some Parameters of Optical Excitations

Investigation of the properties of one-particle light absorption with an external magnetic field applied to a crystal enables one to make a quantitative estimate of parameters of excitons such as $\mathscr{I}^f_{\mathbf{n}\alpha,\,\mathbf{m}\beta}$ and $\mathscr{M}^f_{\mathbf{n}\alpha,\,\mathbf{m}\beta}$. However, the one-particle absorption spectra corresponding to optical transitions into the point $k \simeq 0$ (the center of the Brillouin zone) give no information as to the width of the exciton bands. The main source of information of the exciton dispersion are the spectra of two-particle absorption and luminescence, as it is in these processes that the involvement of whole state takes place. An exciton dispersion is, as a rule, much less than a spin-wave dispersion; however, in a number of cases it turned out to be quite substantial. It is evident when the spin-wave dispersion becomes known from independent experiments, there appears a good opportunity to determine the width of the exciton band proceeding from the width of the observed exciton–magnon absorption and luminescence. Such an estimate of the exciton parameters was, for example, carried out for a two-sublattice antiferromagnetic $RbMnF_3$ ($T_N = 32.6$ K) with a cubic structure of perovskite type [228]. In the expression for the energy of the exciton branch of the spectrum in the nearest neighbors approximation (3.14), the second and last terms describe shifts of the exciton line center of gravity and Davydov splitting, respectively, both of them changing within the magnetic field according to as quare law. In the case of an $^4E_g(^4G)$ exciton it turns out that g-factors of the ground and excited states coincide and $A^f = 2S\mathscr{I}_{\mathbf{n}1,\,\mathbf{n}2}z + 2\mathscr{I}^f_{\mathbf{n}1,\,\mathbf{n}2}z$. For $RbMnF_3$,

$$\gamma(\mathbf{k}) = \tfrac{1}{3}(\cos ak_x + \cos ak_y + \cos ak_z)$$

and

$$\tilde{\gamma}(\mathbf{k}) = \tfrac{1}{3}(\cos ak_x \cos ak_y + \cos ak_x \cos ak_z + \cos ak_y \cos ak_z),$$

where a is the distance between the nearest magnetic ions. The spin-wave dispersion of $RbMnF_3$ is known and adequately described by the expression (3.15) with $S\mathscr{I}_{\mathbf{n}1,\,\mathbf{n}2} = 72\ \mathrm{cm}^{-1}$. The shift of the exciton line center of gravity in a field of 30 T is 3 cm^{-1}, and $\cos\theta = H/2H_E (H_E = 89$ T). Hence, the value of the parameter $z\mathscr{I}^f_{\mathbf{n}1,\,\mathbf{n}2}$ in $RbMnF_3$ for $^4E_g(^4G)$ exciton is equal to 126 cm^{-1}. Using the experimentally found Davydov splitting versus the magnetic field relationship, an estimate is also made of the resonance interaction $z\mathscr{M}^f_{\mathbf{n}1,\,\mathbf{n}2} = 200\ \mathrm{cm}^{-1}$.

The observed shift of the exciton line center of gravity in a magnetic field enables one to make an inference about the value of the exchange field H_E^f too; this field acts on an excited ion from the Mn^{2+} ions side which are in their ground state. As is to be seen from the expression for exciton dispersion (3.13) the shift of the "gravity center" is described by the terms $\Delta\varepsilon^f$ and D^f, which in the molecular field approximation are equal to

$$\Delta\varepsilon^f(H) = g\mu_B H \cos\theta, \tag{3.40}$$

$$D^f(H) = g\mu_B[H_E S - H_E^f(S-1)] \cos(\pi - 2\theta). \tag{3.41}$$

It turns out that for $RbMnF_3$ $H_E^f \simeq 0.6\, H_E$.

Thus, proceeding from the results of the spin-wave spectrum study and from the peculiarities of an exciton line behavior in a magnetic field, it is possible to obtain the values of all the parameters for the nearest neighbors $\mathscr{I}_{n\alpha,\,m\beta}$, $\mathscr{I}^f_{n\alpha,\,m\beta}$, and $\mathscr{M}^f_{n\alpha,\,m\beta}$, which describe a one-particle exciton excitation and a multiparticle exciton–magnon excitation. The next-nearest neighbors have to be taken into account when calculating the energy of excitons, as this account results in the emergence of a dispersion term $\tilde{z}\mathscr{M}^f_{n\alpha}\tilde{\gamma}_{n\alpha}(\mathbf{k})$ which describes the intersublattice exciton dispersion. To determine this term, a study of the exciton–one-magnon light absorption is necessary. If we neglect the exciton–magnon interaction V^{fs}, then the position of the band intensity peak will be determined by the density of the exciton–magnon states $\rho_\mu(\nu)$ modified by the factor $\sin^2 ak_x u_\mu^2(\mathbf{k})$ which gradually changes within the Brillouin zone. The account of the interaction V^{fs} greatly complicates the problem of the maximum determination, but is solved when considering only one singular point $\mathbf{k}_0$ of the state density $\rho_\mu(\nu)$. In separate cases such a simplified approach proves justified, as the number of singular points of the joint exciton–magnon density of states is limited, and, besides, they are additionally separated with respect to the weight factor $\sin^2 ak_x u_\mu^2(\mathbf{k})$. In a cubic two-sublattice antiferromagnetic, the maximum contribution into the exciton–magnon absorption is made only by points of the Brillouin zone which are located on the straight lines passing via the center of the zone and via the centers of the hexagonal faces and satisfying the following condition: $\cos ak_{0x} = \cos ak_{0y} = \cos ak_{0z} = \cos ak_0$,

$$\cos ak_0 = (-1)^\mu \cos^2\theta \frac{z\mathscr{M}_{n1,n2} - S\mathscr{I}_{n1,n2}z/\varepsilon_\mu(\mathbf{k}_0)}{2\tilde{z}\mathscr{M}^f_{n1,n+1,2} - [(S\mathscr{I}_{n1,n2})^2/\varepsilon_\mu(\mathbf{k}_0)]\cos 2\theta}. \tag{3.42}$$

Then the frequency of the absorption maximum is determined by the expression

$$\nu_{max} = E_\mu(\mathbf{k}_0, f) + \varepsilon_\mu(\mathbf{k}_0) + V^{fs}, \tag{3.43}$$

where

$$V^{fs} = (2\cos^2\theta - 1)(\mathscr{M}^f_{n1,n2} - \mathscr{I}^f_{n1,n2})/S.$$

In the case of no external field, the position of singular points coincides with that of the points $L = (\pi/a)(\frac{1}{2}, \frac{1}{2}, \frac{1}{2})$. As the exciton absorption corresponds to the point Γ of the Brillouin zone, the interval between the exciton line and its magnon sideband will be described by the expression

$$\nu_{max} - E(\Gamma^4, E_g(^4G)) = S\mathcal{I}_{\mathbf{n}1,\mathbf{n}2}z - \tilde{z}\mathcal{M}^f_{\mathbf{n}2,\mathbf{n}\pm 12} - \frac{1}{S}(M^f_{\mathbf{n}1,\mathbf{n}2}\mathcal{I}^f_{\mathbf{n}1,\mathbf{n}2}). \tag{3.44}$$

Hence the value of the dispersion part of the exciton energy $^4E_g(^4G)$ in $RbMnF_3$ is found to be equal to $\tilde{z}\mathcal{M}^f_{\mathbf{n}\alpha,\mathbf{n}\pm 1\alpha} = -29\ cm^{-1}$. The validity of the assumption used to find $\tilde{z}\mathcal{M}^f_{\mathbf{n}\alpha,n\pm 1\alpha}$ in $RbMnF_3$ is also confirmed by the fact that the calculated frequencies of the maxima of exciton two- and three-magnon absorption obtained with the help of this parameter are in very good agreement with those experimentally observed.

Information about exciton parameters corresponding to the lowest excited electron state $^4T_{1g}$ of the manganese ions is also given by studying the intrinsic luminescence spectra of the manganese antiferromagnetic compounds. We are going to emphasize that the luminescence has to be intrinsic. The thing is that the main contribution to the luminescence observed in the visible part of the spectrum in many antiferromagnetic compounds of manganese (in particular, in manganese fluorides) is made by the defect luminescence caused by Mn^{2+} ions located near the impurity ions Zn^{2+}, Mg^{2+}, Ca^{2+}, the intrinsic luminescence being extremely week [243]. The energy levels of these "disturbed" Mn^{2+} ions are located below the exciton band formed by the "undisturbed" ions, and at low temperatures these levels trap the electron excitation energy. Besides, a considerable part of the electron excitation energy is trapped by quenching traps. In nominally pure crystals of manganese fluorides, the host luminescence makes up approximately 0.01 of the total intensive luminescence.

Two-particle exciton–magnon transitions play an important role in forming the intrinsic luminescence spectra, as well as the light absorption spectra. There is, however, a considerable difference between exciton–magnon transitions in absorption (3.26) and emission. Whereas in the case of absorption the electron and spin excitations emerge on sublattices with opposed oriented spin moments (α and β sublattices in (3.26)), in the case of luminescence the decay of the exciton ($\boldsymbol{\sigma}^+_{\mathbf{n}\alpha}$) and the creation of the spin wave ($\mathbf{S}_{\mathbf{m}\alpha}$) take place at the same sublattice α. It is important that the exciton and magnon do not coexist, the spin wave emerges after disappearance of the exciton, i.e., they cannot interact. Therefore, the exciton–magnon luminescence spectrum is not deformed by the exciton–magnon interaction, which makes it possible to determine the width of the exciton band if the spin-wave dispersion is known, without making any assumptions concerning the magnitude of the exciton–magnon interaction. On the other hand, if the exciton and magnon dispersions are known, the value of the exciton–magnon interaction can also be determined by comparing the spectra of absorption and luminescence.

Dietz et al. [244] have calculated the form of a magnon sideband in the spectrum of intrinsic luminescence of the MnF_2 crystal. Assuming a zero dispersion of the lowest energy E1 exciton, good agreement was obtained with experimental data. A conclusion was made that the width of the E1 exciton band in this crystal does not exceed 0.5 cm^{-1}. The small value of the E1 exciton dispersion is explained by a strong electron–vibration interaction in the lowest excited electron state of Mn^{2+} ions.

3.3. Multiquantum Excitation of Two-Particle Transitions

3.3.1. Multiquantum Absorption of Light

A multiphonon absorption is a process in which several quanta of energy are simultaneously absored during a single act of absorption. According to the law of conservation of energy, the total energy of quanta in such a process has to be equal to the energy interval between the initial and final states.

A direct computation of probabilities of two-particle transitions in the case of a multiquantum excitation is connected with considerable theoretical difficulties and, to the best of our knowledge, there are no papers in which this problem was solved. The point is that both the probability of a one-quantum pair transition and the probability of the simplest two-photon one-center absorption are calculated in the same (second) order of the perturbation theory, one and the same virtual states of the absorbing medium being actual. Besides, the problem is made more complicated as the exciton formalism is used to describe electron and spin excitations in antiferromagnets. Therefore, to estimate probabilities of multiquantum transitions and to interpret experimental results, use was made of selection rules and conservation laws related to two-particle transitions as well as the results of the multiphoton transitions theory for noninteracting atoms and molecules. Such an approach to multiquantum processes in antiferromagnets is justified by the fact that electron and spin excitations in this medium are Frenkel excitons which, to a certain extent, can be considered to be connected with definite sites in a crystal.

In the framework of the conventional perturbation theory, the probabilities of a system transition from the state $|0\rangle$ into the state $|j\rangle$ with absorption of m quanta of identical frequency is described by the following expression [245]:

$$W_{\mathrm{m}} = \frac{\pi g_{\mathrm{L}}(m\omega)}{2^{2m-1}\hbar^{2m}}$$

$$\times \left| \sum_{g,k,\ldots s,p} \frac{\mathscr{H}'_{ig}\mathscr{H}'_{gk}\cdots\mathscr{H}'_{sp}\mathscr{H}'_{p0}}{[(m-1)\omega-\omega_{g0}][(m-2)\omega-\omega_{k0}]\cdots[2\omega-\omega_{s0}][\omega-\omega_{p0}]} \right|^2 . \tag{3.45}$$

In this expression $g_{\mathrm{L}}(m\omega)$ is the normalized line shape which for an isolated atom or molecule is assumed to be Lorentz-type, $\mathscr{H}'$ is the Hamiltonian which describes the interaction of an electron system with an electromagnetic field. An analysis of expression (3.45) shows that:

(1) the probability of m photons being absorbed by the system in a single act is proportional to the mth power of the exciton intensity;
(2) selection rules for one-quantum and multiquantum transitions differ.

Consider, for example, a crystal with a center of inversion in which the initial state $|0\rangle$ and the final state $|j\rangle$ have the same parity. If only electric dipole interaction is taken into consideration, then the transitions, for which m is an even number, between these states are allowed, and the transitions with an odd m (including one-quantum transitions) are forbidden.

To compare experimental results with theory, consider the parameter σ, the scattering cross section of the multiquantum absorption,

$$\sigma = P/N_V\mathscr{I}, \tag{3.46}$$

where P is the average power absorbed in a unit of volume, N_V is the number of molecules in a unit of volume, and $\mathscr{I}$ is the average power of the incident radiation per unit area. The absorption cross section is related to the transition probability in a simple way:

$$\sigma_{\mathrm{m}} = \hbar\omega W_{\mathrm{m}}/\mathscr{I}. \tag{3.47}$$

Here $\hbar\omega$ is the transition energy. As the cross section of a multiquantum absorption depends upon the intensity of the incident radiation, the parameter δ, independent of the intensity $\delta_{\mathrm{m}} = \sigma_{\mathrm{m}}/\mathscr{I}^{m-1}$, is often used instead of the cross section σ. Cross sections of multiquantum processes are much smaller than the characteristic value of the cross section for a one-photon electric dipole transition (about 10^{-16} cm^2). Even with the pumping intensity $\mathscr{I} \simeq$ 10 MW/cm^2 they do not exceed 10^{-25} cm^2, for example, for a two-photon process, and 10^{-34} cm^2 for a three-photon process [245]. Therefore, a direct measurement of a multiphoton absorption is difficult, and the best method to record such a process is to study the secondary processes, for example, luminescence. Among the great number of antiferromagnetic dielectrics, only managese compounds have a noticeable luminescence in the visible part of the spectrum which is convenient for experimental work. Due to this fact, a multiquantum absorption of light has been studied to date only in antiferromagnetic compounds of manganese.

Experiments have been carried out to investigate the relationship between the integral value of luminescence from the lowest excited electron state of manganese ion $^4T_{1g}(\nu_{\max} \simeq 17{,}000 - 16{,}000\ \mathrm{cm}^{-1})$ and the intensity of a nonpolarized excitation. The source of the luminescence excitation was a neodymium-doped glass laser ($\nu_{\mathrm{ex}} = 9400\ \mathrm{cm}^{-1}$). It was found that the integral intensity of luminescence $\mathscr{I}_{\mathrm{L}}$ of different crystals changes in different ways with an increase in the excitation intensity $\mathscr{I}_{\mathrm{ex}}$ (Fig. 3.7). The luminescence intensity of the $RbMnCl_3$ crystal increases proportionally to the square of the excitation intensity in the whole range of $\mathscr{I}_{\mathrm{ex}}$ values: $\mathscr{I}_{\mathrm{L}} \sim \mathscr{I}_{\mathrm{ex}}^{\mathrm{m}} (m = 2)$; for MnF_2 and $BaMnF_4$ crystals $m = 4$, and for $CsMnCl_3$ and $NaMnCl_3$ crystals the exponent of a power m varies from 2 to 4 with an increase of the excitation

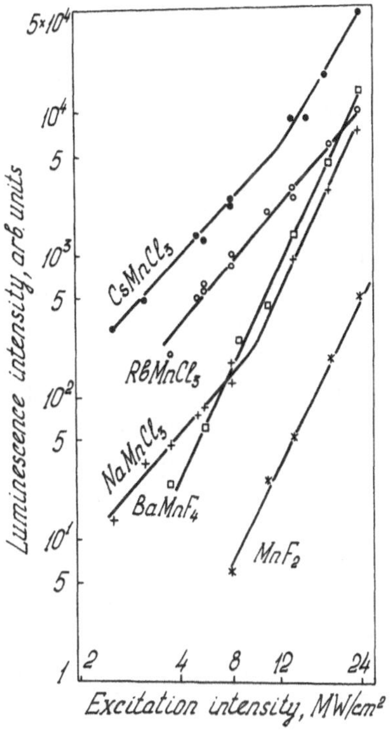

Fig. 3.7. Intensity of luminescence versus intensity of excitation in $RbMnCl_3$, $CsMnCl_3$, $NaMnCl_3$, $BaMnF_4$, and MnF_2 crystals.

intensity (Fig. 3.7). The exponential dependence of the intensity of luminescence on that of excitation testifies that at an excitation intensity less than 10 MW/cm^2 two photons take part in the absorption in $CsMnCl_3$, $NaMnCl_3$, and $RbMnCl_3$ crystals, and at $\mathscr{I}_{ex} > 20$ MW/cm^2 four photons participate in the absorption in $CsMnCl_3$, $NaMnCl_3$, as well as in MnF_2 and $BaMnF_4$ crystals.

Cross sections of multiquantum absorption were measured by comparing luminescence signals within one and the same configuration of the experiment for one-quantum and multiquantum excitations. At temperatures less than T_N and an intensity of excitation 20 MW/cm^2, the cross section of a multiquantum absorption of these crystals makes up approximately 10^{-28} cm^2. In a $NaMnCl_2$ crystal, the cross section of absorption was measured at $\mathscr{I}_{ex} \simeq$ 10 MW/cm^2 at temperatures above and below T_N. In this case, the cross section of absorption increased from 2×10^{-29} cm^2 $(T > T_N)$ to 1.2×10^{-28} $(T < T_N)$. The measured values of the cross section are close to the theoretically predicted one for a two-photon one-center absorption $\sigma_{2ph} \simeq 10^{-25}$ cm^2 [245].

The square law dependence of the luminescence intensity on $\mathscr{I}_{ex}$ and the proximity of the measured value of the absorption cross section to the value σ for two-quantum transitions permitted in electric dipole approximation have made it possible to interpret the absorption of radiation of a neodymium-doped glass laser observed at $\mathscr{I}_{ex} < 10$ MW/cm^2 in $RbMnCl_3$, $NaMnCl_3$, and $CsMnCl_3$ crystals as a two-photon absorption. During this absorption, an electron state of a manganese ion $^4T_{1g}(\nu \simeq 1800 \text{ cm}^{-1})$ became excited. In this case, the parity forbiddenness of an electron transition into the state $^4T_{1g}$ is lifted as two photons take part in one and the same absorption act, and the intercombination forbiddenness is softened, as in the case of a one-quantum excitation, due to the creation of a magnon simultaneously with an exciton (exciton–magnon transition). The decrease of a cross section of a multiquantum absorption in a $NaMnCl_3$ crystal as it is heated above the magnetic ordering temperature is in good agreement with this interpretation, as the temperature increase above T_N is known to reduce the efficiency of the spin forbiddenness removal in the case of an exciton–magnon transition [246].

There are several reasons why it is impossible to explain the processes in which four photons take part in an absorption act using the one-center multiquantum absorption model, i.e., to explain them by the excitation of a manganese ion resulting from absorption at a time of four quanta of a neodymium-doped glass laser radiation. First, according to the energy conservation law for a multiphoton transition, during such a process, an electron state with energy 37,000 cm^{-1} has to be excited in a crystal. There is, however, no such state in the structure of the energy levels of manganese compound crystals. Second, the values $\sigma \simeq 10^{-27}$–10^{-28} cm^2 that have been obtained far exceed the theoretically predicted cross section of four-quanta absorption with one center under conditions of the pumping intensities used: $\sigma_{4ph} \simeq 10^{-43}$ cm^2 [245].

3.3.2. Two-Exciton Absorption of Light in the Case of Four-Quantum Excitation

To interpret the observed four-quantum absorption process, a model was used according to which, under conditions of a powerful optical excitation, a two-exciton absorption of light takes place in antiferromagnetic crystals, the simultaneous creation of each of the two excitons with energy—18,000 cm^{-1} ($^6A_{1g} \to {}^4T_{1g}$ transition) being accompanied by absorption of two photons [247]. Indeed, in spite of the fact that the luminescence intensity versus the excitation intensity relationship is indicative of the four-photon nature of absorption, the value of the cross section of this process which is close to that theoretically predicted for a two-photon absorption ($\sigma_{2ph} \simeq 10^{-25}$ cm^2) testifies to the validity of the model used. A part of the two-exciton absorption spectrum in a MnF_2 crystal at a one-quantum excitation is shown in Fig. 3.8, curve 2 [231]. Comparing curves 1 and 2 one can see that the narrow line in the multiquantum absorption spectrum (at $\gamma = 9230$ cm^{-1}) coincides with the line corresponding to the state of two bound excitons $^4T_{1g} + {}^4T_{1g}(E_{12})$ in a two-exciton absorption spectrum at one-quantum excitation. It is another proof of validity of the suggested model to explain the four-quantum process.

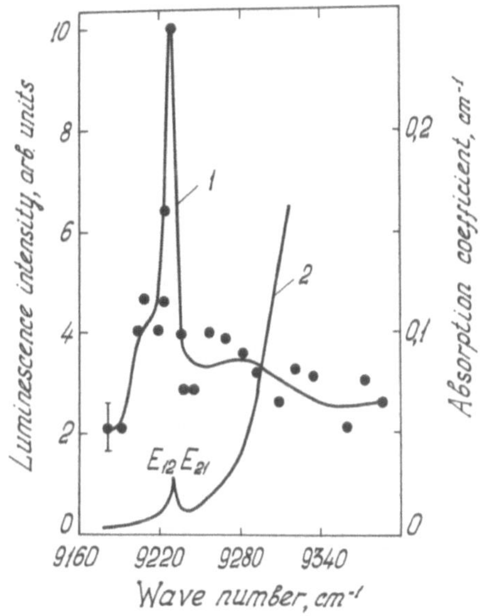

Fig. 3.8. Luminescence excitation spectrum at multiquantum absorption (curve 1) in MnF_2 crystal and a part of two-exciton light absorption at one-quantum excitation (curve 2) (for this curve the values of the energies should be multiplied by four) [247].

The mechanism of two-exciton absorption at four-quantum excitation is efficient particularly for antiferromagnetic dielectrics, as the creation of excitons in sublattices with opposed orientation of spins is possible in them, this being necessary to remove the intercombination forbiddenness. An exciton–exciton interaction is required to make possible such an absorption. For example, in MnF_2 this interaction is about 100 cm^{-1}.

When comparing curves 1 and 2 (Fig. 3.8) we can see an essential difference between one-quantum and four-quantum two-exciton absorption. Whereas at the one-quantum process the coefficient of light absorption at the frequency of a two-exciton bound state E_{12} is 10^{-2} of the absorption in the electron–phonon band peak, in the case of a four-quantum process an absorption at the exciton state frequency is most probable. It is explained by the fact that under the conditions of a one-quantum process the excitation of a pair of exchange-coupled excitons results in the effective removal of the spin forbiddenness only, the parity forbiddenness being partly removed, due mainly to interaction with nonsymmetrical vibrations of the lattice. As for the four-quantum absorption, a two-exciton optical transition becomes permitted in an electric dipole approximation since the creation of each individual exciton is conditioned by absorption of two photons.

The two-exciton four-photon absorption of light in a MnF_2 crystal is peculiar for its unusual dichroism—as the polarization of the excitation radiation is varied, the exponent of the absorption process nonlinearity changes [248].

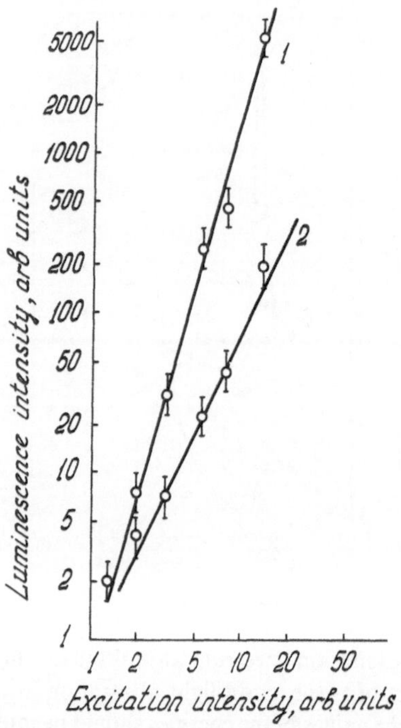

Fig. 3.9. Intensity of luminescence ($\nu \simeq 17{,}000\ \mathrm{cm}^{-1}$) versus intensity of excitation ($\nu \simeq 9400\ \mathrm{cm}^{-1}$) in MnF_2 crystal at $T = 4.2$ K. (1) $\mathbf{E} \parallel [100]$; (2) $\mathbf{E} \parallel [110]$ [248].

As follows from Fig. 3.9, curve 1, at excitation polarization $\mathbf{E} \parallel [100]$ the luminescence intensity of the crystal MnF_2 increases proportionally to the fourth power of the excitation intensity. When $\mathbf{E} \parallel [110]$, the luminescence intensity depends upon the excitation intensity according to a square law (curve 2, Fig. 3.9). Thus, at one excitation polarization a four-photon process takes part, while at another polarization a two-photon absorption is realized. To explain the dependence of the multiquantum absorption order on the polarization of the excitation radiation, it was assumed that at polarization $\mathbf{E} \parallel [110]$ in a MnF_2 crystal the probability of the simultaneous creation of two excitons on opposed sublattices becomes rather small, i.e., excitons are created at the definite sublattice only.

The possibility of such a selective excitation of each of the sublattices in a MnF_2 crystal was demonstrated by inducing magnetization in this crystal with the help of a pulsed one-quantum pumping with polarized radiation at the exciton–magnon transition frequency [239]. The emergence of magnetization at such a pumping is explained by the lifetime of magnons being essentially smaller than that of excitons. Magnetization appears at polarization $\mathbf{E} \parallel [110]$. As the polarization plane of the excitation radiation is turned through 90°, the orientation of the magnetization is changed. As for polarization $\mathbf{E} \parallel [100]$, no magnetization emerges in this case . Hence, at a one-quantum pumping by polarized light with **E** ∥ [110] at the exciton–magnon transition frequency, creation of an exciton and, consequently, of a magnon is possible in a strictly definite sublattice of a crystal, whereas at polarization **E** ∥ [100] and in the case of nonpolarized light, excitons and magnons are created in both sublattices. Such a selective creation of excitons in one sublattice is caused by the fact that in the case of polarization **E** ∥ [110], the sublattices become inequivalent with respect to the electric field of the exciting light. In a MnF_2 crystal this inequivalence is caused by a different arrangement of bonds between ions of manganese and fluoride with respect to **E** ∥ [110].

Thus, it is the impossibility of excitons to become created at both sublattices of a MnF_2 crystal, when using radiation with polarization **E** ∥ [110], which results in a change of the exponent of nonlinearity of the multiquantum absorption process as the excitation polarization is varied.

3.3.3. Influence of Statistical Properties of Light on Two-Exciton Four-Photon Absorption

When calculating the probability of multiquantum transitions (see (3.45)), the electromagnetic field affecting the absorbing system was presented in the form of a classical stationary wave. To characterize such a light beam, it is sufficient to know the average density of energy or intensity which, in general, are functions of coordinates. When studying optical transitions, this approach, however, is rather simplified, and it can be successfully used only for the description of one-quantum transitions when photons are absorbed individually. In the case of multiphoton absorption, m quanta of the electromagnetic field are simultaneously absorbed, as the lifetime of intermediate states at virtual transitions is very small. Therefore, the probability of multiquantum transition depends on the instantaneous intensity of radiation, i.e., it is determined not only by the properties of the absorbing system, but by the properties of the radiation source interacting with the system as well.

Real laser beams used in the experimental studies of multiquantum transitions differ greatly depending on the operation duties of lasers, the number of modes generated, and the amplitude and frequency stability. In the case of such radiation, the instantaneous intensity in an arbitrary space–time point fluctuates and cannot be determined directly. To characterize a light beam, use is made of the average intensity $\tilde{\mathscr{I}}$ which is measured experimentally.

Fluctuations of strength (E) and intensity ($\mathscr{I}$) of optical fields are related to the concept of field coherence. Coherent properties of radiation are described with the help of correlation functions. In particular, the correlation function of the first order is obtained when considering the intensity of light averaged over the period of oscillations in an arbitrary point of a screen in Young's interference. To the accuracy of a constant factor, it is the average intensity of radiation in a space–time point [249]:

$$\langle E^*(\mathbf{r}, 0)E(\mathbf{r}, 0) = \frac{2}{\varepsilon_0 c}\langle\tilde{\mathscr{I}}\rangle. \tag{3.48}$$

The correlation function of the first order normalized to unity is referred to as the first-order degree of coherence. In two space–time points $(\mathbf{r}_1, t_1)$ and $(\mathbf{r}_2, t_2)$ lights is coherent if on superposition it can, in principle, create interferential effects. If $g_{12}^{(1)} = 1$, light is first-order coherent, if $g_{12}^{(1)} = 0$, it is incoherent and is partly coherent at intermediate values of $g_{12}^{(1)}$. For example, for a random radiation

$$g_{12}^{(1)} = \exp(-\gamma|\tau|), \tag{3.49}$$

where $\tau = t_1 - t_2 - (r_1 - r_2)/c$, and γ is the width of the radiation line. Light is first-order coherent at two points if

$$t_1 - t_2 - (r_1 - r_2)/c \ll \tau_0, \tag{3.50}$$

where τ_0 is coherence time.

In the majority of cases it is impossible to calculate the degree of coherence of the mth order; however, its values for two specific cases are known:

(i) For a completely coherent one-mode radiation

$$g^m(r_1 t_1 \cdots r_{2m} t_{2m}) = 1 \tag{3.51}$$

for all possible m, i.e., the classical stationary wave is coherent in all orders.

(ii) For a random light beam in the case when all the space–time points coincide

$$g^m(r_1 t_1 \cdots r_{2m} t_{2m}) = m!. \tag{3.52}$$

The process of the multiphoton one-center absorption of light is a kind of interference experiment in which m photons are simultaneously recorded in one and the same space–time point. To analyze these experiments, use is made of correlation functions of the mth order. Taking into consideration the coherence degree g^m, the probability of m-quantum absorption is written as

$$W_{\mathrm{m}} \sim |\langle \tilde{\mathcal{J}}(t) \rangle|^{\mathrm{m}} g^{\mathrm{m}}. \tag{3.53}$$

An important conclusion follows from (3.52) and (3.53): the probability of an m-quantum absorption of random light by an isolated center exceeds by $m!$ times the probability of m-photon absorption of completely coherent light of the same frequency. The results of the experimental studies of the multiphoton ionization of atoms [250] and of two-photon absorption in isolated centers [251] have confirmed the correctness of the above-mentioned theoretical considerations.

A peculiarity of two-particle transitions discussed in Section 3.3.2, is that each of the interacting excitions created as a result of a multiphoton absorption is "localized" at different sites belonging to sublattices with opposed orientation of spins. When calculating the coherence degree of the mth order, in the case, two space–time points have to be taken into consideration. For coherent light $g^{(m)} = 1$ and for a random source $g^{(m)}$ in this case is unknown. Experimental studies [252] have been carried out to find the influence of the

Table 3.2

Coherent duty (main laser)	Incoherent duty (main laser)	Incoherent duty (additional laser)
1	0.172	0.165

statistical properties of light on the intensity of two-exciton four-photon absorption. The authors used a neodymium-doped glàss giant-pulse laser with a complex resonator, a peculiar feature of which is the possibility of adjusting the degree of the radiation spatial coherence. The time coherence of this laser could be judged by the width of the generation spectrum in different duties of the laser operation. In a coherent duty, the laser generated a single-frequency radiation with spectrum width 10^{-4} nm, this radiation being perfectly spatially coherent [253]. In a noncoherent duty, four to five longitudinal modes were generated (with a large number of transverse modes), and the degree of the laser spatial coherence was substantially smaller.

An additional laser was used to increase the number of longitudinal and transverse modes taking part in the generation. The width of its generation spectrum was about 0.2 nm, and the radiation was spatially incoherent. Given in Table 3.2 are the experimental results characterizing absorption in a crystal MnF_2 in relative units.

The results obtained show that the intensity of two-exciton absorption in the case of four-quantum excitation essentially depends on the degree of the electromagnetic field coherence, the absorption increasing with an increase in the radiation coherence degree. This dependence cannot be explained within the framework of the multiphoton one-center absorption theory. The revealed relationship, between the intensity of a two-particle four-photon transition and the degree of the radiation coherence, is suggestive of the necessity to take into consideration phase relations between the radiation field and quasi particles when considering the probability of such transitions.

3.4. Dynamics of Magnetic Excitons

3.4.1. Transfer of Electron Excitation Energy

Investigations of the luminescence of rare-earth ions dissolved in crystals of manganese fluorides [254–256] have shown that in these compounds there is an effective nonradiation transfer of electron excitation energy from manganese ions to impurity ions, migrating excitons being responsible for the transfer. The Davydov magnetic splitting detected in antiferromagnets and the observed migration of electron excitation energy are the most convincing experimental proof of the exciton nature of excitations emerging in magnetodielectrics during absorption of electromagnetic radiation. To date, a con-

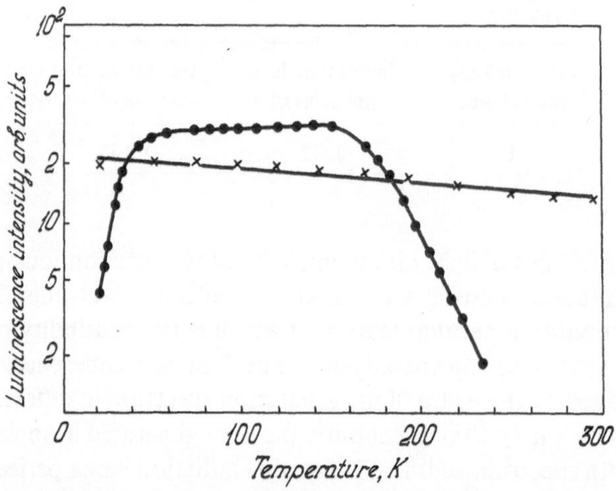

Fig. 3.10. Luminescence intensity of complex Mn^{2+} + radiation defect versus temperature in a $KMnF_3$ crystal. (×) Excitation in the absorption band of the complex; (o) excitation in the absorption band of the matrix [262].

siderable number of studies have already been carried out to investigate experimentally the sensitized luminescence in antiferromagnets [257–260]. A characteristic feature detected in all the experiments is the increase of the impurity luminesence intensity with an increase in temperature in the same interval where peculiarities of matrix luminescence are observed [243, 261].

Discussed below, as an example, is the sensitization of the luminescence of a complex Mn^{2+} ion plus radiation defect [262, 263]. The spectra of the complex luminescence excitation contain both the bands due to light absorption by the matrix, and additional bands related to the intrinsic light absorption by ions of the complex. The shape of the luminescence spectra of the complex does not depend upon the method of their excitations, whereas temperature dependences of their intensity greatly differ (see Fig. 3.10). As follows from this figure, the probability of radiation optical transitions in a complex shows a weak dependence on temperature in the interval 20–300 K. Therefore, the change of the complex luminescence intensity with a variation in temperature, in case the complex is excited via the matrix absorption bands, characterizes the temperature dependence of the efficiency of energy transfer from the matrix to the complex. The increase of the transfer efficiency in the temperature range 20–40 K is caused by the rise of the exciton band population due to "boiling out" of electron excitation from the metastable traps located below the exciton band. The ratio of the exciton band population N_3 to the population of metastable states N_2 is expressed as follows:

$$N_3/N_2 = (P_{23} + P_{21})P_{31}/P_{32}P_{31} = P_{21}/P_{32} + e^{-\Delta/kT}/n. \qquad (3.54)$$

Here, Δ is the energy gap between the exciton band and the trap, n is the concentration of traps, P_{32} and P_{31} are probabilities of transitions from the exciton band into a trap and into the ground state, and P_{21} and P_{23} are probabilities of transitions from a metastable state into the ground state and into the exciton band ($P_{23} = (P_{32}/n)e^{-\Delta/kT}$). As $P_{21} \ll P_{32}$, the population of the exciton band rises with an increases in temperature which results in an increase in the amount of energy transferred to complexes. The transfer mechanism is a nonradiation resonance transition. Radiation processes in the crystals under consideration can be neglected, as the oscillation forces of optical transitions are small. Since the energy is transferred by means of a short-range exchange interaction, and the concentration of impurity ions and the excitation intensity are small, an effective energy transfer is possible only in the case of excitation migration in the crystal.

The migration ability of excitons is characterized by the diffusion coefficient D and the length of diffusion shift $L = \sqrt{D\tau_{ex}}$, where τ_{ex} is the lifetime of excitons. The mobility of excitons corresponding to the lowest excited state of manganese ions in antiferromagnets is small, due to the weakness of the resonance interaction. For example, in a MnF_2 crystal at a temperature of 4.2 K the diffusion coefficient D of excitons is approximately 10^{-8} cm²/s, and in a $CsMnF_3$ crystal $D \simeq 10^{-9}$ cm²/s [264].

When estimating the value of D, use was made of the fact that in antiferromagnetic compounds of manganese the lifetime of excitons in the band and, consequently, the length of their diffusion shift are controlled by the rate of exciton trapping by metastable traps:

$$P_1 = 4\pi D n R_T, \tag{3.55}$$

where P_1 is the rate of exciton decay, n is the concentration of traps, and R_T is the effective radius of trapping.

3.4.2. Character of Exciton Motion

The values of D and the character of exciton motion in different antiferromagnetic compounds of manganese substantially differ. They depend upon the magnitude of the resonance interaction, the exciton–phonon and exciton–magnon bond, as well as upon the dimensionality of the possible motion of excitons.

In the case where the resonance interaction is strong and the interaction with other quasi particles is weak, the electron excitation energy is transferred in a crystal by wave packets which cover the crystal region exceeding the lattice parameter many times. Otherwise, the exciton motion (autolocalized ones) has the character of random jumps from one lattice site to another. This phenomenon takes place when the resonance interaction is weak and the interaction with other quasi particles is strong. In low-dimensional antiferromagnets, for which the exchange interaction integral taken along one or two directions is noticeably greater than the corresponding integral for the orthogonal direaction, there appear characteristic planes or chains of magnetic ions with predominating coupling between their spins. That is the mechanism that controls the existence of planes or chains of the preferential propagation of excitons within a crystal.

The effect the exciton–magnon interaction on the motion of excitons in antiferromagnetic dielectrics was considered in [265]. There it was shown that, depending upon the relation between the probability of the excitation energy resonance transfer inside the sublattice $K^{(0)}$ on the one hand, and the magnitude of the exciton energy modulation due to interaction with magnons Γ_0 and depending on the probability of an intersublattice transfer Γ_1 on the other hand, the motion of magnetic excitations can be either coherent or diffuse. If $\Gamma_0 + z\Gamma_1 \ll K^{(0)}$, then the initially localized wave packet propagates as a superposition of coherent exciton waves. If $\Gamma_0 + z\Gamma_1 \gg K^{(0)}$, the exciton moves diffusely with diffusion constant D which is determined by the probability of an intersublattice transfer Γ_1 and the crystal structure, for example, in a one-dimensional case, $D = \Gamma_1 + 8[K^0]^2/(\Gamma_0 + 2\Gamma_1)$. When the temperature increases in the region $T < T_N$, the diffusion coefficient of noncoherent magnetic excitons has to grow, as the fluctuations of the exciton energy Γ_0 and the probability of an intersublattice transfer Γ_1 increase with the rise in temperature.

The motion of magnetic excitons in crystals with a strong electron–phonon interaction is adequately described within the framework of the theory developed for Frenkel's excitons in molecular crystals [266, 267]. In the case of a strong bond, the time of excitation transference between neighboring sites is greater than the time it takes molecules to shift to new equilibrium positions. These shifts are due to the change of interaction forces between molecules as one of them gets excited. There emerges in a crystal a local deformation of the lattice which is moving together with the excitation. The exciton spectrum changes, too. The new (reconstructed) exciton band becomes narrower than the initial one. The width of the exciton band decreases exponentially with the rise of temperature. In the case when, due to the exciton–phonon interaction, the width of the exciton band becomes less than $\hbar/\tau_e$, where τ_e is the exciton free path time, the motion of such excitons in a crystal has the character of random jumps. With an increase in the crystal temperature, the frequency of jumps and, consequently, the diffusion coefficient of a noncoherent exciton rises from a certain limit value at $T = 0$ up to the magnitude which, in the range of temperatures exceeding the Debye one, changes according to an exponential dependence on T:

$$D \sim \exp(-E_a/kT). \tag{3.56}$$

Here, E_a is the excitation diffusion activation energy.

To determine the character of the exciton motion, use is made of their mobility dependence on temperature. The temperature dependence of their intrinsic or impurity luminescence is used as an indicator to a change in exciton mobility in antiferromagnetic compounds of manganese as the temperature is varied. With an increase in the exciton mobility, the intensity of the intrinsic luminescence drops due to the rise of the probability for the excitation energy to get into extinguishing traps, and the intensity of the impurity luminescence increases due to the rise of the probability of the excitation energy being transferred from matrix ions to impurity ions. Whether the observed luminescence is intrinsic or "defect"—this conclusion is made, as a rule, by comparing the spectra of luminescence and of its excitation. The most convincing proof of the "host" nature of the luminescence of antiferromagnetic compounds of manganese is the overlapping (sometimes, resonance coincidence) of purely exciton bands in absorption (or excitation) and luminescence spectra.

3.4.3. Coherent Excitons

When the electron excitation energy is said to be transferred by wave packets with a wave vector indeterminacy $\Delta k \ll k < 1/a$, where a is the lattice parameter, coherent excitons are meant. For them, the wave vector is a "good" quantum number. The results of kinetic investigations of MnF_2host luminescence are indicative of the fact that exciton states in this crystal can be considered to be, to some extent, delocalized states which are characterized by the quite definite wave vector.

In [244] the exciton lifetime has been measured with different values of wave vectors at the axial pressure 25 kg/mm^2 (the role of the axial pressure being to increase by an order of magnitude the luminescence of the excitons with the lowest values of energy of E1). Use was made of the following characteristic peculiarity of the luminescence spectra of antiferromagnetic crystals: the exciton band in this spectrum corresponds to the annihilation of excitons with $\mathbf{k} = 0$, and the exciton–magnon band is related to the emision decay of excitons with $\mathbf{k}_{max}$. The following exciton decay rate versus exciton wave vector dependence has been found: excitons with greater values of $\mathbf{k}$ decay quicker at $T = 4.2$ K. This dependence indicates that the distributions of excitons over the zone fails to reach its quasi-equilibrium state in the MnF_2 crystal during 3 μs at $T = 4.2$ K.

Even more convincing evidence, testifying that these excitons have a quasi momentum, has been presented in [268] where the relaxation time of $^4T_{1g}$ excitons in MnF_2 within the Brillouin zone (the time of phase memory loss) was measured. The authors used the method of resonance optical pumping into the absorption bands which correspond to exciton and exciton–magnon transitions into the lowest excited state of the Mn^{2+} ion. The relaxation times of excitons over the zone were determined by the rise time of the exciton and exciton–magnon luminescence.

The experimental results are shown in Fig. 3.11. In the first experiment (a) the pumping was made into the absorption band E1, i.e., excitons with $\mathbf{k} = \mathbf{q} = 0$ were created ($\mathbf{q}$ is the wave vector of light). Kinetics of luminescence were studied in the peak of the exciton–magnon band. The rise time of the luminescence was used to measure the interval of time during which there appeared excitons with $\mathbf{k}_{max}$. This interval was about 1 μs. The same time was

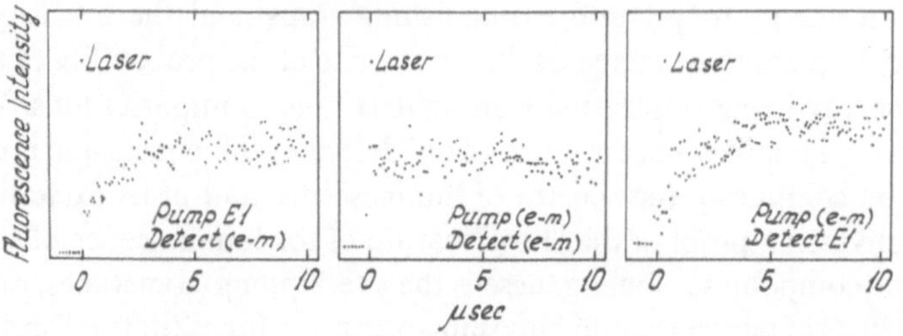

Fig. 3.11. Kinetics of the intensity increase of the host luminescence of MnF_2 under different conditions of its excitation and detection [268].

recorded in the second experiment (c) when excitons with $\mathbf{k}_{max}$ were pumped, and the interval of time was measured during which there appeared excitons with $\mathbf{k} = 0$. At the same time, less than 0.1 μs is required for excitons with quasi momentum $\mathbf{k}_{max}$ to appear in the case of their direct pumping (pumping and detecting of exciton–magnon bands (b)). Consequently, the relaxation time of excitons with $\mathbf{k} = 0$ and $\mathbf{k}_{max}$ over the zone in MnF_2 crystal makes up about 1 μs.

Exciton–exciton scattering plays a dominant role in establishing a quasi-equilibrium distribution of excitons over the zone. At low temperatures, scattering on phonons is not likely to play a noticeable role as wave vectors of phonons make up only 10^{-3} $\mathbf{k}_{max}$, and a great number of steps (stages) are required to scatter excitons from the center of the zone to its boundary (and vice versa). It is probably this fact that explains such a big difference between the values of the time of relaxation of ${}^4T_{1g}$ excitons in MnF_2 which was recorded in the above-mentioned studies [244, 268], as different densities of excitons were created in these investigations.

The realization of the coherent motion of magnetic excitons in MnF_2 is connected with several factors. First, the exciton–phonon bond in this crystal is the weakest (as compared with other compounds of manganese). Second, the exciton–magnon intersection in this case is the weakest also, and third, the values of the intrasublattice and intersublattice interactions are practically equally probable ($K^{(0)} \sim \Gamma_1$). Therefore, the relation $K^{(0)} \approx \Gamma_0 + z\Gamma_1$ seems to hold for manganese fluoride.

3.4.4. Noncoherent Magnetic Excitons

3.4.4.1. Localization of Excitons in Quasi-One-Dimensional Antiferromagnets

The possibility of a magnetic $3d$-exciton migrating along this or that direction is controlled by the spin projection change as the excitation transfers from one lattice site to another. In the case of quasi-one-dimensional antiferromagnets, migration of electron excitation across the chains is hindered, due to very weak exchange interaction between ions in neighboring chains. As for the chains themselves, an antiferromagnetic ordering is established there, so the spins of the neighboring ions have antiparallel orientation. Migration of the exciton along the chain is forbidden, as the transition of excitation from site A to site B would otherwise be accompanied with a great change in the spin projection $\Delta M_s = 2$. Thus, in quasi-one-dimensional antiferromagnets at sufficiently low temperatures $T < T_N$, when spins of neighboring Mn^{2+} ions are oriented strictly antiparallel, the motion of the excitons can be realized by random jumps only. As the mobility of these excitons is extremely low, such electron excitations can be referred to as localized excitations.

Localization of the electron excitation energy in magnetically concentrated compounds of manganese results in some peculiarities in optical spectra [269–271]. Knowledge of these peculiarities is required for the correct interpretation of experimental results in the case where conditions of electron excitation localization (autolocalization) are ambiguous. The optical spectra of a $CsMnCl_3 \cdot 2H_2O$ crystal were compared with the corresponding spectra of a MnF_2 crystal in which free excitons were observed. The whole luminescence observed in the $CsMnCl_3 \cdot 2H_2O$ crystal was interpreted to be a host luminescence. As a crystal is heated, the integral luminescence intensity and the emission time remain practically unchanged up to 300 K, and the peak of the luminescence spectrum shifts into the short-wavelength region. The integral intensity of luminescence being temperature independent up to 300 K testifies, on the one hand, to the absence of any dependence between the probability of nonradiative transitions in Mn^{2+} ions and temperature, and, on the other hand, to the absence of the excitation energy transfer to extinguishing traps. Thus, the mobility of excitons does not change (or remains very low) in a wide temperature range, which includes the temperature at which the correlation between spins of Mn^{2+} ions vanishes in the chain ($T_N = 6.5$ K). The cause of the low mobility of the excitons at temperatures exceeding this is most likely related to a weak intersublattice interaction and a strong electron–phonon interaction in this crystal, as the mobility of excitons was observed to increase in a quasi-one-dimensional antiferromagnetic $CsMnBr_3$ ($T_N = 8.3$ K) at a temperature exceeding 50 K [272]. Impurity Nd^{3+} ions were implanted into this crystal, and the temperature dependence of Mn^{2+} and Nd^{3+} luminescence intensity was investigated in the case of excitation in the manganese absorption band. The intensity of impurity luminescence was found to increase, and the manganese luminescence intensity was found to decrease at a temperature rise above 50 K. In this case it was not the whole energy of manganese excitation that was transmitted to neodymium, the greater part of it was lost, obviously being captured by extinguishing traps.

The main differences between the optical spectra of antiferromagnetic crystals with free excitons and localized excitons are as follows:

(1) Very narrow ($< 1\ \text{cm}^{-1}$) polarized bands correspond to free excitons in luminescence spectra; the free excitons bands coincide both in lumine-

scence and absorption spectra. Similar bands of localized excitons are relatively wide and nonpolarized.

(2) Luminescence spectra of crystals with free excitons are dominated by the emission of ions Mn^{2+} "perturbed" by cation impurities and strongly depend upon the quality of specimens. In crystals with localized excitons the luminescence is intrinsic and its spectrum, to a certain extent, does not depend upon the quality of specimens.

(3) As the temperature rises and approaches T_N, the peak of the wide luminescence band of crystals with free excitons shifts to the long wavelength region of the spectrum, and the integral intensity of emission decreases. In crystals with localized excitons, the emission intensity remains unchanged in a wide temperature range, and the emission peak shifts into the high-frequency region of the spectrum.

3.4.4.2. Crystals with Strong Exciton–Phonon Interaction

In the majority of the antiferromagnetic manganese compounds, there is a strong interaction of excitons with lattice vibrations. A criterion of the force of this interaction is the quantity C which is the average number of vibration quanta created by phototransition. The interaction is considered to be strong if $C > 1$. The magnitude of C can be calculated by dividing half of the Stokes shift energy $\Delta\omega$ by the energy of the phonon ω_{ph} which is the most active in the spectrum formation. As was previously noted, the exciton motion in crystals with a strong exciton–phonon interaction has to be incoherent.

Experimental investigations of luminescence spectra and those of its excitation in manganese chloride crystals have shown that at low temperatures the luminescence from these crystals is host and has a big Stokes shift ($C > 1$). Hence, at low temperatures, the exciton mobility in these crystals is extremely small. With the temperature rise, the integral intensity of luminescence decreases—curves 2, 3, and 4, Fig. 3.12 [273, 274]. Apparently, the exciton mo-

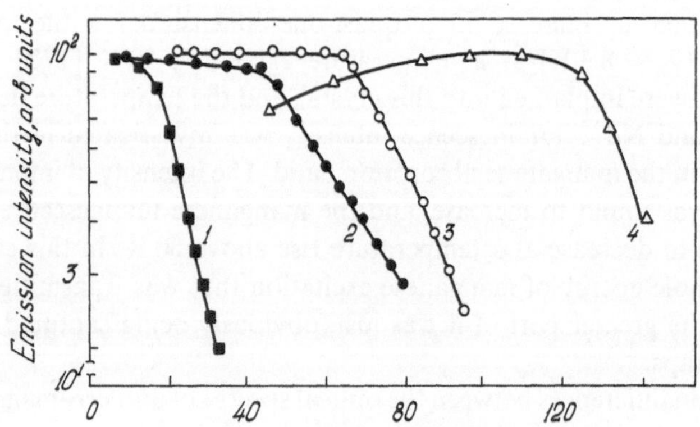

Fig. 3.12. Temperature dependences of the host luminescence intensity of crystals: (1) $BaMnF_4$; (2) $NaMnCl_3$; (3) $RbMnCl_3$; and (4) $CsMnCl_3$ [273].

Table 3.3

Crystal	T_N (K)	$\Delta\omega$ (cm^{-1})	E_a (cm^{-1})	T_A (K)
$NaMnCl_3$	6.5	1200	220	35
$RbMnCl_3$	95	1600	600	60
$CsMnCl_3$	63	1700	800	100

bility increases, a part of their energy being now transmitted to traps. It means that in these crystals the exciton motion has a diffusive character. The temperature T_A, at which a noticeable decrease of the emission intensity is observed, i.e., the temperature at which the exciton mobility increases, does not correlate with the Néel temperature. For example, in $NaMnCl_3$ crystals $T_N \simeq$ 6.5 K, and the exciton mobility increases only when the crystal is heated above 40 K; in other words, magnetic ordering established in crystals with a strong electron–phonon interaction does not affect appreciably the kinetics of magnetic excitons. At the same time, a change in the exciton–phonon interaction strength has a substantial influence on the exciton mobility; the stronger this interaction, the more energy is required to activate the diffusion of excitons, and the higher is the temperature at which a considerable growth of their mobility is observed (Table 3.3). It confirms the thesis that the main cause of incoherent exciton motion in antiferromagnetic manganese chlorides is an exciton–phonon interaction.

The values of the exciton diffusion activation energy presented in the table were calculated from the slope of curves 2, 3, and 4 shown in Fig. 3.12.

3.4.5. Change of Exciton Mobility at Inducing Intersublattice Transitions

As was noted in Section 3.4.1, if the intersublattice exchange interaction exceeds the intrasublattice exchange interaction in antiferromagnetic dielectrics, the induction of intersublattice transitions has to result in an increase of exciton mobility. An increase in the magnetic exciton mobility, due to an increase in the probability of intersublattice transfers with the rise of temperature, was experimentally observed in a quasi-two-dimensional antiferromagnetic $BaMnF_4$. At low temperatures, all the emission from this crystal in the visible spectrum is an intrinsic luminescence. Hence, at these temperatures the exciton mobility is low, i.e., the excitons are practically localized. Their localization is stimulated by peculiarities of the crystal magnetic structure (all the nearest neighboring Mn^{2+} ions have an antiparallel orientation of spins). With the temperature rise, the crystal luminescence intensity decreases (curve 1, Fig. 3.12) [275]. There is a noticeable decrease of the emission intensity and, consequently, an appreciable increase of exciton mobility in a temperature range close to $T \lesssim T_N$, i.e., in that temperature range where the rate of intersublattice transfers is increased.

A resonance transfer of the electron excitation energy in a pair of ions A and B, from sublattices with opposed orientation of spins due to their exchange interactions depending on their spin orientation, is described by expression (3.16). The factor $\cos^2\theta$ characterizes the degree of elimination of the spin forbiddenness on excitation energy transfer as the noncollinearity of sublattices is induced. In particular, in magnetic fields $H > H_{cr}$(H_{cr} is a certain critical field controlled by the crystal anisotropy) $\cos^2\theta = (M/M_0)^2$. Therefore, the probability of intersublattice transfers rises with an increase in the external magnetic field intensity, i.e., with an increase in M. In layered antiferromagnets, in which only a two-dimensional motion of excitons is possible in a collinear state, the dimensionality of exciton motion has to increase as the noncollinearity of sublattices is created.

A change of exciton mobility at the induction of intersublattice transfers with a magnetic field was observed in $NaMnCl_3$ [276] and $CsMnF_3$ crystals [264]. A defect luminescence of a $CsMnF_3$ crystal in a magnetic field was found to increase (Fig. 3.13), this increase not being related to an increase in the probability of radiation transitions in a magnetic field. With an increase in the external magnetic field strength up to 12 T, the intensity of the defect emission increased by 1.4 times.

Since the transfer of the electron excitation energy from the matrix to metastable traps, which are responsible for the defect emission, can only be effected in the process of exciton migration, the rise of the defect emission intensity is indicative of an increase in the exciton diffusion coefficient. An

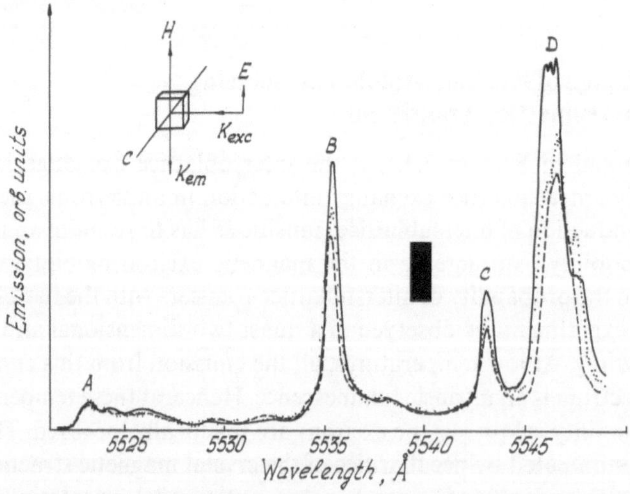

Fig. 3.13. Fragments of luminescence spectra of $CsMnF_3$ at $T = 2$ K for $H = 0$ (---), $H = 7$ T (...) and $H = 12$ T (——). Band A is a pure exciton transition. Band B its magnon sideband + pure electron transition in "disturbed" ion Mn^{2+}. Bands C and D are defect emission [264].

increase in the exciton mobility is caused by an increase in the probability of intersublattice transfers, as the noncollinearity of magnetic sublattices is induced by a magnetic field. In a $CsMnF_3$ crystal the constant of the ferromagnetic exchange interaction of manganese ions responsible for luminescence, $\mathscr{I}_F = 0.22\ cm^{-1}$, and the constant of the antiferromagnetic interaction $\mathscr{I}_{AF} = -4.4\ cm^{-1}$, therefore, indeed, the induction of intersublattice transfers, has to result in an increase in the exciton diffusion coefficient.

In a $NaMnCl_3$ crystal, the dimensionality of the possible motion of excitons is increased in a magnetic field. The magnetic structure of $NaMnCl_3$ is formed by antiferromagnetic packing of ferromagnetic layers containing Mn^{2+} ions. At low temperatures and in the case of a zero magnetic field, only two-dimensional incoherent exciton motion is possible in this crystal, exciton mobility being extremely low up to 35 K. In a magnetic field, $H \simeq 7–8$ T, a $NaMnCl_3$ crystal becomes a saturated paramagnet [277], therefore, a three-dimensional motion of excitons becomes possible in this crystal in the case of such fields. The intrinsic luminescence of the crystal was found experimentally to decrease with an increase in the external magnetic field intensity (Fig. 3.14) which is indicative of an exciton mobility increase. In this crystal, the constant of the ferromagnetic exchange interaction is greater than that of the antiferromagnetic interaction, therefore, inducing intersublattice transfers cannot, as such, result in an appreciable increase in exciton mobility. The observed rise of exciton mobility in a $NaMnCl_3$ crystal in a magnetic field is caused by an increase of the dimensionality of the possible motion of excitons as intersublattice transitions are induced. A connection between the character of the migration of small-radius excitons and the dimensionality of their possible motion was indicated in a theoretical paper [278].

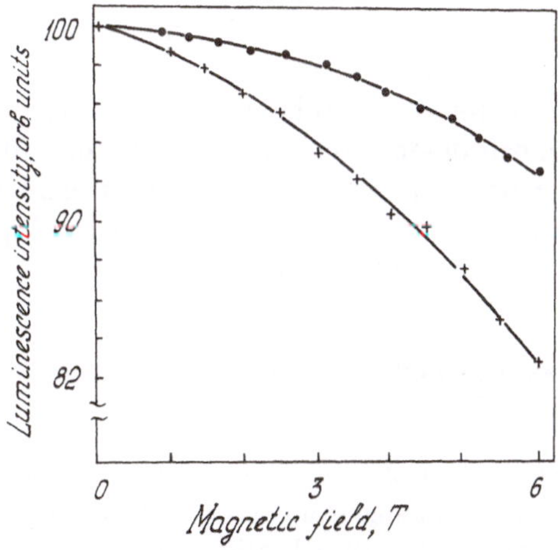

Fig. 3.14. Integral intensity of $NaMnCl_3$ luminescence versus magnetic field strength at different temperatures. (●) $T = 8$ K; (×) $T = 2$ K.

3.5. Radiation Decay of Magnetic Excitons

Radiation decay is a process in which a quantum system passes to a lower energy state, a photon being spontaneously emitted. In antiferromagnetic manganese compounds one-particle (purely exciton) transitions, responsible for the radiation decay of excitons, i.e., for the luminescence of crystals, are weak due to the intercombination forbiddenness. Pair transitions play the main role in a radiation decay of excitons. Pair processes enabled us to explain the appearance of electric dipole bands in the luminescence spectra of antiferromagnets. The probability of an electric dipole moment appearance in a pair of magnetic ions depends upon the mutual orientation of their spins. A method which makes it possible to find this dependence, without calculating the dipole moments themselves, is discussed in [224].

The intensity of pair transitions in the luminescence is proportional to the square of the matrix element module $G_{k_1k_2}$ in the expression (3.29). Each matrix element indicates the character of the dependence of the luminescence intensity of possible types of pair transitions upon the angle between the magnetic sublattices.

The matrix element G_{11} characterizes the process of a "hot" exciton–magnon luminescence, i.e., of a simultaneous annihilation of the electron excitation (exciton) and the spin excitation (magnon) in the opposite sublattices (Fig. 3.2(a)). The frequency of such a transition ν_{em}^h exceeds that of a pure exciton line ν_e by the value of the magnon energy ν_m. The magnetic field dependence of the luminescence intensity is defined by the factor $\sin^4\theta$.

G_{12} is a "cold" exciton–magnon luminescence (the annihilation of an exciton and creation of a magnon at one and the same sublattice (Fig. 3.2)). Its frequency is $\nu_{em}^c = \nu_e - \nu_m$; $\mathscr{I} \sim \cos^4\theta$.

G_{13} is the luminescence in crystals without the inversion center in a pair of ions from opposite sublattices (the exciton is annihilated at one ion, the second ion takes part in the luminescence without getting excited (Fig. 3.2(c))). For this luminescence $\nu = \nu_e$, $\mathscr{I} \sim \sin^2 2\theta$.

G_{33} is a "cold" exciton–magnon luminescence in crystals without the inversion center in a pair of ions from opposite sublattices (the annihilation of an exciton and creation of a magnon take place at one ion, the second ion participates in the luminescence without getting excited (Fig. 3.2(i))). For this luminescence $\nu = \nu_e - \nu_m$, $\mathscr{I} \sim \cos^2 2\theta$.

3.5.1. The Increase of Exciton–Magnon Luminescence in a Magnetic Field

As a photon is emitted in crystals of antiferromagnetic manganese compounds, a Mn^{2+} ion can change from an optically excited state with $M_s' = \frac{3}{2}$ into a ground-electron or electron-vibrational state with $M_s = \frac{5}{2}(\Delta m_s = +1)$, or into a spin-excited ground-electron state with $M_s = \frac{3}{2}(\Delta m_s = 0)$. For the process of photon emission in the first case to be of an electric dipole character, simultaneously with the change in the value of the projection of the first ion spin moment by +1, that of the second ion spin moment projection should change by −1. This is possible in the case when both ions are in the same sublattice. A similar situation is realized in the case of an optical transition in the other sublattice ($M'_s = -\frac{3}{2}$, $M_s = -\frac{5}{2}$), when the spin projection of the neighbor is changed by +1. Therefore, as the noncollinearity of magnetic moments of the opposite sublattices is induced by a magnetic field, the intensity of the exciton–magnon luminescence in the first case has to increase, since the efficiency of a pair process with the participation of ions from the sublattices with opposite orientation of spins becomes noticeable in this case also.

The total intensity of luminescence is described by the following expression:

$$\mathscr{I}_L \sim \sum_{\mathbf{k}} |\pi^\tau(\mathbf{k})|^2 u_k^2, \tag{3.57}$$

where

$$\pi(\mathbf{k}) = \frac{1}{z_F} \sum_{\mathrm{n} \neq \mathrm{m}} \pi^\tau_{\mathrm{n}\alpha,\,\mathrm{m}\alpha} \exp[i\mathbf{k}(R_{\mathrm{m}\alpha} - R_{\mathrm{m}\alpha})]$$
$$+ \frac{\cos^2\theta}{z_{AF}} \sum_{\mathrm{n}\alpha \neq \mathrm{m}\beta} \pi^\tau_{\mathrm{n}\alpha,\,\mathrm{m}\beta} \exp[\mathbf{k}(R_{\mathrm{n}\alpha} - R_{\mathrm{m}\beta})].$$

Here $\pi^\tau_{\mathrm{n}\alpha,\,\mathrm{m}\alpha}$ corresponds to a pair transition with the participation of Mn^{2+} ions from one sublattice, and is proportional to an exchange interaction between ions with a parallel orientation of spins, $\mathscr{M}^f_{\mathrm{n}\alpha,\,\mathrm{m}\alpha}$; while $\pi^\tau_{\mathrm{n}\alpha,\,\mathrm{m}\beta}$ corresponds to a pair transition with the participation of ions from sublattices with opposite orientation of spins, and is proportional to the exchange interaction of ions with an antiparallel orientation of spins $\mathscr{M}^f_{\mathrm{n}\alpha,\,\mathrm{m}\beta}$. As follows from expression (3.57), the magnitude of the exciton–magnon luminescence enhancement effect depends upon the relation between the intra- and intersublattice exchange interactions. It means that information about the relation of these interactions can be obtained by studying the behavior of the exciton–magnon luminescence in a magnetic field.

A change in the exciton–magnon–phonon luminescence of MnF_2 and $KMnF_3$ crystals in a magnetic field was studied experimentally [227]. Phonons taking part in the emission transitions does not affect their probability versus the magnetic field dependence. This follows from [225] where it was shown that the behavior of exciton–magnon and exciton–magnon–phonon light absorption bands in a magnetic field is the same. The integral intensities of the luminescence of crystals was found to increase with an increase in the degree of noncollinearity of their magnetic structures (Fig. 3.15). The experimental dependences are in good agreement with the results of calculation by the formula (3.57) at $\mathscr{M}^f_{\mathrm{n}\alpha,\,\mathrm{m}\beta} \simeq 3\mathscr{M}^f_{\mathrm{n}\alpha,\,\mathrm{m}\alpha}$ for $kMnF_3$ and $\mathscr{M}^f_{\mathrm{n}\alpha,\,\mathrm{m}\beta} \simeq \mathscr{M}^f_{\mathrm{n}\alpha,\,\mathrm{m}\alpha}$ for MnF_2.

In low-symmetry antiferromagnets with a strong anisotropy, the constants of inter- and intrasublattice interactions can differ substantially. If, for ex-

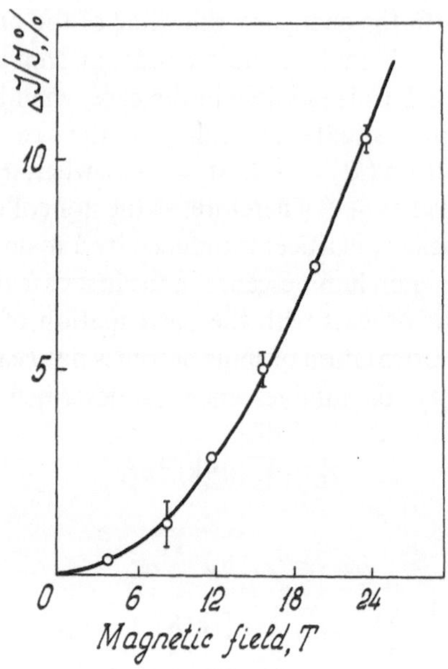

Fig. 3.15. Relative change of integral intensity of $KMnF_3$ luminescence versus magnetic field strength. (o) Experimental points ($T = 20.4$ K, $\mathbf{H} \parallel C_4$); the solid line is the calculation by formula (3.57) [227].

ample, the intrasublattice interaction is much stronger than the intersublattice interaction, the intensity of the exciton–magnon luminescence has to be practically independent of a magnetic field. In the opposite case, with zero magnetic field the exciton–magnon luminescence has to be weak, and the character of the luminescence intensity variation in a magnetic field has to differ from that discussed above. It is exactly this situation that is realized in a quasi-one-dimensional antiferromagnetic crystal $CsMnCl_3 \cdot 2H_2O$ which enabled us to observe a change in a pure exciton transition in a magnetic field.

3.5.2. Enhancement of Exciton Radiation Transition

Results of experimental studies of the $CsMnCl_3 \cdot 2H_2O$ crystal luminescence integral intensity versus the magnetic field strength dependences are presented in Fig. 3.16 [279]. It is remarkable that the character of the $\mathscr{I}_L(H)$ dependence presented differs substantially from that observed in three-dimensional antiferromagnetic crystals (Fig. 3.15). When interpreting the results shown in Fig. 3.16, the following factors were taken into consideration.

(1) In a $CsMnCl_3 \cdot 2H_2O$ crystal, all the optically active ions (Mn^{2+} ions) are shifted from the center of symmetry, therefore, the luminescence at the exciton frequency has an electric dipole character.

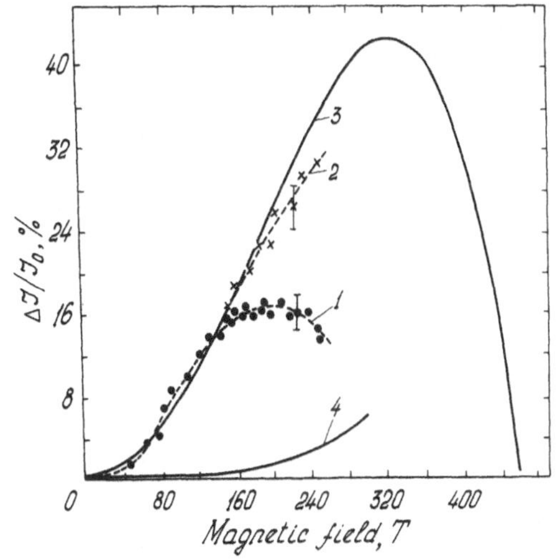

Fig. 3.16. Relative change of the integral intensity of $CsMnCl_3 \cdot 2H_2O$ luminescence versus magnetic field strength: (1, 2) experimental data at stationary and pulsed excitations; (3, 4) exciton luminescence and exciton–magnon luminescence, respectively, (calculation) [270].

(2) Since in this crystal all the nearest magnetic neighbors of Mn^{2+} ions have an antiparallel spin orientation, and the interaction (ferromagnetic) with the next-nearest neighbors is very weak, the exciton–magnon luminescence (at frequencies $\nu_e - \nu_m$), proportional to the exchange interaction of Mn^{2+} ions with parallel oriented spins, has to be very weak in a collinear phase.

(3) As noncollinearity of magnetic sublattices is created, this exciton–magnon luminescence is induced by the interaction with the nearest neighbors. Since the intersublattice exchange interaction in this crystal is much stronger than the intrasublattice interaction, the dependence of the luminescence intensity on that of a magnetic field is described by the expression $\mathscr{I} \simeq \pi^{\tau}_{n\alpha,\, m\beta} \cos^4 \theta$. In a $CsMnCl_3 \cdot 2H_2O$ crystal, $H_E \approx 23$ T, therefore the exciton–magnon luminescence induced by a magnetic field has a noticeable magnitude only in a field with $H > 25$ T.

(4) In the crystal luminescence spectrum at low temperatures a band was observed which shifted with respect to the exciton band into the red region of the spectrum. This band was interpreted as the exciton transition magnon sideband corresponding to the process in which annihilation of the exciton with photon radiation and creation of a spin wave take place at one and the same ion. The probability of such transitions does not depend upon a magnetic field.

To explain the observed enhancement of luminescence by a magnetic field, a mechanism of the electric dipole radiation of light at exciton frequencies is suggested, which is possible only in the case of noncollinear magnetic moments of sublattices for ions shifted from the center of symmetry. According to this model, a dipole moment in induced in a pair of magnetic ions coupled by an exchange interaction, as a result of which one ion changes from an optically excited state into the ground state, a photon being emitted (Fig. 3.2(c)).

The luminescence intensity at the exciton frequency is expressed as follows:

$$\mathscr{I}_e^\tau \sim \langle |P^\tau|^2 \rangle \sin^2\theta \cos^2\theta \sum_{\mathbf{k}} |\pi^\tau(\mathbf{k})|^2 \delta(\mathbf{k}, \mathbf{q}). \tag{3.58}$$

Assuming $\mathbf{q} = 0$ and $\cos\theta = m$, we obtain

$$\mathscr{I}_e \sim m^2(1 - m^2)|\sum_{\delta} \pi^\tau_{\mathbf{n}1,\mathbf{n}1+\delta}|^2. \tag{3.59}$$

The dependence of the exciton luminescence intensity of the angle θ is controlled by the factor $\sin^2\theta\cos^2\theta$, which becomes zero at $2\theta = \pi$ (a collinear antiferromagnet) and at $\theta = 0$ (a saturated paramagnet) and reaches its maximum at $\theta = \pi/4$. The amplitudes of the matrix elements of exciton–magnon and exciton transitions in the case of ions shifted from the inversion center differ by the factor $\sqrt{2S}$ [280]. In the case of Mn^{2+} ions, for whcih $S = \frac{5}{2}$, the intensities of exciton and exciton–magnon luminescence in a magnetic field are related as follows:

$$\mathscr{I}_e/\mathscr{I}_m = 5m^2(1 - m^2)/m^4, \tag{3.60}$$

i.e., up to the field with $H = 1.8H_E$ the exciton luminescence exceeds that of the exciton–magnon.

Plotted in Fig. 3.16 (curves 3 and 4) are theoretical curves of the exciton and exciton–magnon luminescence intensity versus the magnetic field strength. One can see good agreement between the experimental curve which was obtained when the emission was excited by a pulsed source of light, i.e., when the magnetic field effect on the crystal light absorption coefficient was absent (as there was no excitation at the time of the field action) and the theoretical curve describing the exciton luminescence.

Theoretical values of the luminescence intensity in the region of strong magnetic fields are somewhat greater than the corresponding experimental data. This mismatch is explained in the following way: the object of the experimental investigations was the integral emission intensity in which, besides exciton transitions, some contribution is also made by exciton–magnon transitions shown in Fig. 3.2; the probability of these transitions in this range of magnetic fields decreases with the rise of H.

A large difference observed in case the luminescence is induced by a continuous light source (curve 1, Fig. 3.16) is explained by the influence of a magnetic field on the light absorption coefficient, i.e., on the efficiency of excitation. As was noted above, the intensity of the exciton–magnon absorption in antiferromagnets decreases according to the $K = K_0(1 - m)^2$ law. Taking into account the weakening of light absorption in a magnetic field, the theoretical curve (3, Fig. 3.16) will transform into a curve qualitatively similar to curve 1.

3.5.3. Biexciton Annihilation

Besides the magnetic structure of crystals, the exciton–exciton interaction also has a considerable effect on the radiative decay of excitons in antiferromagnets. This interaction is realized in acts of exciton–exciton collisions. The degree of transformations of the luminescence parameters (caused by exciton–exciton) depends upon the magnitude of this interaction and the kinetic properties of excitons. As the exciton–exciton interaction in antiferromagnetic compounds of manganese is a short-range one, i.e., the radius of a "collision" is small, the manifestation of this interaction becomes noticeable only under the conditions of a high concentration of excitons (at high degrees of excitation).

A significant number of interactions between excitons can be expressed only in the case where the exciton lifetimes τ_e is commensurable with their collision time τ_{ee} or is greater

$$\tau_{ee} \geq \tau_{ee} = (8\pi R_0 ND), \tag{3.61}$$

where R_0 is the collision radius, and N is the exciton concentration. A considerable number of collisions for excitons in antiferromagnetic compounds of manganese are possible due to their long lifetimes. Indeed, despite a very slow ($D \sim 10^{-8}$ cm^2/s) migration of excitations in a crystal, the condition is observed, for example, for a MnF_2 crystal at a concentration of excitons of about 10^{17} cm^{-3} [281]. Such a concentration of excitons can be obtained under the pulse laser excitation.

Luminescence studies of antiferromagnetic manganese fluorides MnF_2, $KMnF_3$, and $CsMnF_3$ under the strong one-quantum excitation has shown that in the case of large density excitons, both time and spectral effects are observed [242, 282–283]. Figure 3.17 shows the emission decay character of E1 excitons in a MnF_2 crystal at a high degree of excitations. We can clearly see that at an initial moment after excitation is switched off, when the exciton density is maximum, there is a considerable deviation from the exponential decay law. With an increase of the pumping intensity the decay rate rises at an initial period of time, i.e., the decay rate depends upon the excitation population

$$dN(t) = -\gamma(N)N(t). \tag{3.62}$$

The simplest case corresponding to $\gamma(N) = \gamma_1 + \gamma_2(N)$ gives the following relationship for the rate of the exciton population change:

$$dN(t)/dt = -\gamma_1 N - \gamma_2 N^2. \tag{3.63}$$

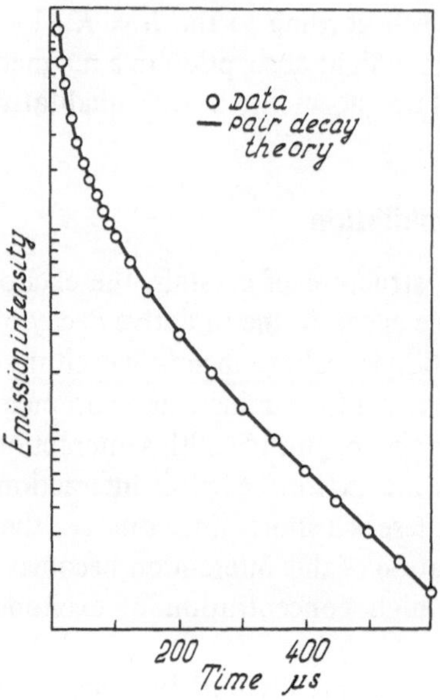

Fig. 3.17. Comparison between E1 exciton decay in MnF_2 crystal at 1.7 K with the theoretical dependence obtained in a biexciton decay model [242].

Here, γ_1 is the sum of rates of the radiation decay and transfer into traps, and γ_2 is the decay rate controlled by the exciton pairs interaction.

To interpret the results obtained, a biexciton decay mechanism was used [242]. The essence of this mechanism is as follows. When a large exciton concentration is created in a crystal, in the case of a collision of two excitons, due to the exciton–exciton interaction, both the electron excitations are accumulated at one site. Then, in the site that has received some additional energy, there is a rapid nonradiation transition into the lowest excited state. Such a relaxation process results in a loss of one exciton at each act of collision. Therefore, an increase in the excitation intensity results in a decrease of the exciton luminescence intensity and to a nonexponential decay of the luminescence. In such a model, a change of the emission intensity with time is described by the following expression:

$$\mathscr{I}(t) = \frac{\gamma_1}{\gamma_2} \frac{\mathscr{I}(0)}{[\gamma_1/\gamma_2 + \mathscr{I}(0)] \exp(\gamma_1 t) - \mathscr{I}(0)}. \tag{3.64}$$

Good agreement of experimental and theoretical results is observed in a MnF_2 crystal at $\gamma_2 \simeq 5 \times 10^{-13}$ cm^3 s, in $KMnF_3$ at $\gamma_2 \simeq 4 \times 10^{-11}$ cm^3 s, and in $CsMnF_3$ at $\gamma_2 \simeq 5 \times 10^{-15}$ cm^3 s. Such a small rate of the biexciton decay in antiferromagnets is caused by a low mobility of excitons. In the case of a $CsMnCl_3 \cdot 2H_2O$ crystal in which electron excitations are localized, no

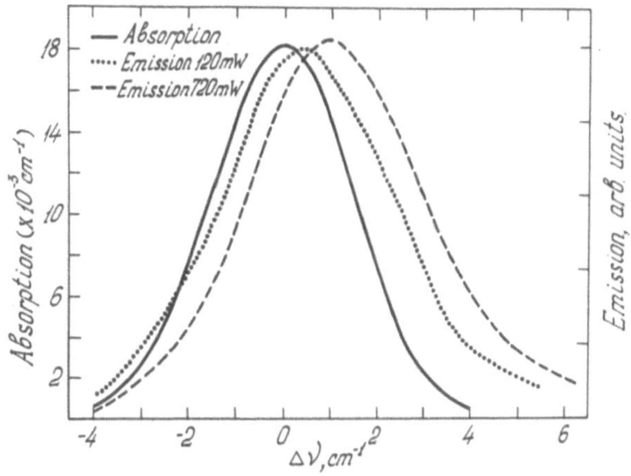

Fig. 3.18. Time-resolved luminescence spectra and exciton lines in the absorption spectrum of $KMnF_3$ crystal at different levels of excitation (120 mW and 720 mW) but at the same delay 300 μs. Excitation is created at 514.5 nm wavelength by a light source with pulse duration of 5 ms [282].

biexciton decay is observed at all up to the exciton concentration $N \simeq 5 \times 10^{19}$ cm^{-3}.

In addition to the nonexponential decay of the emission in the case of a high exciton concentration there is the exciton band broadening and shift to the red region of the luminescence spectrum (Fig. 3.18) [282]. The magnitude of shift and broadening increases proportionally to the concentration of excitons. It should be noted that the revealed shift of exciton bands is not related to the heating of the crystal, as the temperature rise results in a "violet," not "red," shift of the transition frequency.

Like time effects, spectral effects are explained by the interaction among excitons and their collisions. In the case when this interaction is strong enough, a noticeable shift of the luminescence band is observed with the exciton concentration increase, in the case when the interaction is weak there is no shift. In MnF_2 crystals, the interaction between E1 excitons is very weak, whereas the interaction between E1 excitons and E2 excitons, which are 17 cm^{-1} higher on the energy scale, reaches 80 cm^{-1} [232]. When a high concentration of only E1 excitons was created, no shift of the exciton luminescence band was observed, however, when a considerable concentration of E1 and E2 excitons was obtained, a shift was detected [281]. Nevertheless, in an antiferromagnetic $KMnF_3$ crystal, spectral effects were observed when only the E1 exciton concentration was increased. This testifies to the fact that in $KMnF_3$ crystals the interaction of E1 excitons is much stronger than in MnF_2 crystals.

To establish a correlation between the density of excitons and the shift of the E1 exciton emission line, use was made of a simple collision model developed for an atomic spectrum. According to this model which takes into consideration only pair interactions, a change of the shape and frequency of spectral lines of an atom due to its interaction with the neighboring particles is proportional to the concentration of these particles. The experimentally observed linear dependence of the exciton band shift upon the exciton density is in good agreement with the theory.

CHAPTER 4

Magnetic Excitons as a Method of Study of the Magnetic Properties of Antiferromagnets

This chapter is concerned with the application of the spectroscopy of magnetic crystals. The value of the spin wave energy is estimated, and particular crystal magnetic structures are studied on the basis of the crystals light absorption spectra.

4.1. Spectroscopic Studies of Spin Waves

4.1.1. Three-Dimensional Magnetic Crystals. Energy of Magnons at the Boundary of the Brillouin Zone

As was mentioned in the preceding chapter, the exciton–magnon light absorption takes place in a band whose width is equal to the sum of the exciton and magnon bandwidths. Taking no account of the exciton–magnon interaction, whose role in the case of its moderate magnitude is reduced to some redistribution of the intensities within the absorption band, the absorption coefficient turns out to be proportional (3.23) to the total density of the exciton–magnon states $\rho(E(\mathbf{k}) + \varepsilon(\mathbf{k}))$ and to the transition moment which is a smoothly varying function within the Brillouin zone. If the width of the exciton band of the optical transition under consideration is small, compared to that of the magnon band (which is apparently the case for the majority of the transitions in $3d$-magnetic dielectrics), then the form of the "cold" exciton–magnon absorption is indicative of nothing other than the density of the magnon state distribution.

$$\rho(\nu) = \frac{1}{N} \sum_{\mathbf{k}} \delta(h\nu - \varepsilon(\mathbf{k})).$$

The density of the magnon states in three-dimensional crystals is known to have a sharp peak at the boundary of the Brillouin zone. As the values $\mathscr{U}(\mathbf{k})$ and $\pi^{\tau}(\mathbf{k})$ in expression (3.23) are limited ($\mathscr{U}(\mathbf{k}), \pi^{\tau}(\mathbf{k}) \sim 1$), the exciton–magnon absorption will be the most intensive one for the value of the wave vector in the vicinity of the Brillouin zone boundary, if $\pi^{\tau}(k)$ does not cause expression (3.23) to vanish. Thus, by measuring the interval between the pure-exciton line and its "cold" magnon sideband, it is possible in many cases to determine directly the energy of spin waves in separate points at the Brillouin zone boundary with an accuracy up to the exciton–magnon interaction.

The "hot" exciton–magnon absorption is proportional not only to the corresponding dipole moment of the transition and the differential density of states $\rho(E(\mathbf{k}) - \varepsilon(\mathbf{k}))$, but is also dependent upon the thermal population of magnon states, which is described by the Bose distribution function. Disregarding the polarization dependence and neglecting, as above, the exciton dispersion, we obtain the "hot" absorption coefficient to be proportional to the product of the density of magnon states $\rho(\varepsilon(\mathbf{k}))$ and the distribution function. It is clear that at low temperatures, the distribution function "cuts off" the bigger part of the magnon band and distinguishes only the states arranged in the region of small energies, provided, of course, that the density state is not zero at small values of **k**. Therefore, in three-dimensional crystals, these "hot" exciton–magnon bands appear at sufficiently high temperatures at frequencies corresponding to $E(\mathbf{k}) - \varepsilon(\mathbf{k}_{max})$. It will be noted that in this process there is no exciton–magnon interaction (exciton and magnon do not coexist with the exception of the case shown in Fig. 3.2(h)), and the energy intervals between the exciton and exciton–magnon absorption peaks exactly correspond to the energies which display peculiarities in the density of the magnon states.

A study of exciton–magnon transitions in the α-phase of solid oxygen [284] is one of the first examples of using the absorption spectra of antiferromagnets to estimate the spin wave energy at the boundary of the Brillouin zone. The value of the spin wave energy (38 cm^{-1}) found from spectroscopic studies was lately confirmed by the antiferromagnetic resonance data and by thermal properties of α oxygen. One of the latest examples of the above-mentioned investigations is a study of the spin-wave spectrum in $CoCO_3$ [285, 286]. It follows from the spectral position of the "hot" magnon sidebands that the energy of magnons in $CoCO_3$ at 14 K (0.77 T_N) is equal to 25 and 35 cm^{-1} in two points of the Brillouin zone. The presence of both "hot" and "cold" magnon sidebands in the light absorption spectrum makes it possible to determine the exact values of the magnon energy in the whole temperature range, i.e., to study (with a sufficiently direct method) renormalization of the spin-wave spectrum. It was found that when cooling an antiferromagnet $CoCO_3$ from 14 to 1.8 K the magnon energy increases approximately by 3 cm^{-1} to become equal to 28 and 35 ± 1 cm^{-1}. One of these values (28 cm^{-1}) is in perfect agreement with the values of the magnon energies obtained when measuring both the two-magnon infrared absorption [287] and the two-magnon light scattering [288] in $CoCO_3$ crystals, the latter values being 54.7 ± 0.7 cm^{-1} and 58 ± 2 cm^{-1}, respectively. Since the magnon energy with $\mathbf{k} \approx 0$ (according to the same measurements) is equal to 35 cm^{-1}, the spectroscopic investigations indicate that in the spin-wave spectrum of $CoCO_3$ crystals there is also such a **k** direction where magnons practically do not disperse.

As the crystallographic and magnetic structures of the antiferromagnet become more complex, their spin-wave spectrum and the fine structure of the exciton–magnon absorption corresponding to it also become more complicated, magnons of different branches of spin waves and different points at the boundary of the Brillouin zone taking part in this structure formation. It is this complicated structure that is demonstrated by the absorption spectrum of $CsMnF_3$ (a six-sublattice antiferromagnet with two types of Mn^{2+} ions which occupy inequivalent positions in the lattice). Identification of this structure enabled us to find five distinct boundary energies of spin waves in this crystal (34, 38, 51, 60, and 64 cm^{-1}) which correspond to an optical, an acoustic, and four exchange magnon branches [289].

4.1.2. Low-Dimensional Magnetic Crystals. Magnon Energies Inside the Brillouin Zone

Unlike three-dimensional magnetic crystals, the structure of the exciton–magnon absorption in low-dimensional crystals carries information not only about the spin wave energy at the zone, but inside it as well. This phenomenon is related to the difference in a magnon state density distribution: in three-dimensional crystals there are density sharp peaks at the edge of the zone, whereas in low-dimensional crystals the distribution is more uniform. A strong anisotropy of exchange interactions in low-dimensional magnetic crystals results in the following phenomenon: there is practically no dispersion of spin waves along the direction of a weak exchange interaction, and an additional peak appears in the distribution of the state density in the small energy region (Fig. 4.1). Due to this difference in the distribution of the state density the shapes of the exciton–magnon bands in magnetic crystals of these classes differ substantially as well (Fig. 4.2).

One of these low-dimensional antiferromagnets is an orthorhombic two-axis two-sublattice crystal $CsMnCl_3 \cdot 2H_2O$. Its magnetic structure is con-

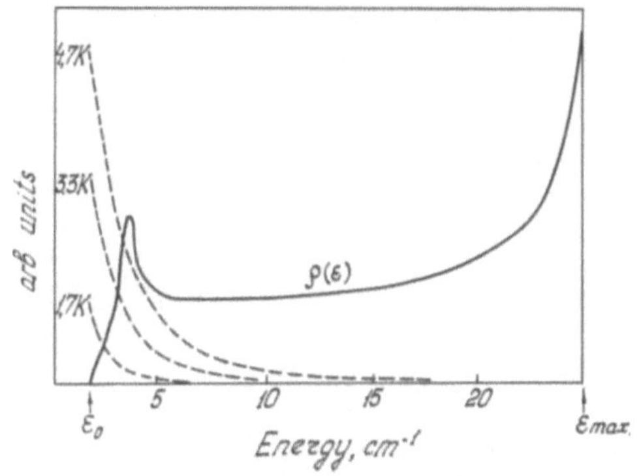

Fig. 4.1. $CsMnCl_3 \cdot 2H_2O$. Magnon density of states $\rho(\varepsilon)$. Solid line—calculation; dashed lines—Bose distribution functions for indicated temperatures [290].

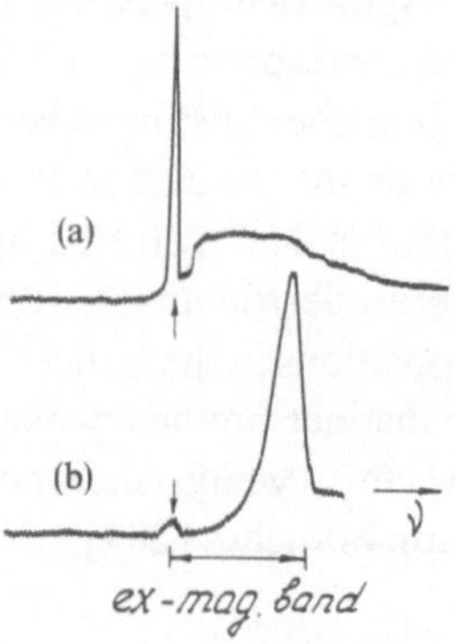

Fig. 4.2. Spectra of the light exciton–magnon absorption by antiferromagnetic dielectric crystals. (a) One-dimensional $CsMnCl_3 \cdot 2H_2O$; (b) three-dimensional $RbMnF_3$ (the beginning and end of exciton–magnon bands are matched; pure exciton lines are marked with arrows) [290].

trolled by the presence of chains of interacting magnetic Mn^{2+} ions. The interchain bond is very weak ($\mathscr{I}/\mathscr{I}' \approx 10^{-2}$) which is the reason for the magnetic properties of the crystal being one dimensional. At $T = 4.89$ K, a completely ordered three-dimensional antiferromagnetic state prevails, but the exchange interactions remain to be of a quasi-one-dimensional character. The exciton–magnon light absorption [290] in the region of the ${}^6A_{1g}({}^6S) \to {}^4T_2({}^4D)$ transition has a pronounced zone character (Fig. 4.3). The figures show that the intensive exciton–magnon absorption starts directly at the exciton line. As the $CsMnCl_3 \cdot 2H_2O$ crystal is a two-axis one, it has two branches of spin waves with a practically common boundary of maximum energies, but with different activation energies ε_0 and with different positions of magnon state density peaks in the small energy region. Such a structure of

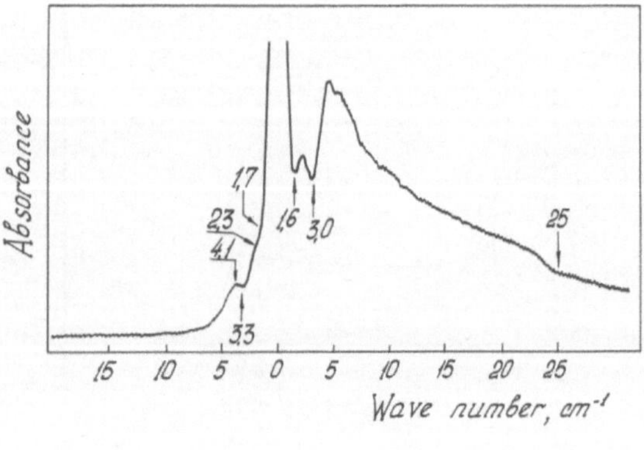

Fig. 4.3. $CsMnCl \cdot 2H_2O$. Fine structure of exciton–magnon light absorption, $E \parallel C$, $T = 1.65$ K [290].

the state densities is reflected in the fine structure of the absorption spectrum in the close vicinity of the exciton line both in the "cold" and "hot" absorption: the position of the peak of the "hot" absorption bands, with respect to the exciton line directly and unambiguously, shows the arrangement of density peaks for each of the two branches of the spin-wave spectrum, and the points at the beginning of a continuous zone absorption in the optical spectrum have to give activation energies for each branch of the spin waves. These points, indicated by arrows in Fig. 4.3, are readily observed in the "cold" and "hot" exciton–magnon absorption spectra. Obtained in this way, values of the energy gaps, 3.3 and 1.7 cm^{-1} at 1.65 K, as well as the temperature dependence of the higher gap value, are in complete agreement with the results of the antiferromagnetic resonance measurements.

As for the exciton–magnon absorption, corresponding to the peak of the density of states in the vicinity of the boundary values of energies, it should be repeated that the absorption coefficient is proportional not only to the magnon state density, but also to the transition moment as well which controls the polarization properties of the absorption. Due to these reasons, the state density peak corresponding to ε_{max} (Fig. 4.1) does not manifest itself in the form of the absorption peak in the experimental spectra. As for the boundary of the magnon state density, a distinct pecularity corresponds to it in the exciton–magnon absorption spectra (Figs. 4.2 and 4.3). Thus, in the case of low-dimensional magnetic crystals, practically all data on the spin-wave spectrum are derived from the absorption spectrum.

Peculiarities of the Spin-Wave Spectrum in the Vicinity of the Brillouin Zone Center

Now let us discuss in more detail the results of studies of the temperature dependence of the intensity of "hot" exciton–magnon transitions of the type shown in Fig. 3.2(d), i.e., the results of studies of the optical excitation formation at one sublattice with the annihilation of a thermally excited magnon at the other. The maximum intensity of such an absorption is reached in external magnetic fields which transform the antiferromagnet into a saturated paramagnetic state ($K \sim \cos^4 \theta$), or into a ferromagnetic state in the vicinity of the Curie temperature. Such a process can take place in some antiferromagnets of the low-dimensional class, in which an exchange interaction along one of the directions is a ferromagnetic interaction and substantially exceeds the antiferromagnetic interaction in another direction. As an example of materials of this class we can mention the metamagnetic haloid compounds of transition metals. As the intensity of "hot" exciton–magnon bands depends upon the thermal population of magnon states, the data on the temperature dependence of the intensity of this absorption have to characterize the peculiarities of magnon dispersion in these crystals. It is from this point of view that the absorption spectra of the following three materials: $FeCl_2$, $CoBr_2$, and $NiBr_2$ were analyzed [291]. The first of these crystals has a large one-ion anisotropy of the "easy axis" type, the second one has an "easy plane", and the last one also has an "easy plane" but a small value. At low temperatures, when only low-energy magnon states are populated (i.e., in the vicinity of the Brillouin zone center) the expression for a spin-wave dispersion of a magnetic of $FeCl_2$-type can be written as follows:

$$E_{sw} \sim \varepsilon_0(1 + \alpha k^2), \tag{4.1}$$

where α is a constant about 1 which depends upon the crystal structure, and ε_0 is the energy gap in the spin-wave spectrum at $k = 0$ whose magnitude depends upon the anisotropy parameter. For a magnetic crystal of $CoBr_2$-type there are two spin-wave modes. Neglecting the terms of high order with respect to k we obtain

$$E_{sw}^{(1)} \sim B_1(1 + \gamma k^2), \tag{4.2}$$

$$E_{sw}^{(2)} \sim B_2 k, \tag{4.3}$$

where B_1 is controlled by parameters of an anisotropic exchange interaction within the plane and of isotropic interplane interaction, and B_2 is controlled to a greater degree by an intraplane anisotropy and to a lesser degree by the isotropic exchange interaction within the plane.

For $NiBr_2$ crystals, at small values of k, the dispersion is of $E \sim k$-type, for large values of k it is of $E \sim k^2$-type. The intensity of "hot" exciton–magnon bands is expressed by (3.23), the thermal population of magnon states being taken into consideration. In the case of a $FeCl_2$ crystal treated as an easy-axis two-dimensional ferromagnet, the temperature dependence of the "hot" exciton–magnon absorption coefficient will be of the following form:

$$K(T) \sim T^2 \exp(\varepsilon_0/k_B T). \tag{4.4}$$

For an easy-plane magnetic crystal with a large anisotropy ($CoBr_2$), if the effect of the zero point on the degree of filling the magnon states is taken into consideration, the intensity of a "hot" one-magnon sideband will vary as T^3.

If we neglect both the interplane exchange interaction and the anisotropy ($NiBr_2$), we shall obtain a square-law dependence on the absorption coefficient on temperature. The introduction of a small anisotropy will lead to a temperature dependence intermediate between T^2 and T^3.

The experiment does show different temperature dependences for the three compounds. Shown in Fig. 4.4 is the dependence of the "hot" exciton–magnon band intensity on temperature plotted in coordinates $\ln K(T) - 2 \ln T$ versus $1/T$. As was to be expected from (4.4), it is a straight line, the slope of which determines ε_0—a gap in the spin-wave spectrum. It follows from this graph that the energy gap is equal to $19 \pm 1\ \text{cm}^{-1}$. It is sknown from neutronographic studies that the gap in the spin-wave spectrum in $FeCl_2$ at 5 K in the center of the zone makes up to $17.2 \pm 0.4\ \text{cm}^{-1}$. Both measurements are in reasonable agreement since the latter value was obtained for 5 K, and the gap drops with temperature rise. In a $FeBr_2$ crystal whose magnetic structure is similar to that of $FeCl_2$, the temperature dependence of a "hot"

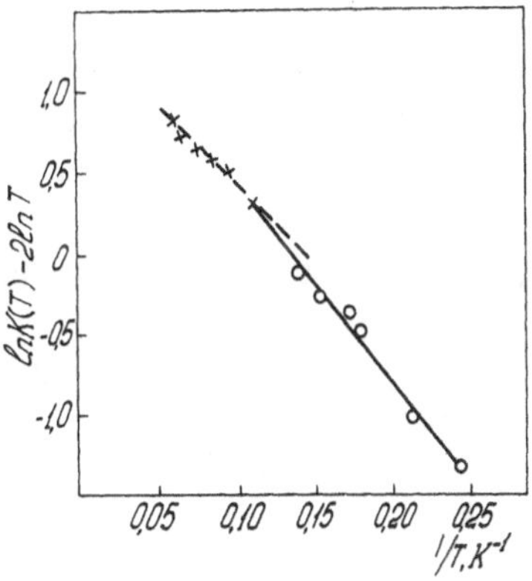

Fig. 4.4. $FeCl_2$. Intensity of the 427 nm band versus temperature [291].

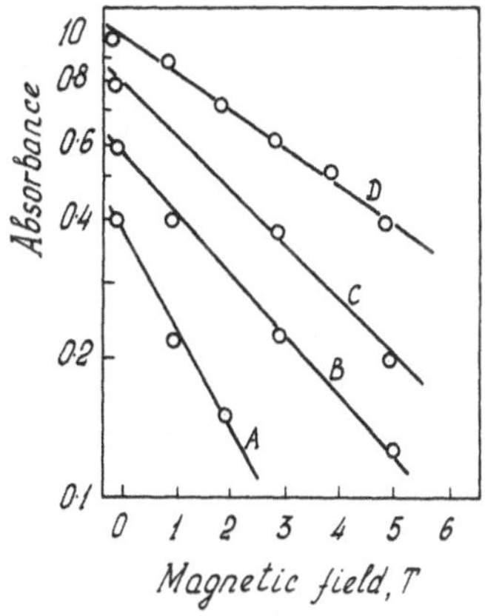

Fig. 4.5. Rb_2CrCl_4. Intensity of the 631 nm band versus the external magnetic field at temperature $T = 2.1, 2.5, 3.2$, and 4.2 K (curves A, B, C, and D, respectively) [293].

exciton–magnon band is also adequately described by expression (4.4) [292]. The gap of the $FeBr_2$ spin wave spectrum derived from absorption spectra 17.0 ± 1.8 cm^{-1} is in excellent agreement with the value 17.7 cm^{-1} that was measured neutronographically. For $FeCl_2$ and $FeBr_2$, expression (4.4) yields figures that agree with experimental data up to $T \approx 0.7\ T_N$. This is indicative of a high density of magnon states in the vicinity of the Brillouin zone center which is typical for a two-dimensional ferromagnet.

In $CoBr_2$, the dependence of the intensity of a "hot" exciton–magnon band on temperature is plotted as a straight line in coordinates $\ln K(T) - \ln T$. The slope of the line gives the power of T. It turns out to be equal to 2.8 which is close enough to the expected 3.

For $NiBr_2$, the power of T is equal to 2, i.e., it confirms the assumption that the plane anisotropy is small. However, in a two-dimensional easy-plane ferromagnetic Rb_2CrCl_4 ($T_C = 57$ K), i.e., in a crystal with the anisotropy of the same type as in $NiBr_2$, the dependence K on T is not a square-law one [293]. At low temperatures, the band intensity decreases much faster than T^2. Such behavior is typical for a magnetic crystal, the spin-wave spectrum which has a gap and the thermal energy become comparable to the gap magnitude. Using (4.4) we obtain that $\varepsilon_0 = 1.4 \pm 0.1$ K. The decrease in the absorption intensity in a magnetic field (Fig. 4.5) confirms this conclusion. The application of a magnetic field shifts the magnon dispersion curve, and the thermal population of magnons changes at a fixed temperature. This, in turn, affects the intensity of the exciton–magnon band. By analogy with (4.4), the influence of the field can be described in the following way:

$$K(H, T) \sim T^2 \exp[-(\varepsilon_0 + g\mu_B H)/k_B T]. \tag{4.5}$$

It follows from this run of measurements (Fig. 4.5) that $\varepsilon_0 = 0.9 \pm 0.4$ cm^{-1}.

Neutron diffraction shows that the nature of the plane anisotropy in Rb_2CrCl_4 crystals is controlled by the Jahn–Teller effect. The gap caused by this effect is small, but it has a noticeable influence on the intensity of a "hot" exciton–magnon absorption at low temperatures. Spectral studies give greater accuracy to the gap magnitude measurement than to neutronographic measurements. Thus, the temperature dependence of the intensity of "hot" exciton–magnon bands, in a low temperature range in compounds with prevailing ferromagnetic exchange, proves to be a very sensitive characteristic to estimate the character of the spin wave dispersion in a magnetic.

4.1.3. Impurity Modes

As was shown above, the peculiarities of the exciton–magnon absorption in magnetic crystals reflect the main singularities in the density of the magnon states of crystals. The introduction of magnetic impurities in magnetic materials causes new states, either resonance or localized, to appear in the spin-wave spectrum of a crystal, this phenomenon also being reflected in the fine structure of the optical spectrum. The introduction of an impurity results in the emergence of additional absorption bands. With an increase in the impurity concentration, the intensity of these bands rises. The absorption spectrum of a $KMnF_3$ crystal doped with Co^{2+} is given as an example [294]. C_4' and C_5 bands are additional bands in this spectrum (Fig. 4.6).

Temperature dependence of the new band frequencies are similar to temperature dependences of exciton–magnon bands in a pure crystal. It confirms the identification of additional bands as exciton–magnon bands which are

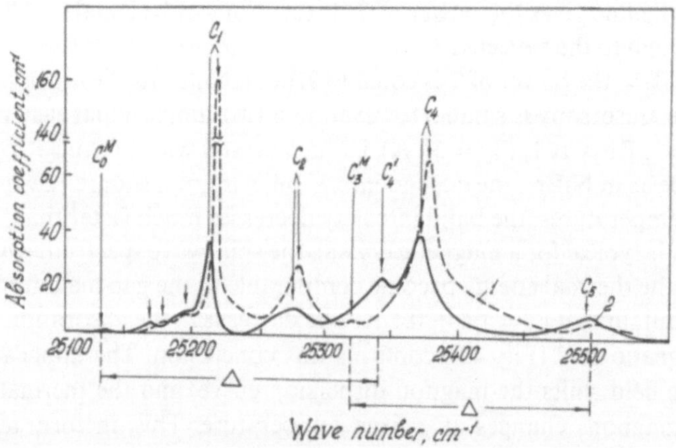

Fig. 4.6. Absorption spectrum of (1) $KMnF_3$ and (2) $KMn_{0.93}Co_{0.07}F_3$ in the region ${}^6A_{1g}({}^6S) - {}^4A_{1g}, {}^4E_g({}^4G)$ transitions, C_0 and C_3 are pure-exciton bands; C_1, C_2, and C_4 are exciton–magnon bands; C_4' and C_5 are additional (impurity) exciton–magnon bands [294].

related to the process at which excited in a crystal they are an exciton with the frequency close to that of an exciton in a pure crystal (at small concentrations of an impurity) and an additional spin-wave localized mode. The interval between pure exciton lines and their additional magnon sideband is 196 cm^{-1} at 20 K, which coincides within the accuracy of the experimental error with the frequency of a localized mode determined later in experiments on neutron scattering.

4.2. Magnetic Excitons as an Instrument to Investigate the Magnetic Configurations of Antiferromagnets

4.2.1. Equilibrium Structures

A possibility of investigating the magnetic configuration of crystals by studying their absorption light spectra stems from the dependence of the polarization properties of pure exciton and exciton–magnon transitions upon the orientation of the magnetic moments of ions, both with respect to the crystallographic axes and to one another. As an example of such study, we shall mention the detection of a spontaneous noncollinearity of sublattice magnetic moments in a $CsMnF_3$ crystal [295, 296], and the determination of the antiferromagnetism vector orientation in a $CoCO_3$ crystal [297].

4.2.1.1. $CsMnF_3$. A Weak Ferromagnet

In an antiferromagnet with a D_{6h}^4 structure, weak ferromagnetism can emerge only in the case of taking into consideration fourth-order invariants (cubic with respect to the vector of antiferromagnetism **l** and linear with respect to the vector of ferromagnetism **m**). It appears at a definite orientation of the antiferromagnetism vector in the basic plane, namely at $\mathbf{l} \parallel [100]$, and is oriented along the hexagonal axis. However, no such moment was detected at the first magnetic and resonance measurements of $CsMnF_3$ crystals. The spectral method proved to be more sensitive. Measurements were carried out in the region of the ${}^6A_{1g} \to {}^4A_{1g}, {}^4E_g$ transition. Two new electric dipole bands C_1'/C_2 (26,016 cm^{-1}/25,020 cm^{-1}) emerge, and gain strength in a magnetic field at frequencies of a pure-exciton doublet C_1'/C_2' connected with the excitation of the centrally nonsymmetrical ions. As was already noted in the preceding chapter, such behavior of the bands is connected with the exciton–magnon mechanism of the type shown in Fig. 3.2(f) (operator $S_{n\alpha}^z \sigma_{m\beta}^-$), which is efficient only in the case where the collinearity of the antiferromagnetic crystal magnetic structure is disturbed and there is no center of symmetry in the pair of interacting ions. The field dependence of the absorption coefficient is governed by the factor $\sin^2\theta\cos^2\theta$ which vanishes in a collinear antiferromagnet and in a saturated paramagnet, and reaches its maximum in the fields with the strength close to that of the exchange fields. In the case of complex magnetic systems, $CsMnF_3$ being one of them, this dependence gets more complicated as this antiferromagnet is not described with one exchange field. The crystal has two types of ions, Mn1 and Mn2, which are in nonequivalent positions not only from the point of view of local symmetry, but also concerning their nearest neighbors–magnetic ions, and magnitudes of exchange interactions in pairs Mn–Mn to this, the angles between the magnetic moments of ions are determined not with the help of the simple relationship $\cos\theta \approx H/2H_E$, but are derived from the following equations:

$$\begin{aligned} H_{E1\text{-}2}\sin(\theta_1 - \theta_2) - H\cos\theta_1 &= 0, \\ H_{E2\text{-}1}\sin(\theta_1 + \theta_2) + H_{E2\text{-}2}\sin 2\theta_2 - H\cos\theta_2 &= 0, \end{aligned} \tag{4.6}$$

where H_E are effective exchange fields on ions of type Mn_i created by neighbors of the ions of type Mn_j; θ_1 and θ_2 are angles of deflection of the magnetizations of sublattices of different types from the collinear state. Due to the nonequivalence of pairs of the nearest neighbors, the dipole moment is written as $P = P_1 \sin 2\theta_1 + P_2 \sin(\theta_1 + \theta_2)$ and the absorption intensity for the case $\mathbf{E} \perp z$ is described by the following expression [298]:

$$\mathscr{K}^{\perp} = \mathbf{E}^2 |P_{22}^y|^2 2S \cos^2 t \sin^2 2\theta_1, \tag{4.7}$$

where P_{22}^y is the projection of the dipole moment of the ionic pair, and t is the angle between the vector $\mathbf{E}$ and the direction of orientation of moments in the basic plane. When comparing experimental data, related to the dependence of the intensity of absorption bands upon the magnitude of the applied magnetic field with the theoretical relationship (4.7), we find perfect agreement which proves the interpretation of the bands C_1'/C_2' to be correct. Hence, the presence of the spontaneous noncollinearity of sublattices will lead to the emergence of the spontaneous nonzero intensity of the doublet C_1'/C_2'. The results of detailed measurements in fields up to 4 T shown in Figs. 4.7 and 4.8 confirm the presence of such noncollinearity. In a small strength field applied in the basic plane, a one-domain structure with the magnetic moment $\mathbf{M} \perp \mathbf{H}$ is

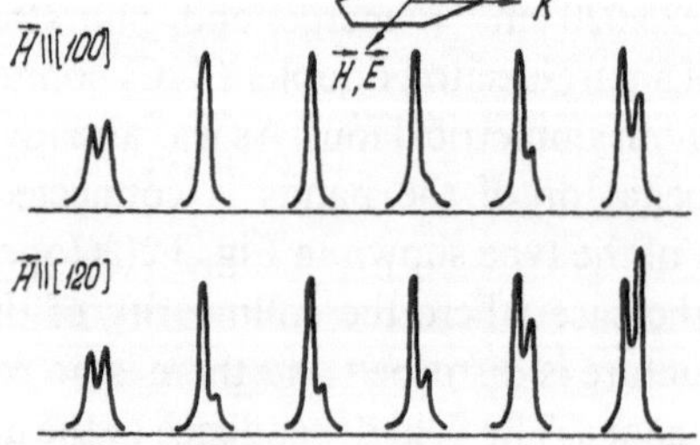

Fig. 4.7. $CsMnF_3$. Densitograms of absorption lines C_1 and C_2 at different orientations and strength of the external magnetic field [296].

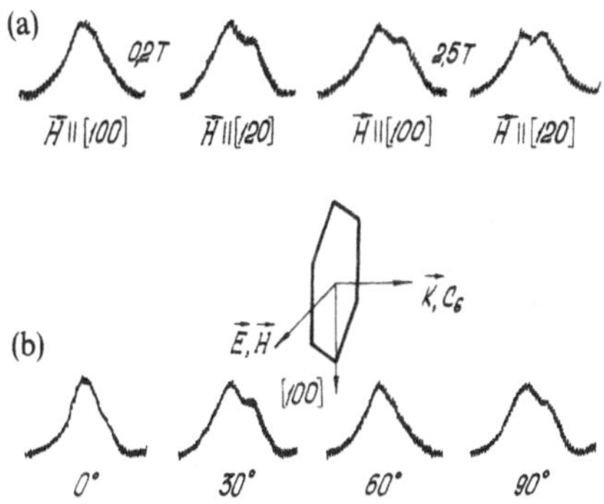

Fig. 4.8. $CsMnF_3$. Oscillograms of absorption lines C_1 and C_2' at different (a) magnitudes and directions of the magnetic field and (b) magnitudes and angles in the basic plane between directions of the field $H = 0.4$ T and the [100]-axis of the crystal [296].

reached, and from a doublet of bands there remains one of them (C_1 at $\mathbf{E} \perp \mathbf{M}$). With a subsequent rise of the field strength, a line C_2' appears and gains intensity in the spectrum. If $\mathbf{H} \parallel [100]$, the line intensity rises from zero, and if $\mathbf{H} \parallel [120]$ it rises from a certain initial value and remains more intensive than in the first case. The spectrum is reproduced with a period 60° (Fig. 4.8(b)). The intensity of C_2' in a weak field $\mathbf{H} \parallel [120]$ is the same as in case $H = 2.5$ T oriented along [100] (Fig. 4.8(a)), i.e., with the antiferromagnetism vector $\mathbf{l}$ being oriented along the [100] axis, the spontaneous noncollinearity of sublattices of centrally nonsymmetrical Mn2 ions is the same as if it is introduced by an external field $H = 2.5$ T at $\mathbf{l} \parallel [120]$. The cant angle θ_1 calculated by formulas (4.6) turns out to be 35°.

Investigations of the polarization properties of light absorption in the exciton doublet region of centrally nonsymmetrical Mn2 ions [299] showed that in $CsMnF_3$ crystals there is a weak uncompensated magnetic moment oriented along the hexagonal axis. The moment $\mathbf{m}_2$ caused by the noncollinearity of spins of centrally nonsymmetrical ions is not compensated for by the moment $\mathbf{m}_1$ caused by the noncollinearity of the spins of centrally symmetrical ions.

4.2.1.2. $CoCO_3$. Orientation of the Antiferromagnetism Vector

For a long time there was no unambiguous answer to the problem of orientation of the antiferromagnetism vector $\mathbf{l}$ in this compound, as the Co^{2+} ion in its ground state has an orbital magnetic moment together with a spin moment which created difficulties in interpreting data on the inelastic scattering of neutrons in $CoCO_3$. Due to spectral studies, information on the orientation of the antiferromagnetism vector in $CoCO_3$ was derived from the polarization dependence of the pure-excitons absorption of this crystal.

The object of the investigation was the magnetic dipole light absorption line corresponding to the transition ${}^4\Gamma_4({}^4F) \rightarrow {}^2\Gamma_4({}^2P)$. The probability of a transition from the ground state $|gr\rangle$ into an excited state φ^f is controlled by the magnitude of the square of the matrix element $P = \langle \varphi^f | h(L + 2S) | gr \rangle$. Omitting the calculation of wave functions φ^f and $|gr\rangle$, we shall present the results:

$$\begin{aligned} P_{\mathbf{h}\|x} &\sim (-i \sin \varphi + \cos \varphi \cos \theta), \\ P_{\mathbf{h}\|y} &\sim (-i \cos \varphi - \sin \varphi \cos \theta), \\ P_{\mathbf{h}\|z} &= 0, \end{aligned} \tag{4.8}$$

where θ, φ are Euler angles in the σ_{ik} transformation.

For definiteness, the y-axis was selected to be oriented along the direction $\mathbf{H} \perp C_3$. Angles θ determine the spin orientation of sublattices $\mathbf{S}_1$ and $\mathbf{S}_2$ with respect to the C_3-axis; it should be mentioned that the absence of magnetism along the C_3-axis means that $\cos \theta_1 = -\cos \theta_2$ or $\theta_2 = \theta_1 + \pi = \theta$ (indices 1 and 2 number the sublattices). The case $\theta = \pm\pi/2$ corresponds to the orientation of $\mathbf{S}_1$ and $\mathbf{S}_2$ in the basic plane, and the case $\theta = 0$ corresponds to the direction along C_3. The difference in angles φ_1 and φ_2 ($\varphi_1 = \pi/2 + \Delta\varphi$, $\varphi_2 = \pi/2 - \Delta\varphi$) determines the magnitude of the noncollinearity of the magnetic moments of sublattices: $\sin \Delta\varphi = (H + H_D)/2H_E$ (where the effective field of the Dzialoshinskii–Moriya interaction is $H_D = 3$ T, and the exchange interaction is $H_E = 16$ T). In low magnetic fields $\varphi_1 = \varphi_2 - \pi/2$, i.e., $\Delta\varphi = 0$. Due to the same reason, it is possible to neglect the exciton line Davydov splitting which is proportional to the square of the crystal magnetization. Then the integral absorption coefficient of the line will be the sum of the contributions from both the Davydov components:

$$K \sim [h_x^2 + h_y^2 \cos^2 \theta] = h^2[\sin^2 t + \cos^2 t \cos^2 \theta], \tag{4.9}$$

where t is the angle between directions $\mathbf{h}$ and $\mathbf{H}$. The expression obtained (4.9) reflects an unambiguous relationship between the dichroism character of the exciton line of the transition ${}^4\Gamma_4({}^4F) - {}^2\Gamma_4({}^2P)$ and the orientation of the antiferromagnetism vector with respect to C_3. In the case when $\mathbf{l} \perp C_3$, $\theta = \pi/2$, the polarization is linear ($\mathbf{h} \| x$, i.e., it is normal to $\mathbf{H}$); otherwise, the line polarization is elliptical (at $\theta = 0$ it is circular). It follows from (4.9) that the dependence of K upon $\sin^2 t$ is a straight line, the slope of which ($\cos^2 \theta$) depends upon the orientation of $\mathbf{l}$ with respect to C_3. Experimental results are given in Figs. 4.9 and 4.10. Under conditions of the zero external magnetic field, cobalt carbonate has a multidomain structure. The ferromagnetic moment of each of the domains is directed along one of the second-order axes. Isotropy in the basic plane of the exciton absorption in a zero field is caused by $CoCO_3$ being multidomain. An averaging by domains oriented along the second-order axes yields $K_{av} \sim (1 + \cos^2 \theta)$, i.e., the line intensity at $H = 0$ is equal to the half-sum of its values in a one-domain structure at $t = \pi/2$ and $t = 0$ (i.e., at $\mathbf{h} \perp \mathbf{H}$ and $\mathbf{h} \| \mathbf{H}$). This conclusion is in agreement with experimental data.

Shown in Fig. 4.10 are data of polarization measurements in a field $\mathbf{H} =$

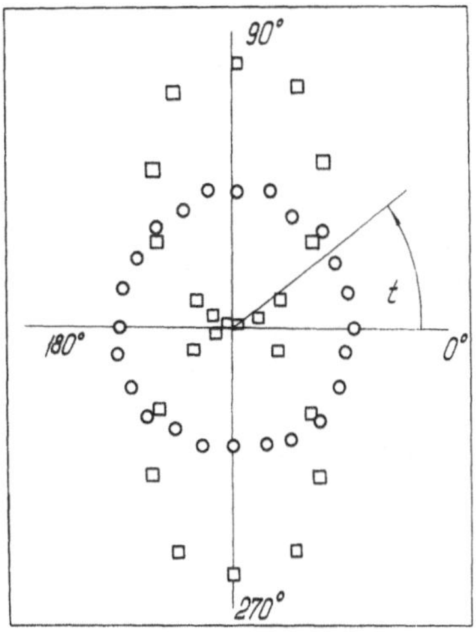

Fig. 4.9. $CoCO_3$. Dichroism of light absorption at 22,860 cm^{-1} frequency (t is the angle between the vector of the magnetic component of linearly polarized light h and the direction of the external magnetic field $\mathbf{H} \parallel C_2$): (o) $H = 0$; (□) $H = 0.1$ T [297].

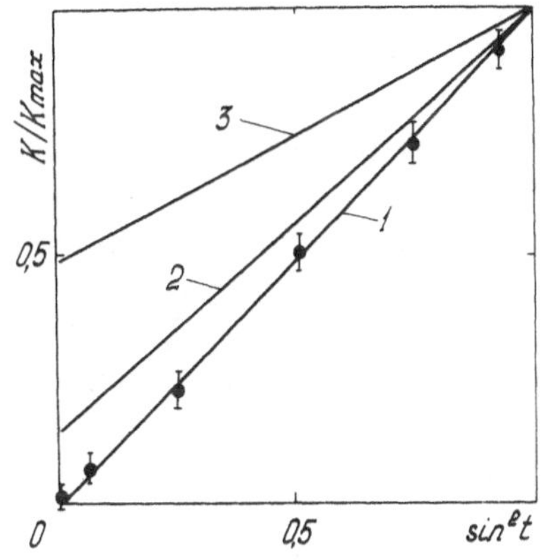

Fig. 4.10. $CoCO_3$. Absorption coefficient of the exciton line ($^4F - {}^2P$) versus angle t at different orientations of vector **l** with respect to C_3. Points are from the experiment, the solid line from calculation $\theta = (1)\, 90°, (2)\, 68°$ and $(3)\, 46°$ [297].

0.1 T (a one-domain specimen). The straight lines correspond to the following values of the angle $\theta = 90°$, $68°$, and $46°$, which were discussed in neutron scattering experiments. Experimental points of spectral measurements are located close to the line corresponding to $\theta = 90° \pm 3°$, confirming thereby that the antiferromagnetism vector in $CoCO_3$ is oriented in the basic plane.

4.2.2. Phase Transitions in an Antiferromagnet with Anisotropy of "Easy-Axis" Type

In antiferromagnets with anisotropy of "easy axis," with application of an external magnetic field which is parallel to the axis of magnetic ordering, the sublattice degeneracy is removed: ions of the sublattice with spins oriented along the field become energetically nonequivalent to ions of the other sublattice with spins oriented against the external magnetic field. In absorption, it results in the splitting of exciton and exciton–magnon bands. For exciton transitions the magnitude of splitting is described by the following expression:

$$\Delta E_e = \pm \mu_B H (g^f M_s^f - g^0 M_s^0). \tag{4.10}$$

For exciton–magnon bands

$$\Delta E_{e\text{-}m} = \pm \mu_B H (g^f M_s^f - g^0 M_s^0) \mp \mu_B H [g^0 (M_s^0 - 1) - g^0 M_s^0], \tag{4.11}$$

for transitions with $M_s^f = M_s^0 - 1$, $\Delta E_{e\text{-}m} = \pm (M_s^0 - 1)(g^f - g^0)\mu_B H$. In the case of compounds with Mn^{2+} ions, $\Delta E_e \simeq 2\mu_B H$ and $\Delta E_{e\text{-}m} \simeq 0$, since $g^f \approx$

g^0. It follows from expressions for the dispersion of exciton and spin-wave energies (Chapter 3) that the linear character of the splitting can be disturbed only by a transformation of the magnetic structure. A shift in the center of gravity of the doublet components which accompanies this transformation will be indicative of the collinearity of moments of the opposite sublattices being disturbed. This disturbance is connected both with the nonequivalence of the Zeeman energies of sublattices and with a change of the molecular field in the case of the "canting" of moments. All this enables us to study phase magnetic transitions by means of a spectral method. Since in antiferromagnets the intersublattice migration of magnetic excitons is hampered, the spectral method makes it possible to follow the behavior of each individual magnetic sublattice. This possibility is an obvious advantage of the spectral method.

4.2.2.1. Spin-Flop Transition

In the case of spin-flop in an antiferromagnet, to indicate such a phase transition it is sufficient to observe a qualitative change in the absorption spectrum of the antiferromagnet. It is even not necessary to concretize the fine structure of the absorption—the very fact that changes are observed in it is of prime importance: the spectra before and after the transition prove to be substantially different. Should a spatially nonuniform state of the crystal emerge during the transition, for example, an intermediate state with a periodic alteration of domains of the initial and spin-flop magnetic phases [300], the observed spectrum will be a superposition of the spectra of both the simultaneously coexisting phases.

MnF_2. In accordance with what was said above, in a magnetic field **H** which is strictly oriented along the C_4-axis of manganese fluoride, in the light absorption spectrum of the latter (transition ${}^6A_{1g} \to {}^4T_{1g}({}^4P)$), a linear splitting of the pure exciton line is observed. As for the field with strengthens above the critical value $H_c = 9.3$ T, here the splitting abruptly disappears as magnetic sublattices have already occupied energetically equivalent positions with respect to the magnetic field (they jump into a plane normal to **H**), i.e., there again emerges a sublattice degeneracy as in the case of no magnetic field. In a field **H**, tilted from the C_4-axis over the angle $\varphi_c > 30'$, the sublattices gradually turn and the splitting of the exciton line changes just as gradually (Fig. 4.11). With **H** being strictly oriented along C_4, spin-flop in MnF_2 takes place as a phase transition of the first order. This is confirmed by investigations in the region of the optical transition ${}^6A_{1g} \to {}^4T_{2g}({}^4D)$: first, the spectral position of the absorption lines does change jumplike—there is no absorption at intermediate frequencies (between the initial and final positions of the lines) [301]; second, in a narrow range of magnetic fields in the vicinity of $H_c(\Delta H = 0.25$ T) a coexistence of spectra of two phases is observed [302]. There arises the question of the position of the antiferromagnetism vector **l** in the crystal basic

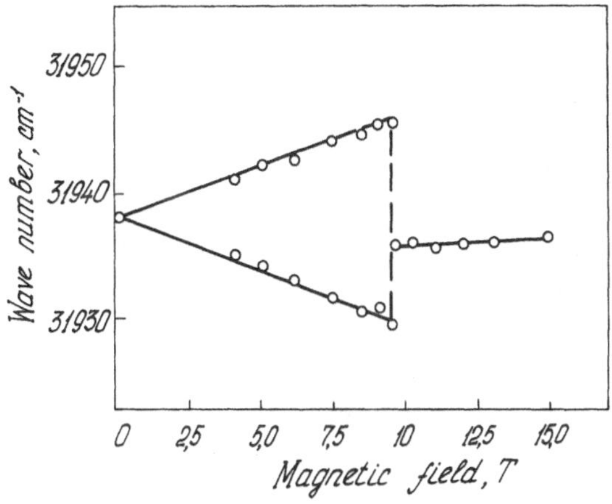

Fig. 4.11. Splitting of absorption band $\nu = 31{,}938\ \text{cm}^{-1}$ versus the external magnetic field strength, $T = 14$ K [303].

plane in the case of the first-order transition. The orientation of **l** at $H > H_c$ is governed by the sign of the constant f of the fourth-order anisotropy $fl_x^2 l_y^2$: the orientation $\mathbf{l} \parallel C_2$ ([100]) corresponds to the case $f > 0$, and the orientation $\mathbf{l} \parallel C_2'$([110]) corresponds to the case $f < 0$. However, it is impossible to determine the sign of the constant f from the data on the magnetization of MnF_2 because the magnitude of anisotropy in the basic plane is small and the spin-flop phase is divided into domains. A possibility of determining the orientation of **l** is provided by the spectral method: an exciton line splitting versus the **l** orientation relationship is detected for the $^6A_{1g} \rightarrow {}^4T_{2g}(^2P)$ transition (see Section 4.3 below). The line is split in the case where the antiferromagnetism vector component in the basic plane $\mathbf{l}_\perp$ is oriented along C_2'. There is no splitting if $\mathbf{l}_\perp \parallel C_2$. It is the absence of the line splitting in the case of a strict orientation of **H** along C_4 which signals the vector **l** in the spin-flop phase being oriented along the axis [100] [303].

$NiWO_4$. The spectral method was used to detect the spin-flop transition in $NiWO_4$ [304] (the spatial structure $P2/c$, $T_N = 67$ K)—a crystal which is described by a thermodynamic potential of a two-axis antiferromagnet. The axis of the sublattice spontaneous magnetization z does not coincide with any of the crystallographic axes, and lies in the plane ac normal to the crystal monoclinic axes $b(C_2)$ at the angle 15° with respect to the direction c. The object of investigation was a pure exciton absorption band of the transition $^3F \rightarrow {}^1G(\nu = 18{,}930\ \text{cm}^{-1})$ in an external magnetic field (Fig. 4.12) oriented in a different way in the ac-plane. An abrupt transition of the spectrum in the field $\mathbf{H} \parallel \mathbf{z}$ was observed at $H = 18$ T. The range of angles (between the directions of **z** and **H**) in which an abrupt transformation of the magnetic structure is observed is conspicuously broad. It makes up approximately 30° in the ac-plane. As for the deviation from this plane, it shall not exceed $\pm 1°$. Such

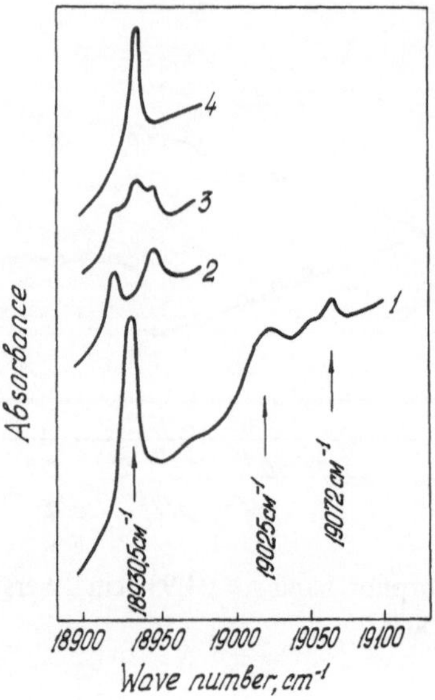

Fig. 4.12. Densitograms of a section of the $NiWO_4$ absorption spectrum in magnetic fields of different strength are: (1) 0 T, (2) 19.77 T, (3) 19.80 T, and (4) 19.83 T [304].

an anisotropy of the phase transition is caused by this crystal being a two-axis crystal. When the angles have the above-mentioned values the magnetic sublattices gradually turn, which is exactly what they are expected to do in the case of the second-order phase transition when the angle φ is above the critical value. The interval of an abrupt transformation of the magnetic structure does not exceed 0.03 T. Three lines are simultaneously observed in this interval in the spectrum, i.e., the absorption spectra of two phases are superimposed. Consequently, the phase transition goes through an intermediate state which proves that the phase transition in $NiWO_4$ in this case takes place as the first-order transition.

$CsMnCl_3 \cdot 2H_2O$. The absorption spectrum transformation during spin-flop proceeds in a somewhat different way in a quasi-one-dimensional $CsMnCl_3 \cdot 2H_2O$ [305], which is also a two-axis antiferromagnet with the spontaneous magnetization directed along the **b**-axis. The direction of magnetization is oriented in a spin-flop phase along the intermediate **C**-axis. So far, because of relatively small critical fields, attempts to resolve the doublet splitting of the absorption lines in the subcritical range of fields have failed. However, in $CsMnCl_3 \cdot 2H_2O$ during the spin-flop transition frequency shifts of absorption bands are observed; these shifts emerge due to the difference between the directions of the antiferromagnetism vector **l** in an antiferromagnetic phase ($l_{\parallel}$) and in a spin-flop phase ($l_{\perp}$). Figure 4.13 shows changes in the spectrum of $CsMnCl_3 \cdot 2H_2O$ in the region of the optical transition ${}^6A_{1g}({}^6S) \rightarrow {}^4T_2({}^4D)$. The first and last spectra are spectra of practically pure phases $l_{\parallel}$ and $l_{\perp}$,

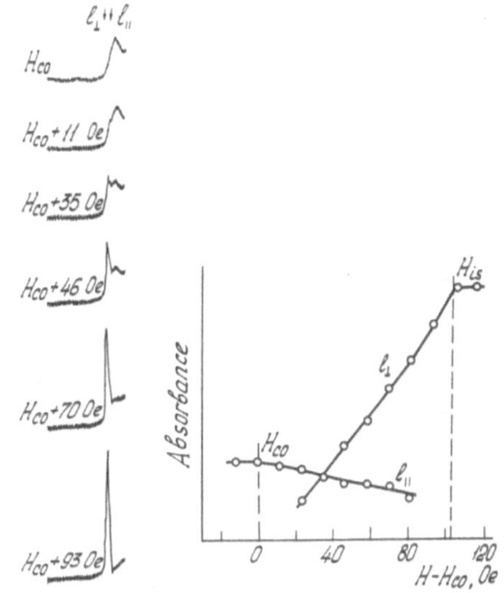

(a) (b)

Fig. 4.13. $CsMnCl_3 \cdot 2H_2O$. (a) Pseudosplitting of the 26,736 cm^{-1} exciton line of ${}^6A_{1g}({}^6S) - {}^4T_2({}^4D)$ transition at sin-flop $\mathbf{H} \parallel \mathbf{b}$. (b) Change of intensities of the absorption lines of $l_\parallel$ and $l_\perp$ phases in the region of the intermediate state [305].

and the intermediate pseudosplitting represents a simultaneous existence and gradual replacement of the initial phase spectrum with that of the spin-flop phase. Proceeding from the maximum intensity of the bands, it is possible to determine (Fig. 4.13(b)) the interval of the magnetic fields in which an intermediate state exists ($\Delta H_{is} = 8.5 \cdot 10^{-3}$ T). With an increase in the angle ψ (between the directions **b** and **H** in the bc-plane) a gradual decrease of the doublet splitting is observed which reflects a reduction in the orientation jump of the vector **l**. The moment of the doublet disappearance ($\psi \simeq 15'$) determines the angular boundary of the region of the intermediate state existence, i.e., the critical angle ψ_c(sin $2\psi_c = \rho/\delta$, where ρ is the magnetic anisotropy constant of the term $(M_{1y}^2 + M_{2y}^2)/2$ in the expression for the energy density of a two-axis two-sublattice antiferromagnet, and δ is the exchange constant). At angles $\psi \geq \psi_c$, the phase transition proceeds by means of a smooth rotation of the antiferromagnetism vector. There is no coexistence of the spectra of two phases in the spectrum, and the transformation of a spectrum of one type into the other is taking place continuously (Fig. 4.14).

Fe_3BO_6. Of special convenience is the spectral method of investigating phase transitions in multi-sublattice systems, as it enables one to study the motion of each individual sublattice which was demonstrated [306] when studying a weak ferromagnetic orthorhombic Fe_3BO_6 crystal ($T_N = 508$ K). At temperatures less than 415 K, the spins of iron ions in this crystal are oriented along

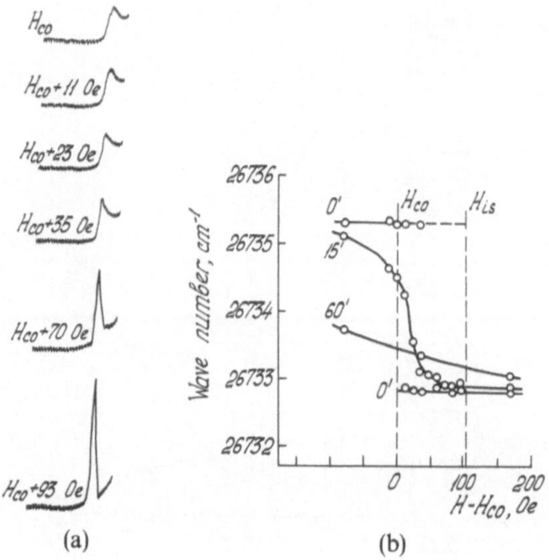

Fig. 4.14. $CsMnCl_3 \cdot 2H_2O$. (a) Change of exciton line frequency observed in the region of spin-flop transition in the case of exact orientation of the magnetic field along $b(\varphi = 0)$ and in the case of the tilted magnetic field ($\varphi > \varphi_c$). (b) Continuous transition of the $l_\parallel$ phase spectrum into the $l_\perp$ phase spectrum at $\psi = 15'$ [305].

the [001] direction. In the (001) planes the interaction is ferromagnetic, and between the planes it is antiferromagnetic. In an elementary cell there are twelve iron ions, eight of which have octahedral surrounding of one type (Fe(I)), and the remaining four of the other type (Fe(II)), these types differing from each other by the site symmetry. Each type of ions of iron forms its own sublattice. In a magnetic field **H** || [001] a spin-flop transition takes place ($H_c = 15.7$ T). At $H > H_c$, the crystal magnetization is directed along [001]. The difference of the crystalline field on ions results in Fe(I) and Fe(II) being optically excited at somewhat different frequencies. To research was carried out in the region of an ${}^6A_{1g} \to {}^4T_{1g}$ transition. The intersublattice exchange in the crystal is small, so excitons of each type migrate in a magnetic sublattice of a definite spin orientation (at any rate, the splitting of excitons controlled by the intersublattice transition energy is less than 0.3 cm^{-1} in a zero field).

In **H** fields which are almost parallel to [001] (Figs. 4.15 and 4.16) the following pattern, characteristic of a spin-flop transition, is observed in the absorption spectrum: first both exciton lines are split into doublets, and then the splitting is flopped. An analysis of the deviation of the splitting from linearity and of the shift of the doublet center of gravity shows that at 4.2 K the spin re-orientation is a continuous process. During this process, a residual splitting is observed for one of the absorption bands (exciton B), whereas for the second band (exciton C) there is no splitting. Such behavior of the absorption bands is indicative of the different orientation of the ferromagnetic moment of sublattices of two different types: for a B sublattice this orientation is

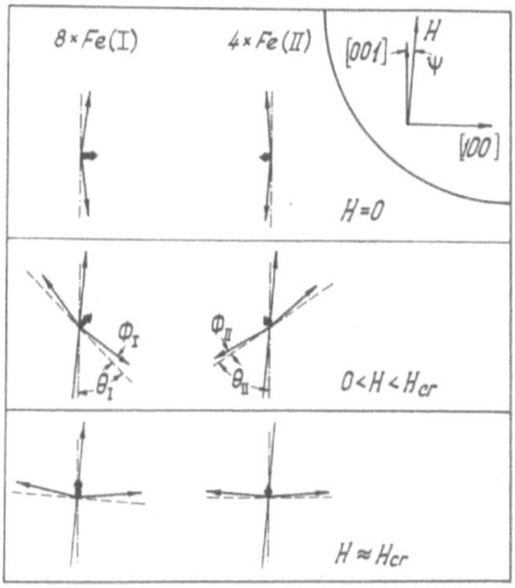

Fig. 4.15. Fe_3BO_6. Schematic diagram of a spin structure for $T = 416$ K in magnetic fields of increasing strength, Φ_I, Φ_{II}, θ_I, θ_{II}, and ψ are angles of spin canting, spin rotation, and magnetic field tilting, respectively [306].

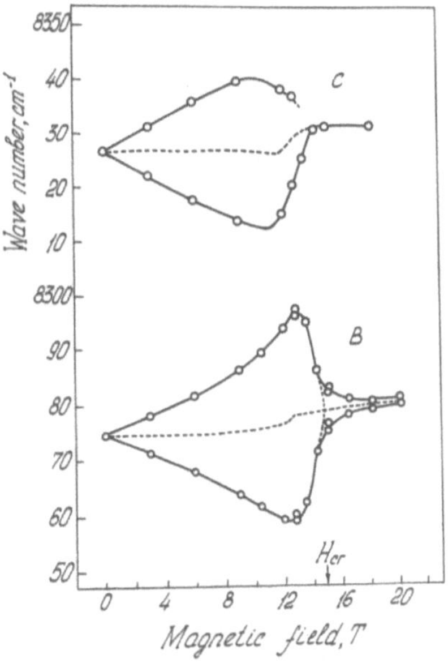

Fig. 4.16. Frequencies of lines B and C versus a magnetic field in the case $\mathbf{H} \parallel [001] \pm 5°$ [306].

practically along the direction [001] (because of the anisotropy), and for a C sublattice it is along **H**. The presence of the B exciton band in all magnetic fields implies that the ferromagnetic moment $\mathbf{m}_B$ connected with this splitting turns clockwise during the phase transition under the geometrical conditions shown in Fig. 4.15, whereas the C exciton ferromagnetic moment $\mathbf{m}_c$ turns counterclockwise. This fact, in its turn, makes it possible to state that ferromagnetic moments of ions Fe(I) and Fe(II) are oriented along the direction [100] antiparallelly. This statement is also in agreement with a shift in the center of gravity of B and C exciton bands into the opposite sides in fields $\mathbf{H} \parallel [100]$. This shift, in the case of Fe_3BO_6, is mainly controlled by the Zeeman term. For fields $H > H_c$, angles Φ (Fig. 4.15) show a weak dependence on the field. A linear dependence of the shift in the center of gravity (Fig. 4.16) is indicative of this fact. On the basis of the data on the splitting of lines in a field $\mathbf{H} \parallel [001]$ and on their shift at $\mathbf{H} \parallel [100]$, the angles were determined to be equal to 7.5° and −4.7° for ions in the Fe(I) sublattice and in the Fe(II) sublattice, respectively.

4.2.2.2. Phase Transition into a Noncollinear Magnetic Phase

A two-sublattice one-axis tetragonal antiferromagnet in the case of the Dzialoshinskii–Moriya interaction is known [154] to have its magnetic

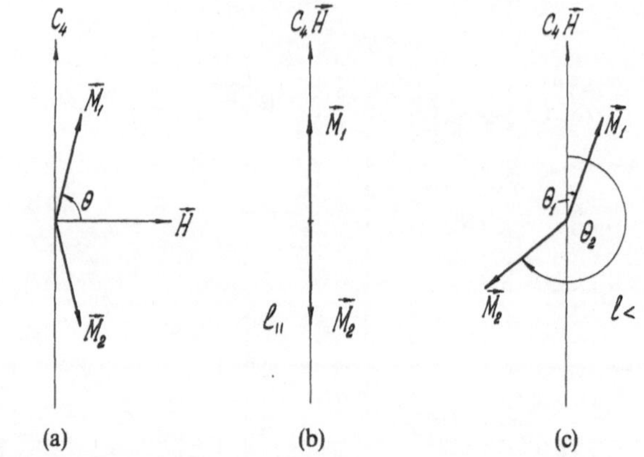

Fig. 4.17. Equilibrium configurations of the moments of sublattices of a two-sublattice tetragonal antiferromagnetic at different orientations of the external magnetic field.

structure transformed in an external magnetic field **H** oriented along the easy C_4-axis. This transformation differs from a spin-flop: the initial configuration $l_\parallel$ with the antiferromagnetism vector $\mathbf{l} \parallel C_4$ becomes a configuration $l_<$ at which the magnetic moments of the opposite sublattices are oriented at different angles with respect to the C_4-axis (Fig. 4.17).

The spectral study results of a tetragonal antiferromagnet FeF_2 (crystallographic structure D_{4h}^{14}, $T_N = 78.4$ K) in the fields up to 30 T are presented in Fig. 4.18 [155]. Light absorption was studied in the region of a ${}^5T_{2g} \to {}^3T_{1g}$ transition. It is to be seen from the figure that in fields $\mathbf{H} \parallel C_4$ with strength above 22 T, the linearity of the splitting is disturbed and the doublet center of gravity is shifted into the long-wave region, i.e., the magnetic structure is transformed into a noncollinear phase. The phase transition from a configuration $l_\parallel$ to $l_<$ can be the transition of both the first and second order, depend-

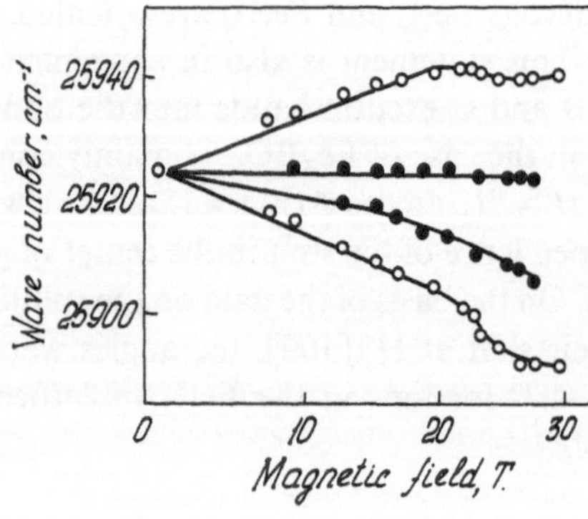

Fig. 4.18. Splitting of the exciton–magnon absorption band $\nu = 25{,}924$ cm^{-1} in the external magnetic field, $T = 14$ K, (o) $\mathbf{H} \parallel C_4$, (•) $\mathbf{H} \perp C_4$ [155].

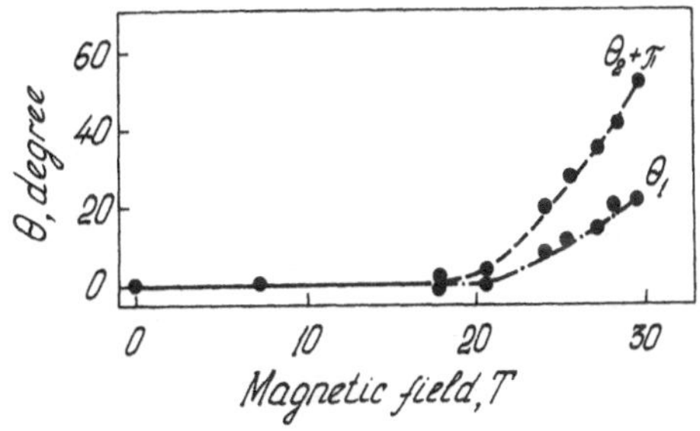

Fig. 4.19. FeF_2. Orientation of the magnetic moments of sublattices ($\mathbf{M}_1$ and $\mathbf{M}_2$) versus external magnetic field strength, $\mathbf{H} \parallel C_4$, $T = 14$K [155].

ing upon the relationship between the value H_D (effective field Dzialoshinskii–Moriya interaction) and parameters H_E and H_A. In the case $H_D < H_D^{cr}$, the first-order phase transition takes place, in the case $H_D > H_D^{cr}$ the transition is of second order. The magnitude of H_D^{cr} is estimated to be equal to $H_D^{cr} \approx \frac{1}{2}(H_C H_A)/(H_E + H_A)$. Using the values of H_A and H_E which are known for FeF_2, viz. $H_A = 20$ T and $H_E = 54$ T, we find that $H_C = 45$ T and $H_D^{cr} \approx 6$ T. It can be concluded from the data of a spectral experiment (Fig. 4.18) that, probably, the transition in FeF_2 is the second-order phase transition.

To determine the equilibrium configuration of magnetic moments at $H > 22$ T, one has to consider the energy dispersions of exciton states and spin waves, in which the nonequivalence of the Zeeman energies of ions from the opposite sublattices are taken into account. These dispersions contain parameters which characterize statical and resonance interactions among magnetic ions (see Section 4.1). These parameters are determined from the experimental data on absorption band splitting in fields $H \perp C_4$ [228]. Shown in Fig. 4.19 is a numeric estimate of a change in the magnetic moment orientations of sublattices FeF_2 at the phase transition $l_{\parallel} \to l_{\langle}$. For example, in a field 33 T, $\theta_1 = 12°$, and $\theta_2 = -27°$. This configuration corresponds to the value of the magnetization projection on the C_4-axis equal to 0.08 M_0, where M_0 is the full moment of the sublattice. Measurements of magnetic susceptibility also confirmed the presence of such a phase transition in FeF_2.

4.2.2.3. Metamagnetic Phase Transition

Characteristic changes in the absorption spectrum of an antiferromagnet in the case of a metamagnetic phase transition are, on the whole, similar to the changes typical for a spin-flop transition: a linear doublet splitting of light absorption lines before the transition, and unsplit lines after the transition. There are, however, certain difference in the absorption spectrum. For example, in the case of a spin-flop the absorption line frequency becomes close to that of the line in a zero field, whereas, in the case of a metamagnetic transition, the line frequency is close to the frequency of that component of the split doublet which corresponds to light absorption in the sublattice with a mag-

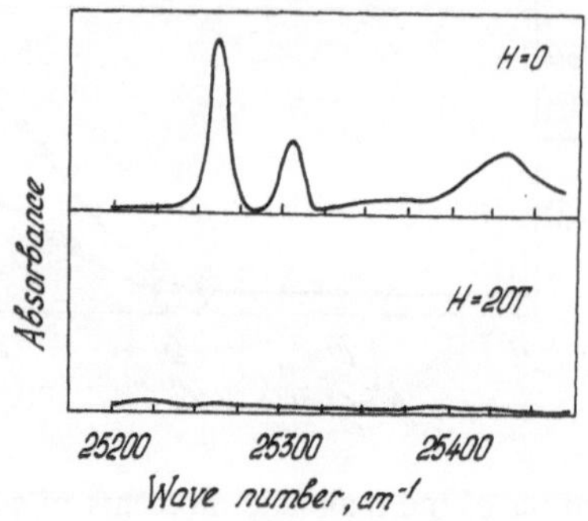

Fig. 4.20. $FeCO_3$. Decreasing of exciton–magnon bands of light absorption in a strong magnetic field, $T = 4.2$ K [225].

netic moment oriented along the field. These changes in the spectral positions of absorption lines are related to the Zeeman energies of magnetic ions in an external magnetic field before and after transitions. Besides, some changes in the spectral position of the lines can be caused by changes in the magnetic ion interactions, both with one another and with the ligands surrounding them. Comparing the two phase transitions under consideration, one can notice an even more substantial difference in "cold" exciton–magnon band intensities: in the case of spin-flop the intensity of bands remains practically unchanged, whereas in the case of a metamagnetic transition their intensity drops practically to zero (Fig. 4.20). The above listed peculiarities observed in the absorption spectra of an antiferromagnet enable an easy identification of a metamagnetic phase transition. This transition was studied by the spectral method, for example, in antiferromagnetic ferrous carbonate [307]. For this crystal, it was found that there is a complete correlation between its spectral and magnetic properties. A phase transition in $FeCO_3$ in pulsed magnetic fields has a number of peculiarities: the phase transition region covers 3 T (14.8–16.8 T) which by far exceeds demagnetization fields; at the beginning and at the end of the transition, magnetization changes with a jump, and in intermediate fields this change is gradual; the transition is accompanied by magnetocaloric effects. All these peculiarities are reflected in the pure-exciton and exciton–magnon absorption of light.

4.3. Photoinduced Magnetic Effects

If a crystal magnetic ion is photoexcited to a high energy state (this excitation being accompanied by a considerable change in its magnetic moment, the symmetry of the state, the parameters of exchange interactions or magnetic anisotropy), then it is sometimes convenient to consider this ion as a non-equilibrium magnetic substitutional impurity. It is this approach that makes it possible to explain a number of photoinduced magnetic effects.

4.3.1. $ErCrO_3$. Spin-Reorientation Phase Transitions

In $ErCrO_3$, as in a number of other rare-earth orthochromites, the direction of the light axis of magnetization is very sensitive to the action of external parameters—the temperature or magnetic field. A spin-reorientation phase transition is controlled by all types of interactions in a crystal (Cr–Cr, RE–RE, RE–Cr; here RE means "rare earth"). Therefore, any magnetic substitutional impurity affects the critical temperature of a phase transition in $ErCrO_3$. The temperature of a spin-reorientation phase transition in $ErCrO_3$ is $T_c = 9.4$ K. A field-induced phase transition is a first-order transition, and the coexistence of magnetic phases is observed in fields from 0.08 to 0.22 T. In an optical spectrum, this manifests itself in the coexistence of two absorption bands with wavelengths 804.4 and 805.0 nm [308], which correspond to a ${}^4\mathscr{I}_{15/2} \to {}^4\mathscr{I}_{9/2}$ transition in the Er system (the first band is typical of a low-temperature phase, and the second of a high-temperature phase) (Fig. 4.21).

If an $ErCrO_3$ crystal is illuminated by a laser pulse at the wavelength $\lambda = 798.3$ nm, the crystal goes completely into a high-temperature phase and then gradually relaxes to its initial state. The development of the phase transition in time is shown in Fig. 4.22. The time constant of the phase transition from a low-temperature phase into a high-temperature phase is about 6 μs. Dynamics of a spectral change in the case of laser pumping (a shift and broad-

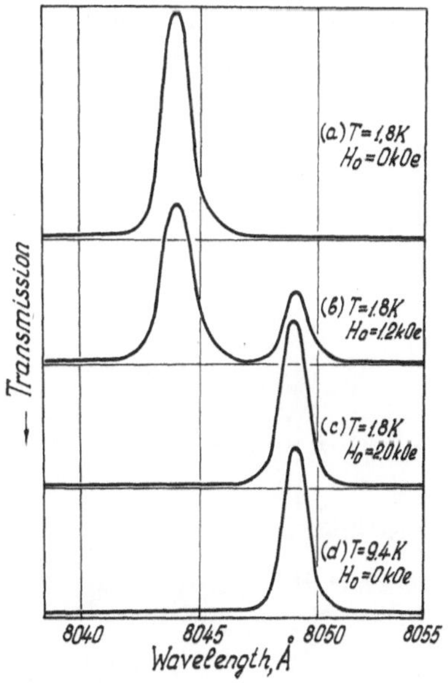

Fig. 4.21. $ErCrO_3$. Change of Er^{3+} absorption lines (${}^4J_{15/2} - {}^4J_{9/2}$ transition) with temperature change and magnetic field change [308].

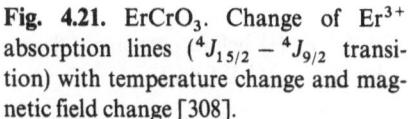

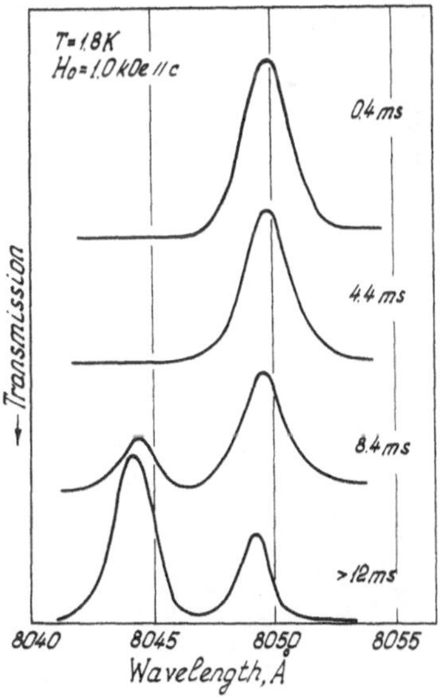

Fig. 4.22. $ErCrO_3$. Changes of the Er^{3+} absorption spectrum induced by a laser pulse at 798.3 nm with duration 20 ms [308].

ening of absorption bands) exclude the thermometric nature of the photo-induced effect observed, and the only reasonable explanation of this effect is as follows: the phase transition is affected by photoinduced ions as if they were a magnetic impurity, as a change of the magnetic moment of an ion Er^{3+} in a $ErCrO_3$ crystal is rather great in the case of the ion photoexcitation.

4.3.2. $EuCrO_3$. Origin of a New Magnetically Ordered Phase

A photoinduced magnetic effect of another type was observed in a $EuCrO_3$ crystal [309] whose ions Eu^{3+} are nonmagnetic in the ground state $^7F_0$. The absorption spectrum of $EuCrO_3$ in the region of the electron-vibration $^4A_2 \rightarrow {}^4T_2$ band of a chrome subsystem considerably differs from the spectrum of other orthochromites whose rare-earth ions are magnetic: in the case of $EuCrO_3$, a "giant" red shift of the absorption edge of this band is observed (Fig. 4.23). This phenomenon is explained as follows: the phonon energy in $EuCrO_3$ turns out to be close to that of the lowermost magnetic state $^7F_1$ of Eu^{3+} ions, a possibility being thereby created of the excitation of a complex consisting of a Cr exciton surrounded by a "mantle" of optical phonons and low-excited ions Eu^{3+}. The energy of a Cr–Eu exciton is decreased with respect to that of a Cr exciton typical for orthochromites with magnetic rare-earth ions, at the expense of the reduction of the energy of the exchange bond among Eu^{3+} ions belonging to this complex. The exchange bond among them is realized via Cr ions (an f–d–f exchange). In the case of a powerful optical pumping, the reduction of energy in Eu ions inside the clusters coupled by the f–d–f exchange can reach the magnitude of the difference of energies of states

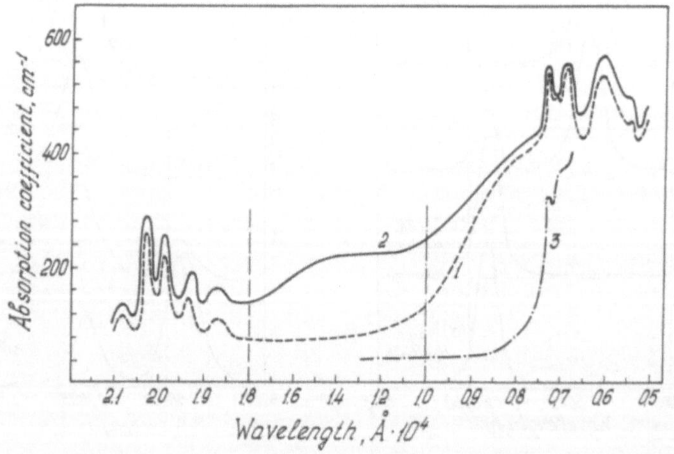

Fig. 4.23. $EuCrO_3$. Absorption spectra at $T = 4.2$ K: (1) before pumping; (2) after optical pumping; (3) $YCrO_3$ absorption spectrum at $T = 245$ K [309].

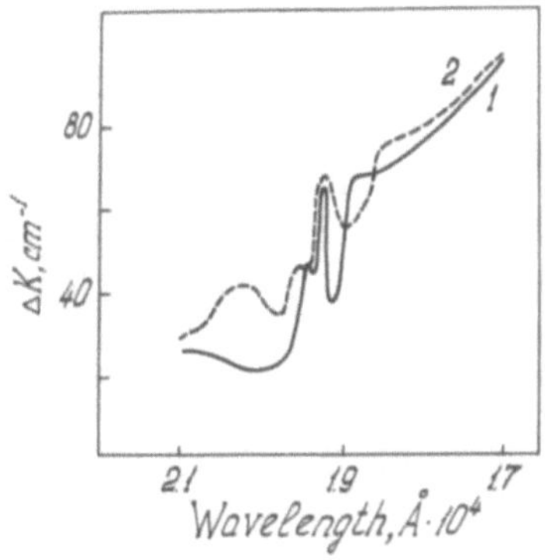

Fig. 4.24. $EuCrO_3$. Difference of absorption coefficients before and after pumping in the region 2100–1700 nm: (1) specimen $2 \times 2 \times \ \times 0.18$ nm, maximum pumping; (2) specimen $2 \times 2 \times 0.9$ mm, less powerful pumping [309].

$^7F_1$ and $^7F_0$, and the crystal can go over to a stable state with a magnetically ordered Eu subsystem. The maximum reduction of energy at the expense of an f–d–f exchange takes place in the case of an antiferromagnetic ordering of Eu^{3+} ions. Such behavior of an $EuCrO_3$ crystal in the case of optical pumping at the wavelength $\lambda = 532$ nm was registered both with the help of magnetic measurements and spectral investigations.

Since Eu ions inside macroscopic domains of the new phase have to be at the energy level $^7F_1^*$ shifted down with respect to the level $^7F_1$ by the magnitude $7b_0$ (where b_0 is the energy of the f–d–f exchange), there must be new lines $^7F_1^* \to {}^7F_6$ (Fig. 4.24) in the absorption spectrum instead of, for example, lines $^7F_0 \to {}^7F_6$ (a region about 2000 nm). The new lines are shifted with respect to the initial ones by the magnitude $\Delta = 7b_0 - \varepsilon = 250\ \mathrm{cm}^{-1}$. As the optical pumping power used was not adequate to bring the whole crystal to a new phase, the absorption spectrum of a $EuCrO_3$ crystal is practically a mixture of spectra of different phase, and the lines of the $^7F_0 \to {}^7F_6$ transition keep being observed after the pumping too.

The presence of a new phase in the crystal has an impact on other regions of the $EuCrO_3$ absorption spectrum as well. The $^7F_0$ state of the ions at the interface boundaries is excited with respect to the $^7F_1^*$ state, "hot" processes thereby becoming possible: during excitation of a Cr–Eu exciton, there is a simultaneous transition of Eu^{3+} ions from the $^7F_0$ state into the $^7F_1^*$ state. This results in an additional red shift (after the one induced by optical pumping) of the edge of the absorption band $^4T_2$. As to the long-life metastable clusters of positive energy with respect to $^7F_0$, their presence in the vicinity of interface boundaries makes possible the excitation of a Cr–Eu exciton with simultaneous relaxation of these clusters into the $^7F_1^*$ state and, consequently, enables the appearance of a new absorption band shifted with respect to the edge of the $^4T_2$ absorption band on the magnitude equal to $8[(\varepsilon - 4b_0) + \Delta(^7F_0 \to {}^7F_1^*) \simeq 3000\ \text{cm}^{-1}$ (for clusters with the longest lifetime, $n = 8$) (Fig. 4.23).

The effects, discussed above, in a $EuCrO_3$ absorption spectrum, which emerge during optical pumping, are stable and disappear only in the case when the crystal is heated up to 120 K when the longrange magnetic order is destroyed.

4.3.3. MnF_2. Photoinduced Single-Ion Magnetic Anisotropy

In the first two examples of photomagnetic effects, a change in exchange interactions of powerful optical pumping was sufficient. However, an essential difference in the magnetic behavior of an excited ion and the matrix of a crystal is possible in special cases, even with noninteracting photoinduced ions under conditions of the usual linear spectral investigations. Indeed, if the ground state of a magnetic ion is totally symmetrical (e.g., a $^6A_{1g}(^6S)$ state of the Mn^{2+} ion), and the photoinduced state is a lower symmetrical one, then this photoinduced ion becomes much more "sensitive" to the crystallographic structure. The role of some terms in its Hamiltonian can noticeably grow, especially in magnetic fields which lower the symmetry of the antiferromagnet and lead to the phase transition.

Antiferromagnetic manganese fluoride (space group of symmetry D_{4h}^{14}, $T_N = 68$ K) is exactly the crystal which has the above-mentioned peculiarities. The spin orientation of a photoexcited ion (a substitutional impurity) is governed both by intra-ion interactions and the fields created by surrounding ions. In the first approximation, the reverse disorientating influence of the photoexcited ion on its nearest neighbor can be neglected.

The Hamiltonian of a photoinduced ion can be written as follows

$$\mathscr{H}_a = \mathscr{I}^f \sum_{i=1}^{g} (\mathbf{S}_{\alpha' i}^0 \mathbf{S}_\alpha^f) + \mathscr{I}^{f\zeta} \sum_{i=1}^{z} \mathbf{S}_{\alpha' i} \mathbf{S}_\alpha^{f\xi} + B(\mathbf{S}_\alpha^\xi)^2$$
$$+ (-1)^{\alpha+1} 2D \mathbf{S}_\alpha^f \mathbf{S}_\alpha^{f\eta} - \mu_B g^f \mathbf{H} \mathbf{S}_\alpha^f. \qquad (4.12)$$

Here, α and α' are numbers of the sublattices ($\alpha, \alpha' = 1, 2; \alpha \neq \alpha'$). $\mathbf{S}_\alpha^f$ is the spin of the photoexcited ion, and $\mathbf{S}_{\alpha' i}^0$ is the spin of its nearest unexcited ith neighbor from the opposite sublattice, the number of these neighbors in the case of the crystallographic structure of rutile is eight; ξ, η, and ζ are three mutually perpendicular axes, the ζ-axis coinciding with the fourth-order crystallographic axis C_4 and the ξ-axis being selected along the second-order axis C_2. The first term in the sum (4.12) corresponds to the isotropic exchange interaction of the excited ion with unexcited ions, the second term corresponds to the anisotropic exchange interaction, the third term describes the single-ion one-axis anisotropy, and the fourth term describes anisotropy in the basic plane; the signs "+" and "−" before these terms reflect the fact that the structures of the ligands in the F–Mn–F complex which belong to ions from the opposite sublattices are turned over 90°; the last term of the sum (4.12) describes the Zeeman interaction of an excited ion with an external magnetic field **H**. The positive sign of the sum $\mathcal{J}^f + \mathcal{J}^{f\zeta}$ corresponds to the antiferromagnetic ordering of an excited ion with respect to its nearest neighbors in the subcritical region ($H < H_C$). An analysis of the equilibrium orientation of the magnetic moments of photoinduced ions shows that if in the subcritical region of fields ($H < H_C = 9.5$ T) the difference of ion energies from the opposite sublattices has a purely Zeeman character, then at $H > H_C$ the difference in energies for Mn^{2+} ions in a low-symmetrical photoinduced state can be caused by a noticeable anisotropy in the basic plane, since in the case of rutile structures the easy directions in the basic plane for spins from the opposite sublattices are perpendicular to one another. The presence of such a structure brings to noticeable values of anisotropy phenomenological constants in the basic plane for highly symmetrical states of a magnetic ion (and first of all, for the ground state).

The orientation of the antiferromagnetism vector **l** along the crystallographic axis [110] corresponds to the minimum energy of anisotropy for ions from one sublattice and to the maximum energy of anisotropy for ions from the other sublattice, the opposite one. In case **l** is oriented along [100], ions from the opposite sublattices become equivalent from the energy point of view, and their moments are oriented at equal angles with respect to the C_4-axis. Their components $\mathbf{S}^f_\perp$ deviate in the basic plane from the direction [100] through the angles about $2DS^f/\mathcal{J}^f zS^0$ which are opposite in sign and equal in value (Fig. 4.25). In the second extreme case when $\mathbf{l} \parallel [110]$, the vectors $\mathbf{S}^f_\perp$ of the moments from the opposite sublattices turn out to be antiparallel and oriented along the same axis [110]. With respect to the C_4-axis the spins $\mathbf{S}^f$ are oriented at different angles. Such orientation corresponds to the maximum difference of energies of ions. A contribution to the difference of energies is made by the term $D(S^f)^2(\sin^2\theta_1 + \sin^2\theta_2)$ where θ_1 and θ_2 are angles between the C_4-axis and the spins of photoexcited ions from the first and second sublattices, respectively. Hence, both in the first and second cases there is a difference in the orientation of spins of excited and unexcited states, and this difference turns out, mainly, to be proportional to the ratio of the anisotropy energy in the basic plane and the energy of exchange between an excited ion and its unexcited neighbors. Therefore, the photoinduced ions brought to a low-symmetry state have to behave in MnF_2 from a magnetization point of view as a typical substitutional impurity.

Such a photoinduced anisotropy of Mn^{2+} ions was found in the study of an exciton line of a $^6A_{1g} \rightarrow {}^4T_{1g}(^4P)$ transition [310] (Fig. 4.25). Indeed, as follows from the above consideration, the observed anisotropy in the splitting of this line at $H > H_C$ is indicative of an anisotropy of an excited state in the basic plane. This effect is connected with an excited state, not with the ground one. A proof of this is the experimentally established fact that no such anisotropy is observed for the lines of an $^6A_{1g} \rightarrow {}^4T_{1g}(^4G)$ transition. The investigation was carried out in a magnetic field which was not strictly oriented along

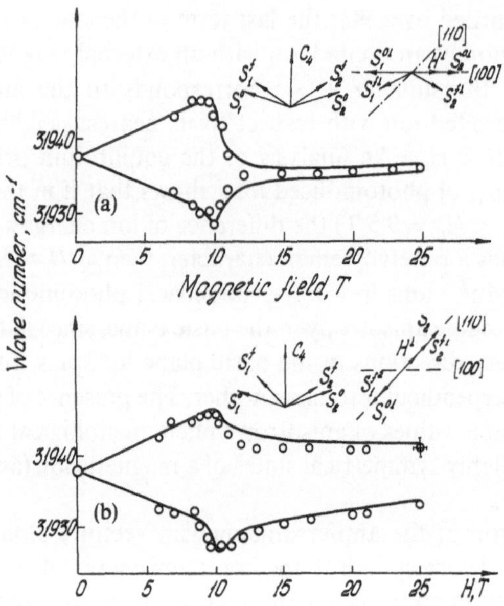

Fig. 4.25. Splitting of exciton absorption line $\nu = 31,\ 938\ \text{cm}^{-1}$ $({}^6A_{1g} - {}^4T_{1g}({}^4P))$, $\psi = 3°$: (a) $H_\perp \parallel [100]$; (b) $H_\perp \parallel [100]$. Solid line is calculation. Shown in inserts is the orientation of magnetic moments of ions in ground and excited states with respect to the C_4-axis and components of moment in the basic plane [310].

the C_4-axis, but was tilted to it at a small angle. The small tilt of the field ($\psi \approx 3°$) in the experiment was meant to fulfil the only function of setting the direction to the projection of the antiferromagnetism vector $\mathbf{l}_\perp$ in the basic plane in fields $H > H_C$ ($\mathbf{l}_\perp \parallel \mathbf{H}_\perp$).

The dependence of the line splitting upon the orientation of $\mathbf{H}_\perp$ corresponds to the tetragonal symmetry of MnF_2 and confirms the dominant role of anisotropy in the basic plane. Comparing the experimentally observed splitting with the results of calculation, one can obtain the exchange integral $\mathcal{J}^f \simeq 1\ \text{cm}^{-1}$ and the parameter of anisotropy in the basic plane $D = 3.33$ cm^{-1}. A considerable difference (up to 20°) in the spin orientations of the excited and ground states corresponds to this value of Dzialoshinskii–Moriya photoinduced anisotropy.

Chapters 3 and 4 cover practically all aspects of the spectroscopy of optical magnetic excitons, the only exception being the problems related to light scattering on magnetic dielectrics. Light scattering as well as its absorption and luminescence make it possible to obtain important and comprehensive information about magnetic exciton properties. However, since it is a large and, in many respects, independent branch of spectroscopy, it should be considered elsewhere.

Among the problems which require further development we would attribute the nonlinear spectroscopy of magnetism. We mean, first, the necessity to develop a theory to describe the multiquantum excitation of two-particle transitions in antiferromagnets both in the field of coherent and random radiation sources. Second, of considerable interest is the experimental study of dynamics of bound two-exciton states with positive and negative bond energy, as well as of a bound exciton–magnon state. These states can be created both at one-quantum and multiquantum excitation, therefore, both visible light and infrared lasers can be used to create them. The problem of creating a considerable nonequilibrium population of magnons at the boundary of the Brillouin zone under the conditions of a power optical pumping at the frequencies of exciton–magnon transitions also remains urgent today.

Among the problems related to the dynamics of magnetic excitons that require attention, we would mention the investigation of crystals in which broad zones (i.e., strong resonance interaction) correspond to luminescence actual excitons. In this case, it is possible to observe a coherent motion of excitons. Nickel compounds, in which intensive zero-phonon lines are observed, belong to eventual objects of such investigations.

Further experimental and theoretical efforts are required to obtain insight into the processes which are magnetic analogues of polaron effects. It seems likely that it is such a "magnetic polaron" that emerges in manganese fluoride at the optical excitation of magnetic ions in the region of an ${}^6A_{1g} \rightarrow {}^4T_{1g}({}^4P)$ transition: since the spin orientation of the photoexcited ion itself considerably differs from that of nonexcited ions [110], the motion of such an optical magnetic exciton has to be accompanied with a "magnetic polarization" (a local disturbance of the magnetic order) of the surroundings.

As for the spectral technique of studying the magnetic properties of dielectrics, the promising fields of its wider application can be, for example, investigations of fluctuations in the vicinity of phase transitions or microspectroscopy of domain walls. Application of this technique in such investigations can provide a number of principally new possibilities as compared to conventional magnetic measurements.

CHAPTER 5

Delocalization of Impurity Excitations in Antiferromagnetic Insulators

The present chapter is concerned with crystals characterized by a small amount of a substitutional impurity of concentration $c \ll 1$. These crystals are of great interest both theoretically and experimentally. In particular, they are a suitable subject for the analysis of localized and delocalized impurity states which are fundamental for a number of physical effects. For the most part, we shall concentrate on antiferromagnets (AFM), as the major body of theoretical and experimental work has been done for this type of crystal.

The properties of impurity excitations observed earlier in antiferromagnets with impurity concentrations $c \ll 1$ were well fitted by the model of localized oscillations. The results in agreement with that theory have been covered in [311, 312, 313]. However, experimental results obtained on cobalt carbonate and cobalt fluoride could not be accounted for in the framework of the local oscillations model. In cobalt carbonate, doped with a few tenths of a percent of iron carbonate, splitting was found in the frequency-field dependences of antiferromagnetic resonance and impurity resonance instead of their expected interception [314, 315]. In cobalt fluoride, doped with a few hundredths of a percent of manganese fluoride, there was found an appreciable amplification of the impurity absorption line intensity as it approaches the antiferromagnetic resonance line in a magnetic field [200].

A complete explanation of these experimental results required, along with the ideas of delocalized impurity oscillations [316, 317], an account of the interaction between impurities via host ions, and was first done in [318–327]. We would like to emphasize that the impurity concentrations are low, $c \ll 1$, so that direct interaction between impurities can be neglected in the analysis of almost all effects; hence, the idea of taking into account the interaction between impurity centers separated by several lattice constants is far from trivial, especially in view of the fact that, usually, the electron wave function of the impurity ion is localized in its immediate vicinity.

By its nature, a doped crystal is a disordered physical system, the impurity being randomly distributed in the crystal. Accordingly, the questions that we are going to analyze here can be approached in terms of the general theory for such systems [328] developed with the use of the Green function formalism.

However, this is not the only way to describe impurity-containing crystals. Whenever possible, we will make use of simpler alternative approaches, emphasizing the physical aspects of the problem.

The results obtained in earlier papers related mainly to two cases, differing in the position of the characteristic frequencies of the impurity excitations. In the first case, these frequencies lie far from the intrinsic excitation band of the host crystal, while in the second case they are within this band. Hence, in the first case, the wave functions of the impurity excitations are localized near the crystal defect, i.e., the excitations remain local and cannot propagate over macroscopically large distances. In the second case the excitations are not localized. They manifest themselves through "resonant" perturbations of the host crystal excitations, in particular, through changed densities, a state of the intrinsic magnons, and different bandwidths and frequencies of the magnons for some values of the wave vector as compared with the pure crystal.

Both localized modes and resonant perturbations have been observed using different crystals and experimental techniques (e.g., optical spectroscopy, neutron scattering, heat conduction, and Mössbauer effect measurements) to study phonon, electron, and magnetic excitations.

Most frequently, the experimental results were described in terms of simple theoretical models, like the cluster model and the Ising model, which yield satisfactory results as far as frequency calculations are concerned. Yet some of the parameters (e.g., bandwidths) cannot be described without resorting to more sophisticated theories employing the Green function formalism. The Green function technique has also proved to be an adequate tool to describe the properties of highly concentrated crystals ($c \approx 0.3$) characterized by the existence of two fully-fledged (i.e., characterized by a quasi momentum) spin wave (magnon) bands, the impurity band, and the intrinsic band [329, 330]. The description of such systems involves an account of the direct interaction between the impurity centers.

The earlier theory was good for the description of experiments in those cases where the frequency of the impurity excitation was sufficiently far from the spin wave band of the crystal. When the frequency approached either the bottom or the top of the band [314, 200, 331] then the theory failed to provide correct predictions. The first to describe some properties of crystals with a similar arrangement of the energy levels pertaining to the intrinsic and impurity excitations was Rashba [332–334]. He stressed the importance of the energy separation between the levels. He analyzed the coupling of impurity and exciton states in molecular crystals and showed theoretically that the impurity excitation becomes delocalized, involving the environment rather than the impurity molecule alone, as the impurity level approaches the exciton band edge. The impurity-induced absorption was characterized by the polarization forced by the intrinsic absorption. This type of interaction between the impurity molecule and the crystal should result either in an increase or a decrease of the impurity line intensity, depending on whether the impurity line is close to the $\mathbf{k} = 0$ or $\mathbf{k} \neq 0$ edge of the exciton band (here $\mathbf{k}$ is the exciton wave vector). The effect was observed experimentally soon afterwards [335–337].

Rashba [334] made use of the perturbation method to develop a consistent theory for the anomalously large impurity-induced absorption. As has been shown in that paper, if the separation between the energy levels of the host and the impurity is much larger than the resonant interaction energy of the solvent molecules determining the exciton bandwidths, then the impurity-induced absorption leads to the appearance of an excitation that is practically completely localized near the impurity molecule. The effects are more interesting when the separation is of the same order as the exciton bandwidth. In that case the excitation is no longer localized near the impurity molecule but invovles the neighboring molecules of the host. Hence, the latter begin to influence essentially the impurity-induced absorption. In case the level separation is much less than the bandwidth, the excitation is fully concentrated on the host molecules, embracing an area of a much larger size than the lattice constant.

Rashba and Gurgenishvili [323] described the "transfer" of the oscillator force from intrinsic to impurity oscillations occurring in semiconducting crystals.

A breakthrough in our understanding of the delocalization mechanism for impurity excitations is associated with the papers by Ivanov, Loktev, and Pogorelov [319–327] (hereafter referred to as ILP). Most of the earlier results were obtained either in the single-impurity approximation or with only the terms linear in the concentration retained. The ILP papers took the terms quadratic-in-the-impurity concentration into account, thus allowing for the interactions within all conceivable pairs of impurities. The calculations can be carried out if the impurity level lies near the renormalized edge (i.e., with the substitution of host ions by impurity ions taken into account) of the intrinsic oscillation band. The linear approximation is not valid here, at least near singular points of the spectrum. The various self-consistent approaches that have been developed to analyze the case of high impurity concentrations, like the coherent potential approximation, face a number of difficulties if the effects due to the interaction of impurities are treated consistently. The methods possessing asymptotic accuracy, e.g., the "optimum fluctuation" technique, prove inapplicable just in those regions of the spectrum which are of greatest physical interest.

ILP have developed a technique for the description of elementary excitation spectra which is based on the expansion of the single-particle diagonal Green function in complexes of interacting impurity centers. The group expansion takes into account only pairs of interacting impurities. The approach has allowed clarification of the mechanism of spectrum rearrangement in a crystal with impurities, including cross rearrangement and formation of a quasi gap.

The impurity band is often understood as a set of energy levels associated with several impurity centers which are coupled by an extremely weak long-range interaction [317, 318]. Such a band is characterized, in particular, by a density of states caused by the fluctuating potentials of individual impurity centers and the lack of periodicity in their arrangement. By using the term "full-fledged impurity band" we wish to emphasize the reality of the energy versus quasi-momentum dependence that exists in such a band, as a result of which the impurity oscillations are no longer localized. They become coherent and capable of propagating over macroscopic distances in the crystal. The impurity excitations become quasi particles, to be called "impuritons."

The term "impuriton" was first introduced by Andreev and Lifshitz [338] in connection with low-temperature properties of quantum crystals that contained impurities and were characterized by a small interaction energy between the atoms (particularly in crystalline helium). With a sufficiently large amplitude of the zero-point oscillations, the impurity becomes delocalized and can move practically freely through the crystal or over macroscopic distances. The effect can be described in terms of an excitation, or an impuriton, whose possible states are classified according to the values of the quasi momentum $\mathbf{k}$. The impuriton energy $\varepsilon(\mathbf{k})$, which is a function of the quasi momentum, assumes all possible values in a band of width $\Delta\varepsilon$ proportional to the tunneling probability of the impurity. At finite temperatures the impuriton undergoes collisions with other excitations in the crystal. When the collision frequency increases, the impuriton is able to come to equilibrium with the lattice in the period of time spent by the impurity in the vicinity of some fixed site. Then we speak of a localized impurity. The term "magnetic impuritons" will be applied to such impurity excitations in a magnetic crystal that are delocalized and become coherent over a macroscopic area of the crystal.

Antiferromagnets have proved best suited for investigating magnetic impuritons, since by mere application of a magnetic field one can easily change the energy separation between the impurity level and the edge of the intrinsic magnon band. Thereby the energy structure of the crystal impurity band can be altered and the magnetic oscillations of the impurity transferred from the highly localized to the coherent state.

To understand the results obtained while studying collective impurity excitations and to estimate their novelty, we should first consider localized states and present delocalization concepts in terms of the Anderson and Lifshitz models. Then we should proceed to the main subject paying special attention to the studies of the energy spectrum and the absorption line intensity.

5.1. Localized States (Single-Impurity Approximation)

The condition for the impurity excitations to be localized is that they should be sufficiently separated in energy from excitations of the host matrix. As noted before, localized states currently are rather well known, both experimentally and theoretically. The review papers of Weber [311] and Cowley and Buyers [312] contain vast lists of references. We shall present the principal theoretic results following the monograph by Izyumov and Medvedev [313].

5.1.1. Theory

To begin with, consider a crystal with a single substitutional impurity center. In view of the translational symmetry, the eigenenergies $\varepsilon_{\mathbf{k}}$ and eigenfunctions $|\mathbf{k}\rangle$ of the single-particle Hamiltonian $\mathscr{H}_0$ satisfying the Schrödinger equation

$$\mathscr{H}_0|\mathbf{k}\rangle = \varepsilon_{\mathbf{k}}|\mathbf{k}\rangle \tag{5.1}$$

are characterized in the perfect crystal by the quasi momentum $\mathbf{k}$. However, in the presence of a local perturbation V due to an impurity, it proves more convenient to make use of the so-called site representation, whose basis is given by the functions $|n\rangle$ localized at the lattice sites, viz.,

$$|n\rangle = \frac{1}{\sqrt{N}}\sum_{\mathbf{k}} e^{-i\mathbf{k}\mathbf{R}_n}|\mathbf{k}\rangle. \tag{5.2}$$

Here N is the number of lattice sites in the crystal and $\mathbf{R}_n$ the position vector of the site characterized by the function $|n\rangle$ (for simplicity, we consider a single band).

With the perturbation V, the eigenfunctions of the Hamiltonian $\mathscr{H} = H_0 + V$ can no longer be characterized by the quasi momentum. However, the system can be described completely with the aid of the operator $(E - \mathscr{H} - i\delta)$. In the site representation it takes the form of a matrix which is the Green function for the Schrödinger equation and has the elements

$$G_{nm}(E) = \left\langle n\left|\frac{1}{E - \mathscr{H} - i\delta}\right|m\right\rangle, \qquad \delta = 0^+. \tag{5.3}$$

Using the operator identity

$$\frac{1}{E - \mathscr{H} - i\delta} = \frac{1}{E - H_0 - i\delta} + \frac{1}{E - H_0 - i\delta}V\frac{1}{E - \mathscr{H} - i\delta},$$

we arrive at

$$G_{nm}(E) = G^0_{nm}(E) + \sum_{pl} G^0_{np}(E)V_{pl}G_{ln}(E), \tag{5.4}$$

where $V_{pl} = \langle p|V|l\rangle$ is the matrix elements of V taken in the site representation and G^0_{nm} is the Green function for the perfect crystal, viz.,

$$G^0_{nm}(E) = \frac{1}{N}\sum_{\mathbf{k}}\frac{\exp[i\mathbf{k}(\mathbf{R}_n - \mathbf{R}_m)]}{E - \varepsilon_{\mathbf{k}} - i\delta}. \tag{5.5}$$

Formally, (5.4) can be written as

$$G = G^0 + G^0VG, \tag{5.6}$$

whence its formal solution is

$$G = \frac{1}{1 - G^0 V} G^0, \tag{5.7}$$

or, in a different form,

$$G = G^0 + G^0 V \frac{1}{1 - G_0 V} G^0. \tag{5.8}$$

The spectrum of elementary excitations is defined by the equation

$$D(E) \equiv \det|1 - G^0(E) V| = 0. \tag{5.9}$$

Thus, generally it would be necessary to calculate the determinant of a $N \times N$ matrix, which is technically impossible. However, since V is a quasi-local perturbation decreasing rather rapidly with the distance from the impurity, the matrix V_{nm} may be approximately substituted by some truncated matrix. The latter should correspond to a perturbation involving just a few coordination spheres closest to the impurity, as was first noted by Lifshitz [339, 340]. Hence, the matrix V_{nm} has the effective dimension $n_0 \times n_0$, where n_0 is the number of atoms involved in the perturbed sphere.

First consider a perturbation that can be regarded as localized within the atomic volume of the impurity, i.e., in the site representation the perturbation matrix V_{nm} should have the form

$$V_{nm} = u_0 \delta_{n0} \delta_{0m} \tag{5.10}$$

(the impurity atom has been assumed to be at the site $n = 0$). For this localized perturbation, (5.4) takes the form

$$G_{nm} = G^0_{nm} + \frac{u_0 G^0_{n0} G^0_{0m}}{1 - u_0 G_0}, \tag{5.11}$$

where

$$G_0 \equiv G^0_{00} = \frac{1}{N} \sum_{\mathbf{k}} \frac{1}{E - \varepsilon_{\mathbf{k}} - i\delta}. \tag{5.12}$$

Equation (5.11) allows us to find the density of states $g(E)$ for the spectrum of single-particle excitations. It is related to the Green function by means of the well-known formula

$$g(E) = \frac{1}{\pi N} \operatorname{Im} \operatorname{Sp} G(E), \tag{5.13}$$

where $1/N$ is the normalization for the density of states per one atom. Indeed, we have

$$\operatorname{Sp} G = \sum_n G_{nn} = \operatorname{Sp} G^0 + \frac{u_0}{1 - u_0 G_0} \sum_n G^0_{00} G^0_{0n}. \tag{5.14}$$

Now it is necessary to make use of the identity relating Green functions of the perfect crystal (which can be derived from (5.5)), viz.,

$$\sum_l G^0_{nl} G^0_{lm} = -\frac{d}{dE} G^0_{nm}. \tag{5.15}$$

Substituting (5.14) and (5.15) into (5.13) we arrive at

$$g(E) = g_0(E) + \frac{1}{\pi N} \operatorname{Im} \frac{d}{dE} \ln[1 - u_0 G_0(E)], \tag{5.16}$$

where $g_0(E)$ is the density of states in the perfect crystal. It might be useful to separate explicitly the imaginary part of (5.16). Noting that

$$G_0(E) = \operatorname{Re} G_0(E) + i \operatorname{Im} G_0(E), \tag{5.17}$$

where

$$\operatorname{Re} G_0(E) = \frac{1}{N} P \sum_{\mathbf{k}} \frac{1}{E - \varepsilon_{\mathbf{k}}} \tag{5.18}$$

and

$$\operatorname{Im} G_0(E) = \pi g_0(E) = \pi \frac{1}{N} \sum_{\mathbf{k}} \delta(E - \varepsilon_{\mathbf{k}}), \tag{5.19}$$

we obtain

$$g(E) = g(E) - \frac{1}{N} \frac{u_0^2 \operatorname{Re} G_0'(E) g_0(E) + u_0[1 - u_0 \operatorname{Re} G_0(E)] g_0'(E)}{[1 - u_0 \operatorname{Re} G_0(E)]^2 + [\pi u_0 g_0(E)]^2}. \tag{5.20}$$

The prime denotes here differentiation with respect to E. Let us analyze the formula in some detail. The second term in (5.20) has a maximum at the point $E = E_0$ where

$$1 - u_0 \operatorname{Re} G_0(E) = 0. \tag{5.21}$$

To write an approximate form of (5.20) near the maximum, $\operatorname{Re} G_0(E)$ should be expanded in powers of $E - E_0$, viz.,

$$\operatorname{Re} G_0(E) = \frac{1}{u_0} + \operatorname{Re} G_0'(E_0)(E - E_0) + \cdots \tag{5.22}$$

Assuming $g_0(E)$ to be a slowly varying function over the interval $E - E_0$, we can write for $g(E)$

$$g(E) = g_0(E) + \frac{1}{N\pi} \frac{\Gamma_0}{(E - E_0)^2 + \Gamma_0^2}, \tag{5.23}$$

with

$$\Gamma_0 = -\pi g_0(E)/\operatorname{Re} G_0'(E_0). \tag{5.24}$$

As can be seen, $g(E)$ has a sharp peak if Γ_0 is small. Further, it is necessary to distinguish between two cases:

(a) E_0 lies within the quasi-continuous spectrum of the perfect crystal; and
(b) E_0 lies out of that spectral region.

Case (a) is characterized by a peak in the density of states of width Γ_0. Physically, it can be treated as a virtual resonant level E_0. As shown below, the states of the system corresponding to these energies have a tendency towards localization of the wave function near the impurity atom.

In case (b) the function $g_0(E)$ is zero in the vicinity of $E = E_0$; therefore the limit for (5.23) is Dirac's delta $\delta(E - E_0)$, i.e., the density of states for $E > E_{max}$ is given by the formula

$$g(E) = \frac{1}{N}\delta(E - E_0). \tag{5.25}$$

This implies a discrete level in the system.

To better comprehend the conditions for the realization of this or another situation, let us represent graphically the results obtained. Figure 5.1 shows typical cases for the graphical solution of (5.21). Here $g_0(E)$ is the density of states in the perfect crystal, and Re $G_0(E)$ can be expressed in terms of $g_0(E)$ (according to (5.18)), i.e., in the form of the dispersion relation

$$\text{Re}\, G_0(E) = P\int_{-\infty}^{\infty} \frac{g_0(E')\, dE'}{E - E'}. \tag{5.26}$$

This integral allows us to analyze qualitatively the trend of Re $G_0(E)$ compared to $g_0(E)$. It might be useful to note the analogy with the formula for the electric potential if $g_0(E)$ were understood as a charge density and E as the separation between the observation point and the "center of gravity" of the charge.

As can be seen from the figure, (5.21) may fail to have solutions for some values of u_0 (case 1). Case 2 corresponds to two solutions of (5.21) (points $E_0^{(v)}$ and E_0') which both lie within the spectral band of the perfect crystal, i.e., both are virtual levels. However, there is an essential distinction between the two levels that can be seen in the figure as well, if one recalls (5.24) for the level width. Indeed, the derivative of Re $G_0(E)$ is negative for one level and positive

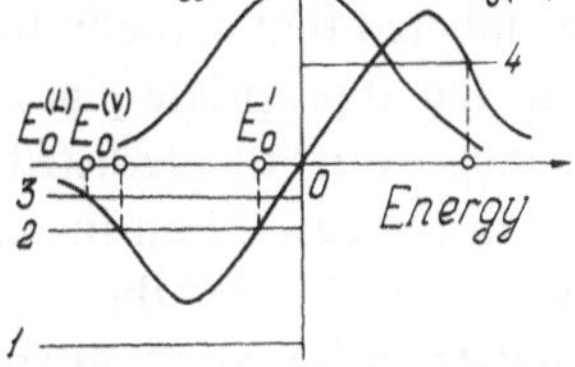

Fig. 5.1. Graphical solution of (5.21) for the position of a local or resonant level. The horizontal lines indicate different magnitudes of the perturbation in units of $1/u_0$ [313].

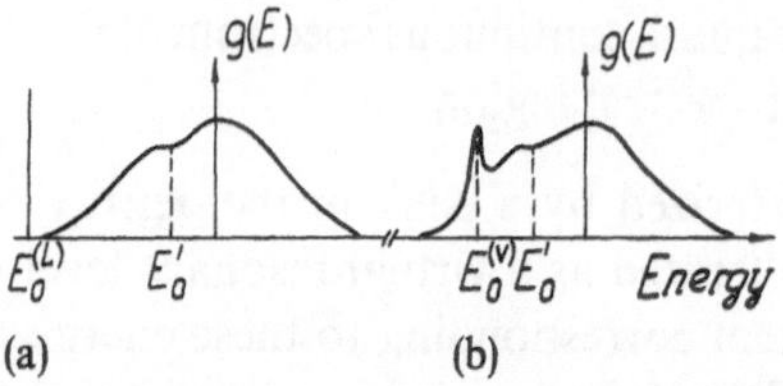

Fig. 5.2. The density of states $g(E)$ in a crystal with an impurity atom, $u_0 < 0$: (a) local level; (b) virtual level [313].

for the other. This means an increased density of crystal states at the point $E_0^{(v)}$ and a reduced density of states near the second point, where the second term in (5.23) becomes negative. Yet the corresponding extremum in the density of states should not be a pronounced minimum, since we have for the situation of Fig. 5.1 $|\Gamma_0(E_0')| \gg |\Gamma_0(E_0^{(v)})|$, in view of $g_0(E_0') \gg g_0(E_0^{(v)})$. As $|u_0|$ is increased, the virtual level can be seen to approach the bottom of the band. At some critical value it can leave the band to make a local level (case 3 in Fig. 5.1). This would also correspond to some point near the center of the band with a very large Γ_0. With $u_0 > 0$, similar situations should occur, resulting in the formation of virtual levels near the "top" of the band, as well as of local levels above the band. Case 4 corresponds to a virtual level. The densities of crystal states corresponding to two values of u_0 in the perturbed crystal (cases 2 and 3 in Fig. 5.1) are shown in Fig. 5.2.

We have stated above that the positions of local and virtual levels are governed by (5.21). In fact, this equation determines all energy levels of the system considered. Indeed, taking (5.18) into account it can be rewritten as

$$\frac{1}{u_0} = \frac{1}{N} \sum_{\mathbf{k}} \frac{1}{E - \varepsilon_{\mathbf{k}}}. \tag{5.27}$$

The quasi momentum $\mathbf{k}$ assuming consecutively N values, this is an algebraic equation of order N; hence it should have N solutions, which are the energy levels of the system. The presence of an impurity atom results in either a local or a virtual level. This is a level lying outside the intrinsic band of the crystal in the former case and within the band in the latter.

In the following sections we will use the term "virtual level" rather often, although it should be remembered that actually there exists no virtual level. What is in fact observed at the appropriate place is either an increase or a decrease in the density of states owing to a redistribution of the eigenenergies in the band. For example, in the case of an increased density of states the following pattern can be seen. For $E < E^{(v)}$ the shifted energy levels lie above the corresponding levels of the quasi-continuous spectrum of the perfect crystal, while for $E > E^{(v)}$ they are below them; hence, near $E^{(v)}$ the density of states is increased as compared to the perfect crystal. Obviously enough, the effect is proportional to $1/N$.

Now we will consider the case of a quasi-local perturbation V involving several nearest coordination spheres around the impurity. This is just the situation that is characteristic of magnetic materials. The perturbation matrix in the site representation actually has the size $n_0 \times n_0$, where n_0 is the number of atoms in the crystal which are involved in the perturbed sphere (including the impurity atom itself). Let the perturbation possess the point symmetry of the initial perfect crystal (this is a natural assumption for substitutional impurities). Then (5.6) can be analyzed in terms of group theory. The perturbation symmetry can be taken into account most simply with the aid of a unitary matrix [341]. A unitary transformation U corresponding to the perturbation V is introduced. The matrix of V apparently should be a $n_0 \times n_0$ matrix. It changes the set of localized functions $|n\rangle$ ((5.2)) corresponding to the sites lying in the perturbation sphere, into sets of symmetrized linear combinations $|\alpha, R(\alpha)\rangle$, which transform as bases of irreducible representations of the point group. Here α labels the irreducible representation according to which the symmetrized combination transforms. Apparently, such a combination can be composed of functions localized at atoms which:

(i) belong to the same coordination sphere around the perturbation center; and
(ii) are coupled through symmetry operations of the point group considered.

The symbol $R(\alpha)$ is used to enumerate the symmetrized combinations which transform according to the representation α. In fact, $R(\alpha)$ indicates how many times the state α is encountered among the states owing to the localized functions of the atoms in the perturbation sphere.

Thus, the matrix elements $U(l, \alpha, R(\alpha))$ of the unitary transformation U are given by the equation

$$|\alpha, R(\alpha)\rangle = \sum_{l=0}^{n_0-1} U(l, \alpha, R(\alpha)|l\rangle, \tag{5.28}$$

where the functions $|\alpha, R(\alpha)\rangle$ are supposed to be orthonormal. The symmetrized combinations of $|\alpha, R(\alpha)\rangle$ corresponding to irreducible representations of the point group can be obtained through standard procedures of group theory, after which the unitary matrices U can be written easily.

The set of n_0 functions $|0\rangle, |1\rangle, \ldots, |n_0 - 1\rangle$, which are localized at the sites within the perturbation sphere, realizes a n_0-dimensional representation of Γ of the point group of the perturbation. This representation is reducible, hence can be decomposed into a complete set of irreducible representations of the group. Each individual representation Γ_μ can be included in the representation Γ several times. Denoting the multiplicity by n_μ we can write this statement symbolically as

$$\Gamma = \sum_\mu n_\mu \Gamma_\mu. \tag{5.29}$$

The values of n_μ can be readily found if the characters of the group operations are known for both representations, i.e., Γ and Γ_μ. Equation (5.29), combined with the invariance of the trace of the matrix under unitary transformations, yields for the characters of the representation

$$\chi(R) = \sum_{\mu} n_{\mu} \chi_{\mu}(R), \tag{5.30}$$

where $\chi(R)$ is the character of the reducible representation Γ and $\chi_{\mu}(R)$ is that of the irreducible representation Γ_{μ}. Taking consecutively all classes of conjugate group elements we can obtain from (5.30)the necessary set of equations for Γ, as the number of classes in a group is equal to the number of its irreducible representations. The characters of the irreducible representations can be taken from any known monograph on group theory, whereas the characters of Γ can be calculated straightforwardly, by applying the group oeprations R to the initial set of localized functions within the perturbation sphere.

The set of n_{μ} is sensitive to the number of the coordination spheres after which the perturbation is truncated. Normally, not all of the irreducible representations of the group are used for the first coordination spheres within the range of the perturbation V. As long as new spheres are included, new irreducible representations come into play, until the set is exhausted and some of the earlier representations are included several times.

By decomposing Γ into irreducible representations, the formulae of the theory are greatly simplified. In particular, the determinant of (5.9) specifying the density of states splits into several factors corresponding to the irreducible representations of the point group that are implemented within the perturbation sphere. Further, the contribution to the single-particle Green function of a quasi-local perturbation can be represented as a sum of contributions, each corresponding to an irreducible representation of the crystal point group. Ultimately, these procedures make it possible to make a complete analysis and obtain specific results. Yet the calculations become very involved even taking into account only the first nearest neighbors. Meanwhile, the perturbation due to an impurity may decrease but very slightly with distance, thus involving a large number of coordination spheres. Similar quasi-local perturbations require some approximate method for calculating the density of states.

Some techniques can be suggested when the magnitude of the perturbation is much larger at the impurity site than at the neighboring sites and other, more remote points. Then the solution of a strictly local problem in which the perturbation is localized at the impurity site can be taken as a zeroth-order approximation. The remaining part of the perturbation is assumed to be a correction term, i.e.,

$$V = V_0 + V',$$

with $V_{0nm} = u_0 \delta_{n0} \delta_{0m}$ and $|V'_{nm}| \ll |u_0|$. It can be shown that, if the solution E_0 for the strictly local perturbation has been found and happens to lie in the quasi-continuous spectral band, and if E_0 corresponds to a resonant state, then the long-range part of the perturbation just acts to shift the resonance level and probably split and broaden it.

In case it does not seem possible to represent the local perturbation as a sum of the contributions V_0 and V', it might be useful to introduce a V' that would not be localized at the impurity site alone but rather involve a few initial coordination spheres, and a V_0 that would be a small perturbation in other spheres. The analysis shows that such a V' results in minor corrections to the energies of the resonance level due to V_0.

Consider in more detail one of the simpler cases, specifically a cubic ferromagnet consisting of N atoms of the same kind, characterized by a spin S. We will assume the Heisenberg model of ferromagnetism and employ the nearest neighbor approximation for the exchange interaction in the crystal (the corresponding exchange integral is denoted as I). The energy spectrum of single-particle spin excitations of such a system represents a band of N quasi-continuous levels, whose positions are specified by the energy-momentum relation $\varepsilon_{\mathbf{k}}$. The corresponding states are plane waves. Let the substitutional impurity atom have a spin S' and the impurity-host matrix exchange interaction integral be I'. With $I' > 0$ the "ferromagnetic" impurity is aligned, in the ground state of the impurity spin, parallel to the magnetic ordering direction of the host matrix. With $I' < 0$ (an "antiferromagnetic" impurity) the alignment is antiparallel.

First consider the simpler case of a "ferromagnetic" impurity. The problem can be solved with the use of symmetry considerations and a double-time Green function. The results are as follows. The crystal allows oscillations of the s-, p-, and d-modes (in the notation of Koster and Slater [342, 343]); the corresponding representations are Γ_1, Γ_{15}, and Γ_{12}. Γ_{15} and Γ_{12} are a three-dimensional and a two-dimensional representation, respectively, whereas Γ_1 is a one-dimensional representation. However, Γ_1 is involved twice, since the basis of this representation can be formed from localized states of either the zeroth or the first coordination sphere. According to the Landau–Lifshitz nomenclature, these are the representations A_{1g}, F_{1u}, and E_g. In other words, the seven-dimensional representation Γ of the group O_h, produced by the localized functions of the impurity atom and its nearest neighbors, splits into the irreducible representations A_{1g}, F_{1u}, and E_g, of which the first is taken twice. This fact can be symbollically written as

$$\Gamma_{sc} = 2A_{1g} + F_{1u} + E_g \tag{5.31}$$

(where sc stands for "simple cubic lattice"). Accordingly, the d-mode excitations are triply and doubly degenerate, with the impurity spin failing to participate. The s-mode excitations are denoted and s_0 and s_1 [344–347]. The first refers to excitations concentrated mainly at the impurity, and the second to those concentrated primarily at its nearest neighbors. Depending on the value assumed by the parameters $\varepsilon = I'/I - 1$, $\rho = I'S'/IS - 1$, and $\sigma = S'/S$, the oscillations can lie either within or out of the spin wave band. In the first case they are virtual, in the second local oscillations. The local oscillations that are possible in the simple cubic lattice belong to the s-, p-, and d-modes, with the p-mode level being triply degenerate and that of the d-mode doubly degenerate. The p-mode local levels appear for $I'S'/IS > 5.76$ and those of the d-mode for $I'S'/IS > 6.41$; resonant levels appear for

$$4.09 < I'S'/IS < 5.76\,(p\text{-mode}), \qquad 2.83 < I'S'/IS < 6.41\,(d\text{-mode}).$$

Thus, the appearance of either resonant or local p- and d-mode levels is conditioned by the presence of a strongly bound (in the sense of the exchange interaction) impurity.

The s-mode oscillations show a more complicated behavior as they are not controlled by the parameter ρ alone, but by ε and σ as well. Virtual s-mode levels can appear both near the bottom and near the top of the spin wave band. For each fixed value of σ, local s-mode oscillations begin at a specific value of I'/I, which is the higher, the lower σ (i.e., the value of ρ must be sufficiently high). For $I'S'/IS \ll 1$ there is a virtual s-level near the bottom of the spin wave band, whose energy can be calculated from the approximate formula

$$E_s^0 \approx 2I'Sz \tag{5.32}$$

and the width from

$$\Gamma_s \approx \pi(S'/S)(E_s^0)^2 g_0(E_s^0), \tag{5.33}$$

where $g_0(E_s^0)$ is the density of states of the perfect crystal for $E = E_s^0$.

Taking the perturbation in the second coordination sphere into account would result in the appearance of new representations, Γ_{25} and Γ_{25}' (both three dimensional), and their related high-energy states. The s-, p-, and d-mode states would become somewhat shifted but not altered essentially.

Excitation of a local or a resonant p-state requires an energy of the order of $2ISz$, which is about equal to the bandwidth, whereas the energy necessary to excite the s-mode oscillation at the impurity site (i.e., s_0) is of the order of $2I'Sz$. This can become arbitrarily low as I' is decreased. However, an s-mode oscillation can also be excited at atoms of the first coordination sphere (i.e., s_1). In this case, the energy required is of the order of $2ISz$ (i.e., once again about the bandwidth), which amount is spent to overcome the coupling to the host matrix atoms.

Until now we have considered crystals with a single impurity atom. The situation does not change basically for finite low-impurity concentrations ($c \ll 1$). The factor $1/N$ in the expression for the density of states should be replaced by c.

Consider the structure of the spin wave band in a ferromagnetic crystal with an impurity. Figure 5.3 shows computed densities of states for a ferromagnetic crystal with a simple cubic lattice for $c = 0.01$. It can be seen in the figure, in particular, that case 1 is characterized by a sharp resonant peak of $g(E)$, corresponding to the contribution of the s-mode oscillations. Indeed, the corresponding parameters meet the "weakly bound impurity" condition of (5.32). The condition is not satisfied, however, in cases 2 and 3, and therefore the respective maxima are broad, with the density of states varying monotonically off the maxima.

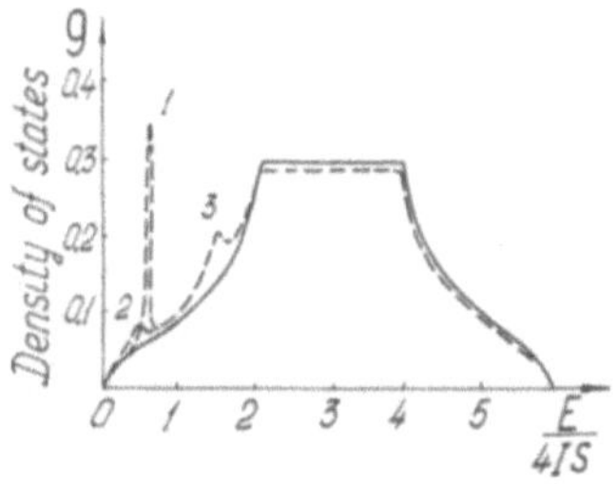

Fig. 5.3. The density of states of a ferromagnetic crystal with $c = 0.01$, for various values of the parameters: (1) $I'/I = 0.2$, $S'/S = 0.25$; (2) $I'/I = 0.2$, $S'/S = 0.4$; (3) $I'/I = 0.5$, $S'/S = 0.25$. The solid line shows the density of state g_0 of the perfect crystal [313].

Figure 5.4 corresponds to such a combination of the parameters ε, ρ, and σ that the impurity levels lie near the top of the spin wave band. Virtual levels of the p- and d-modes and a local s-level exist here.

Until now we have analyzed the case of a ferromagnetic impurity. If the impurity is antiferromagnetic (i.e., $I' < 0$), the situation does not change essentially for the p- and d-mode oscillations. However, the conditions for the appearance of local and reasonant s-mode levels are strongly different from the case $I' > 0$. The solution involves negative poles of the Green function for spin wave excitations in the crystal. Physically, this means the existence of a nonuniform state of the ferromagnet. A localized level with th energy

$$E_s^{(-)} \approx 2|I'|Sz$$

can appear in the spin wave band of the crystal.

In the case of a nonmagnetic impurity in a magnetic crystal, neither resonant nor local p- or d-states appear. The density of states increases near the bottom of the band. However, the integral over the entire spectrum should yield a lower number of states as the nonmagnetic atoms do not participate.

Finally, let us consider some results for an impurity atom placed in a crystal with antiparallel sublattices (taking for the sake of generality, $S_1 \neq S_2$). Obviously, the structure of the spin excitation spectrum depends on the sub-

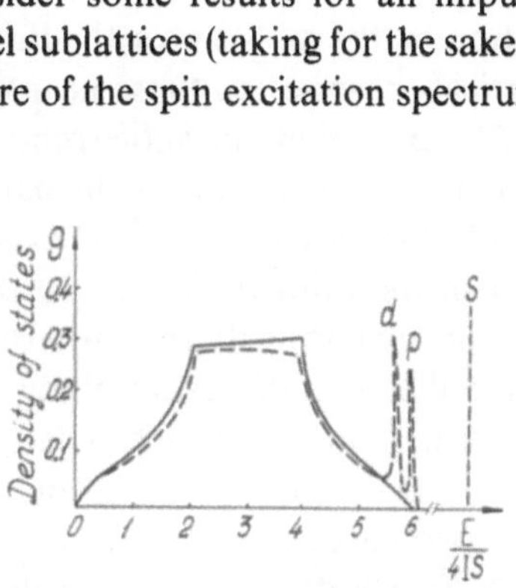

Fig. 5.4. The density of states g of a ferromagnetic crystal with $c = 0.01$, for $I'/I = 2.75$ and $S'/S = 2$ [313].

lattice in which the impurity atom occurs. Specific results can be obtained by means of numerical calculations only. However, in the particular case where the bottom of the band of the optical spin wave branch is much higher than the acoustic branch, the structure of the crystal spectrum near the bottom of the acoustic band is very similar to that for a crystal with a single magnetic sublattice. For example, with a low exchange coupling of the impurity ion to the host matrix, there is a resonant s-level near the bottom of the acoustic band if the magnetic moment of the impurity is oriented parallel to the spontaneous magnetic moment of the crystal. If the impurity moment is oriented antiparallel to the spontaneous crystal moment, the level is local and corresponds to a localized excitation. For the case of low coupling (i.e., $|I'| \ll |I|$) the energy can be expressed analytically, viz.,

$$E_{1s}^{(-)} \approx 2|I_1'|zS_2.$$

5.1.2. Experimental Data and Their Interpretation

We will now discuss the possibility of observing the impurity oscillations in an experiment. The neutron scattering technique allows one to identify the oscillation mode (symmetry), frequency, and wave vector. This is the richest information provided by any experimental method. Unfortunately, the technique suffers from as yet low resolution and experimental complexity. In this respect infrared (IR) and far infrared (FIR) spectroscopy, Raman scattering, and fluorescent technique have an advantage. However, in the case of single-particle excitations the latter methods allow observations of only the A_{1g} (i.e., s_0 and s_1) modes, while in the two-particle case (i.e., one photon excites two magnons at a time) only the $s_0 + p_0$, $s_0 + d$, and $s_0 + f$ modes (the latter for rutile structures) are observed. Raman scattering allows observation of the $s_0 + s_1$ mode as well.

The p-, d-, and f-mode oscillations are often called shell modes as the impurity atom does not participate. As is well known, single-particle processes provide information on the modes near the Brillouin zone center (i.e., $\mathbf{k} = 0$), and on two-particle processes on those near the zone edge (i.e., $\mathbf{k} = \mathbf{k}_{max}$). Note that impurity excitations in antiferromagnetic crystals can lie below the bottom of the magnon band, inside the band, or above the top of the band. In the first case they are called "gap" excitations, in the second "band" excitations, and in the third optical exciations. By applying some external forces (e.g., a d.c. magnetic filed), we can provoke mode conversion.

As an illustration, we will present the results of some experiments aimed at observing impurity modes in antiferromagnets. Dietz et al. [348] performed experiments on luminescence, Raman scattering, and FIR absorption to observe single-particle and two-particle local optical modes in a MnF_2 crystal with a small amount of NiF_2. The data obtained are given in Table 5.1.

The frequencies, calculated with the use of Green functions, are in excellent agreement with the experiment; the cluster and the Ising model calculations

Table 5.1. Local mode symmetries and frequency values for $MnF_2 + Ni^{2+}$.

Mode	s_0	s_0^2	$s_0 + d_{xy}$	$s_0 + d_{xz}$ $s_0 + d_{yz}$	$s_0 + f$
Frequency (cm^{-1})	120.4	254.1	164.5	167	167.8

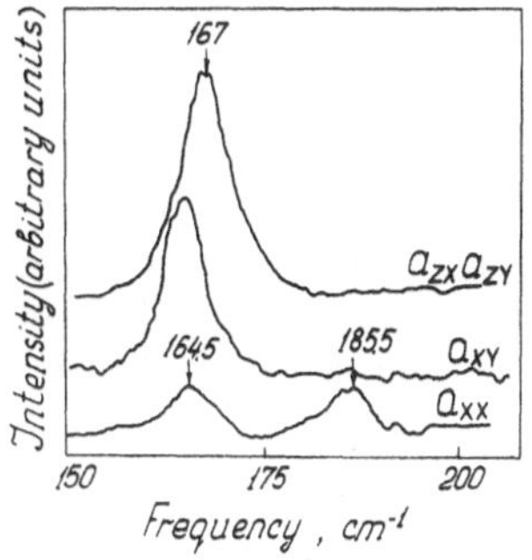

Fig. 5.5. Measured Raman spectra over the frequency range 150–200 cm^{-1} of a MnF_2 crystal doped with 1–2% Ni (temperature 6 K, different polarizations). The intensity ratios for the different peaks are only approximate. The lines shown near 165 cm^{-1} are observed only in Ni-doped crystals and at temperatures below 67.3 K (T_N). The weak scattering near 185 cm^{-1} has not been understood [349].

[311] show fair agreement too. Shown in Fig. 5.5 is the Raman scattering spectrum for the same crystal. The $s_0 + d$ (167 cm^{-1}) and $s_0 + f$ (164.5 cm^{-1}) impurity modes are clearly seen. The line shape has been satisfactorily calculated by Thorpe [350], who employed the Green function technique. However, he obtained a linewidth too low by a factor of two as compared with the experiment.

Figure 5.6 shows the neutron scattering intensity versus wave number in a $Mn_{0.95}Co_{0.05}F_2$ crystal (below we will use a shorter form of notation, viz.,

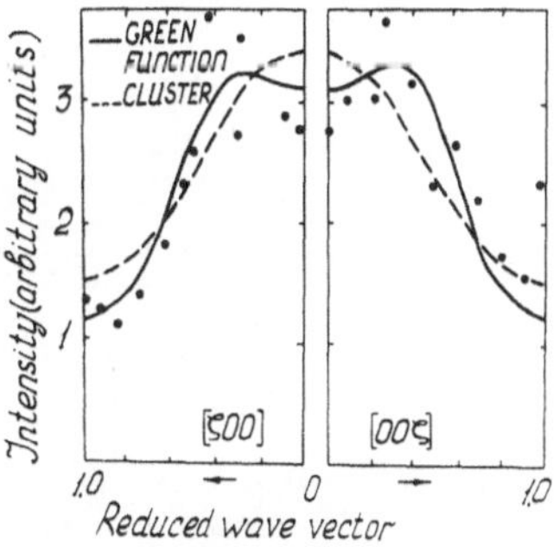

Fig. 5.6. The neutron scattering intensity in MnF_2 + 0.05 Co [330]. The measurement accuracy is about ±10% [311].

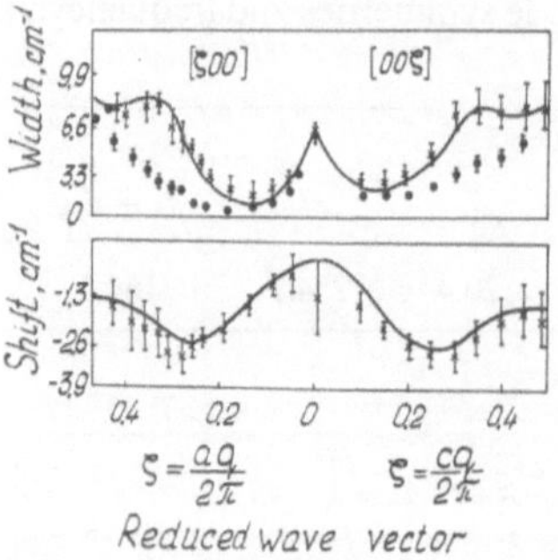

Fig. 5.7. The frequency shift and spectral width of magnons in MnF_2 + 0.05 Zn as compared with the pure crystal. The experimental data [351] are full width at half maximum and frequency shift of neutron groups in the doped sample (crosses) and in pure crystal (full circles). The solid lines are calculations with the Green function technique.

MnF_2 + 0.05 Co). The neutrons are scattered by the local mode s_0, which has a frequency of 118 cm^{-1}, and thus lies above the top of the spin wave band. As can be seen in the figure, calculations by the Green function method yield better results than those in the cluster model.

An example of intrinsic spin wave states resonantly perturbed by a non-magnetic impurity is given by the 43 cm^{-1} mode in MnF_2 + 0.05 Zn. Figure 5.7 shows linewidths and magnon frequency shifts in this crystal as functions of the wave number. As can be seen for $\zeta \approx 0.35$ some features appear in the curves which did not exist in pure MnF_2. Figure 5.8 illustrates line shapes of the neutron groups and the good agreement with the theory. The same excitation has been observed in fluorescence measurements [352].

Finally, we give an example of a "band" mode where the impurity-induced excitation lies in the spin wave band. The frequency shift and line width for such a mode (frequency 50.6 cm^{-1} [311]) are shown in Fig. 5.9. The calcula-

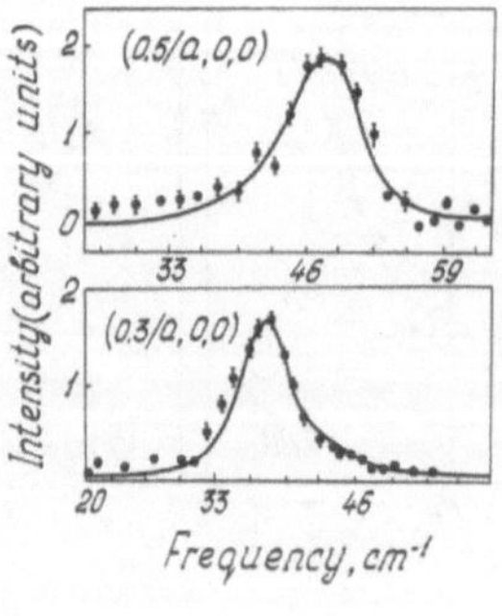

Fig. 5.8. The line shape of neutron groups in MnF_2 + 0.05 Zn [311].

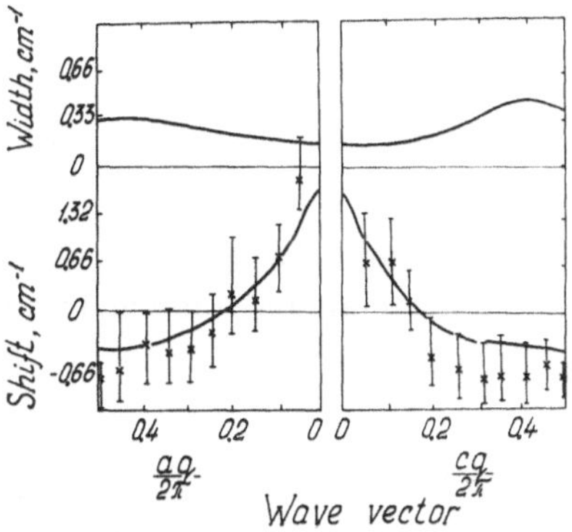

Fig. 5.9. Spectral widths and shifts for the "band" modes of MnF_2 + 0.05 Co. The crosses correspond to measured points and the solid lines are theoretical [311].

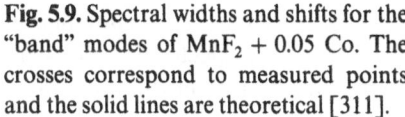

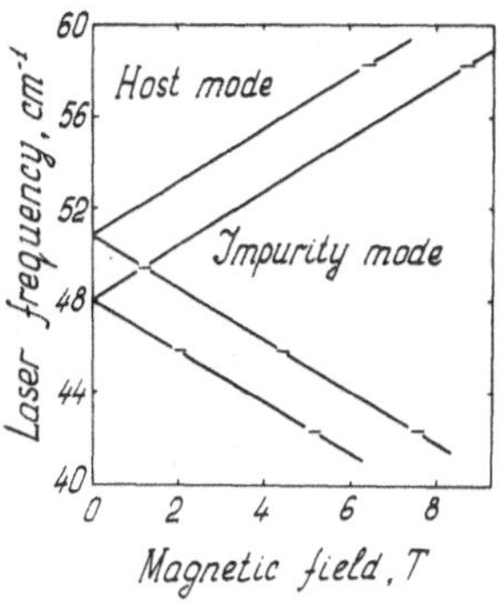

Fig. 5.10. Resonant field H versus laser frequency for the host (AFMR) and the impurity mode in $FeF_2 + 10^{-4}$ Mn [354, 355]. The bars indicate experimental points.

tion has been done by the Green function method, taking into account the complex level structure of the Co^{2+} ions (specifically, the mixing of the ground state doublet with high-energy excited states owing to the molecular field acting on the Co^{2+} ions).

Separation of a band impurity mode out of the magnon band has been observed by Borovik-Romanov and Meshcheryakov [314] in a $CoCO_3$ + Fe crystal by variation of the applied magnetic field. Naumenko and Pishko [353] observed, in $CoCO_3 + 10^{-4}$ (Fe, Mn, Ni), the separation of three band modes. Besides the 1.5 cm^{-1} excitation of Fe^{2+} they also measured excitation of Mn^{2+} and Ni^{2+} near 14.3 and 22 cm^{-1}, respectively. Gap modes lying outside the magnon band near its bottom are greatly enhanced at the expense of the modes of the host matrix. The effect disappears when the modes enter the magnon band. For this reason, with low impurity concentrations such modes have been detected deep in the band in just only one experiment [354, 355], owing to the high sensitivity of the laser spectrometer. The crystal was FeF_2 + Mn, the frequency 58.25 cm^{-1} (1.75 THz), the results are shown in Fig. 5.10.

At this point we shall give a brief description of the linewidth for this mode, following the paper by Thayamballi and Hone [356]. It is appropriate to do so in this section as the description does not include the concept of delocalized impurity excitations, the band mode being strongly localized and showing a much stronger coupling to magnons than the impurity–impurity coupling. The central idea of the paper was to take into account the orthorhombic and the dipole anisotropy terms in the Hamiltonian of the crystal that mix the local and the magnon mode. Without allowance for the anisotropy terms these modes are not coupled, being orthogonal in symmetry. The problem is solved in the framework of the standard Green function approach. The Hamiltonian of the pure crystal has the form

$$\mathcal{H} = \mathcal{H}_e + \mathcal{H}_Z + \mathcal{H}_{ua} + \mathcal{H}_{oa} + \mathcal{H}_d, \tag{5.34}$$

where the terms represent, respectively, Heisenberg's exchange interaction, the Zeeman terms, the uniaxial and the orthorhombic anisotropy, and the dipole interaction. With $\mathbf{H} \parallel z$ the three first terms retain the rotational symmetry with respect to the z-axis; therefore they commute with the net z-component of the spin,

$$\mathcal{H}_e + \mathcal{H}_Z + \mathcal{H}_{ua} = \sum I_{ij}\mathbf{S}_i\mathbf{S}_j - \gamma H_0 \sum S_i^z - D \sum (S_i^z)^2,$$

where S is the spin of Fe^{2+}. The principal exchange term corresponds to the interaction between second-nearest neighbors at sites i and j. The interaction with the two nearest neighbors located at the z-axis is neglected, for it is weak in antiferromagnetic crystals of the rutile type. The weak orthorhombic anisotropy in the $x - y$ plane can be represented as

$$\mathcal{H}_{oa} = -\frac{E}{2S} \sum_i \eta_i [(S_i^x)^2 - (S_i^y)^2], \tag{5.35}$$

where $\eta_i = +1$ if the subscript i relates to the sublattice with spins up and $\eta_i = -1$ for the sublattice with spins down, owing to the $\pi/2$ difference in the orientation of the ligand environment (fluorine "dumbbells" for Fe^{2+} ions from the different sublattices), that is, the orthorhombic anisotropy is local rather than macroscopic. The magnetic dipole interaction has the form

$$\mathcal{H}_d = \tfrac{1}{2}\hbar\gamma^2 \sum [\mathbf{S}_i \cdot \mathbf{S}_j - 3(\mathbf{S}_i \cdot \hat{\mathbf{r}}_{ij})(\mathbf{S}_j \cdot \hat{\mathbf{r}}_{ij})] r_{ij}^{-3}. \tag{5.36}$$

An important alteration in the spin dynamics of the pure crystal, which occurs due to these terms breaking the symmetry, is not revealed in the spectrum but rather in the appearance of new nonzero correlation functions $G^{++}(\mathbf{q}, \omega)$ and $G^{--}(\mathbf{q}, \omega)$. These contain the spectral density for fixed spin-flipping directions where it did not exist before. This allows establishing the decay channel of the local mode. Further, a single substitutional impurity Mn is introduced, which is characterized by a different spin and different coupling constants compared with the iron ion it replaces. However, the total average exchange field is changed only slightly. Besides, with these Heisenberg interactions, S_z is still conserved; hence, they do not affect the results of our prime interest.

A further step in the paper is to apply the perturbation method. The results are shown in Fig. 5.11. The contribution to the linewidth owing to the dipole term is of the order of the squared dipole coupling constant, times the density of states. The orthorhombic coupling constant is much larger than the dipole constant, but its contribution to the linewidth is reduced by two factors. First, it is a quantity of local origin, hence the corresponding perturbation is associated either with the local excitation or the continuum. Second, the sign of E alternates for the two sublattices and no net orthorhombic anisotropy exists. If the impurity feels the same anisotropy as the ion it replaces, then the alter-

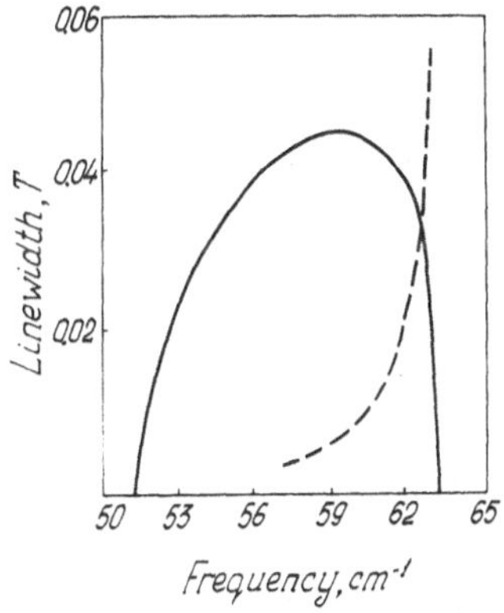

Fig. 5.11. Calculated linewidths of band modes as functions of their frequencies. The solid line shows the contribution owing to an orthorhombic anisotropy and the dashed line that of the dipole interaction [356].

ation in sign can result in a total cancellation of the effect due to terms that might contribute to the linewidth of the local mode. Thus, a finite contribution to the linewidth arises from the difference between the orthorhombic anisotropy for the impurity Mn ion and that for the replaced Fe ion. The predicted frequency dependence of the linewidth of the local mode follows exactly the spectral dependence of the density of states described by $G_{00}^{-+}(\omega)$.

Estimated linewidths of the above-mentioned 58.25 cm^{-1} band mode (1.75 THz) in $FeF_2 + 10^{-4}$ Mn are about 0.005 T, while the measured vlaue is 0.3 T. (Note that the latter value is already corrected for contributions due to radiative damping and the impurity–impurity interaction, which have been substracted from the real measured linewidth. To make the correction possible, the measured values were extrapolated to zero sample volume and concentration $c = 0$.) The authors believed that the discrepancy between theory and experiment was due to spin-lattice relaxation of the local mode. Some clarity perhaps could be achieved through measurements of the linewidth of the local mode as a function of the applied field.

Order of magnitude linewidth estimates for the "band" mode in CoF_2 + Mn can be obtained from the formula [357]

$$\Gamma_{zm} \simeq \tfrac{1}{2} m_+^2 [(\omega_0 + g'\mu H)^2 - \Omega_-^2(0, H)]^{1/2}, \tag{5.37}$$

where m_+ is the coupling constant between the upper Zeeman level of the impurity and the lower Zeeman level of the host matrix (i.e., the low-frequency antiferromagnetic resonance mode); Ω is the coefficient of the quadratic term in $\omega(\mathbf{k})$ for the magnon band [viz., $\omega(\mathbf{k}) = \omega(0) + (4\pi)^{-2/3}\Omega(a\mathbf{k})^2$], whose magnitude is roughly equal to the magnon bandwidth; ω_0 is the impurity excitation frequency for $H = 0$ and g' the corresponding g-factor; $\Omega_-(0, H)$ is the frequency of the low-frequency magnon for $\mathbf{k} = 0$ and the magnitude of H sought for. The numerical values obtained experimentally are [358] $m_+ =$

$17.5\ cm^{-1}$; $\Omega = 45\ cm^{-1}$; $\omega_0 + g'\mu H = 15\ cm^{-1}$ and $\Omega_- = 33\ cm^{-1}$, from which we have $\Gamma_{zm} = 4.6\ cm^{-1}$ for $H = 14$ T. With $H = 10$ T, Γ_{zm} becomes $\Gamma_{zm} = 3.6\ cm^{-1}$. If expressed in terms of units of H (with $g' \simeq 1$), there linewidths are, respectively, 10 and 8 T, i.e., substantially different from the result of Thayamballi and Hone [356]. Presumably, the disagreement can be ascribed to the difference in the orthorhombic anisotropy constant, which is one order of magnitude higher in CoF_2 than in FeF_2.

With the aim of explaining the concentration dependence of the linewidth shown by the gap impurity mode, Hone and Wiecko [359, 360] have taken into account the impurity–impurity coupling due to the weak overlap of their wave functions. The appropriate summation has been performed over all possible pairs of impurities. Allowance has also been made for the applied magnetic field, which brings the magnetic excitation closer to the magnon band, thereby increasing the wave function overlap. Fair agreement has been obtained with tentatively measured concentration depedences of the linewidth at 45 cm^{-1} for $FeF_2 + c$ Mn crystals.

A Hamiltonian similar to (5.34) (however, without allowance for the weak anisotropy of dipole origin) has been used by Loktev and Pogorelov [361] to calculate intensities of impurity-induced absorption lines in CoF_2 + Mn. In the Introduction we have already mentioned the abnormally high intensity of an impurity line near 28 cm^{-1} that was discovered in this material [200]. As the magnetic field was increased, the Zeeman component of the impurity mode, shifting towards higher frequencies, increased in intensity even more. Loktev and Pogorelov [361] were the first authors to suggest a correct explanation of the effect. They considered the frequency range where the single-impurity approximation was valid, i.e., where the "radius" $r_{imp} = a[\Omega/|\omega_0(H) - \omega_2(H)|]^{1/2}$ of the impurity state is less than the mean separation $r = ac^{-1/3}$ between impurities (for more details, see Section 5.3). Here a is the crystal lattice parameter, $|\omega_0(H) - \omega_2(H)|$ is the energy separation between the impurity level and the magnon band edge, and Ω has the same meaning as before. To be specific, it was assumed that the impurity spin is parallel to the spins of its "own" sublattice in which it is substituted. The single-ion anisotropy of the impurity spin was neglected, since the ground state of the Mn^{2+} ion is the s-state. The results of a quantum-mechanical analysis with the use of the Green functions are as follows:

(a) Taking into account the low-symmetry orthorhombic anisotropy reduces the spin, $S_\alpha < 3/2$. In weak fields the spin behaves as

$$S = S_0 + (-1)^{\alpha-1}\tfrac{1}{2}\chi g_{Co}H,$$

where α is the sublattice number, $\chi_\parallel$ the longitudinal susceptibility, g_{Co} the g-factor of Co^{2+} in the crystal field, and $S_0 = 1.13$.

(b) The effective g-factor of Co^{2+} is

$$g_1 = g_{Co}(1 - \tfrac{1}{2}\chi_\parallel)$$

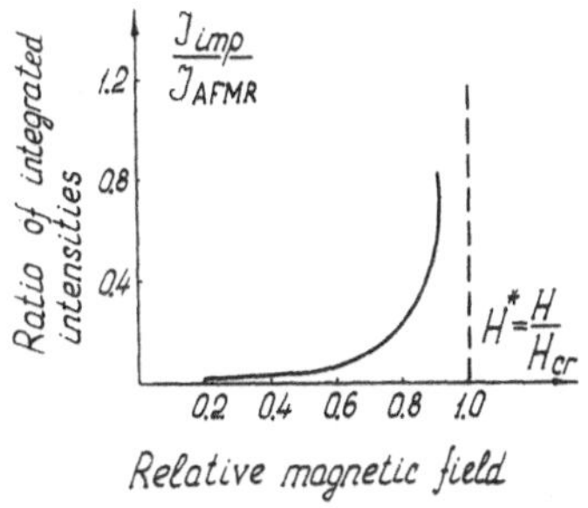

Fig. 5.12. The integrated intensity ratio of the impurity line to the AFMR line as a function of the (relative) applied magnetic field (H_{cr} is the field corresponding to $r_{imp} = \bar{r}$) [324].

and that of Mn^{2+}

$$\tilde{g}_1 = g_{Mn} - \tfrac{1}{2}\chi_{\parallel} g_{Co}$$

with $H \ll [(S_0 \pm a)^2 + 3b^2]^{1/2}$, the levels of the impurity and the host matrix ions are split linearly and symmetrically (here $a = D/Iz$ and $b = E/Iz$).

(c) As the impurity level approaches any of the AFMR modes, the impurity excitation increases in intensity. With $c \ll 1$ and assuming $\omega_0(H) = \omega_0 + \tilde{g}_1 H$, we can write for the integrated impurity line intensity

$$\mathscr{I}_{imp} \propto c[\omega_0(H) - \omega_0 - (g_1 + \tilde{g}_1)H]^{-2}.$$

This relation is plotted in Fig. 5.12.

As we have seen, the orthorhombic anisotropy leads to a resonant interaction of the impurity and the intrinsic excitations near the bottom of the band. Moreover, it has occurred that a large anisotropy can result in an interaction of the upper Zeeman impurity level (for $\mathbf{H} \parallel c_4$) with both Zeeman levels of the host matrix. However, both the above equation for $\mathscr{I}_{imp}$ and Fig. 5.12 show an infinite increase of intensity near H_{cr}. Such an effect, quite naturally, is not observed in the experiment. The fact is that the single-impurity approximation employed in the analysis is not valid in this region. It becomes necessary here to take into account the impurity–impurity interaction via the magnons of the host matrix (see Section 5.3).

The intensity of impurity excitations in FeF_2 + Mn and MnF_2 + Fe was also considered by Rezende [362]. In the first crystal he analyzed a gap mode which approached the bottom of the magnon band as H increased. In the second crystal an optical mode approached the top of the magnon band. The analysis was again performed with the use of Green functions. However, the initial Hamiltonian contained neither the orthorhombic, nor the dipole anisotropy term. Instead, the exchange interaction between the impurity atom and its nearest (I'_1) and text-nearest (I'_2) neighbors was taken into account, with the usual assumption $|I'_2| \gg |I'_1|$.

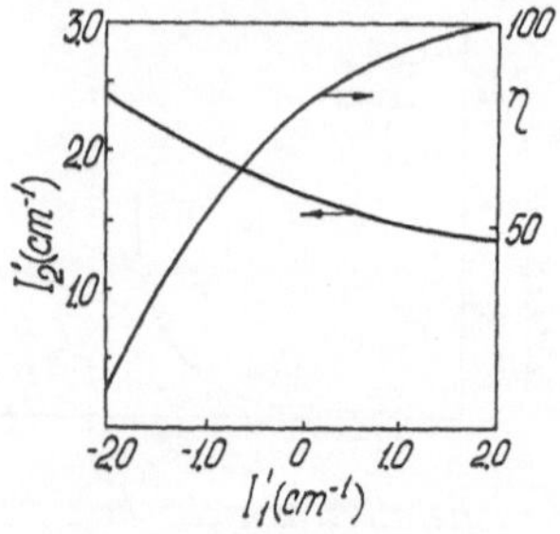

Fig. 5.13. The dependence of the impurity–host interaction parameter I'_2 (antiferromagnetic) on the magnitude of the nearest neighbor parameter I'_1 (ferromagnetic if positive), fitted to match the theoretical frequency of the local mode s_0 in FeF_2 + Mn with 50.27 cm^{-1}. The variation in the relative intensity η of the Raman scattering is also shown.

Values of the intensity enhancement coefficient η and the relative magnitudes as I'_1 and I'_2 providing agreement with the measured frequency of the impurity mode s_0 in FeF_2 + Mn (i.e., $\omega_s = 50.27$ cm^{-1}) are given in Fig. 5.13. The enhancement coefficient was the ratio of the local mode intensity to that of the magnon, with the latter normalized to the impurity concentration c. The choice of this value for the analysis was not accidental, as the emission observed in Raman scattering experiments may depend on the impurity and host spins in very different ways. From the experimental knowledge of the enhancement coefficient Rezende evaluated I'_1 and I'_2 to be 0.2 cm^{-1} and 1.79 cm^{-1}, both of antiferromagnetic type. His expression for the intensity as a function of the energy separation between the modes (hence, as a function of H) is roughly the same as the results of Loktev and Pogorelov [361]. This dependence is shown in Fig. 5.14, where Λ and η are, respectively, the increase of the IR absorption and the relative Raman scattering intensity.

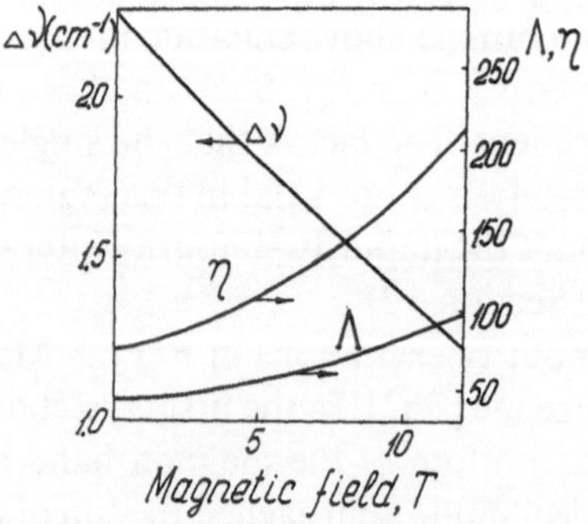

Fig. 5.14. The frequency separation $\Delta\nu$ between the host and the impurity mode and the intensity of the local mode s_0 in FeF_2 + Mn as functions of the magnetic field applied along the c-axis (impurity concentration tending to zero). η and Λ are, respectively, the relative Raman intensity and the amount of enhancement in the infrared absorption.

Despite the apparent similarity shown by the two dependences, there are two reasons for being doubtful about Rezende's interpretation of the effects involved. Certainly, taking the interaction with the nearest neighbors taken into account can effect the coefficient of enhancement in the impurity mode, as this interaction changes the nature of the delocalization for the impurity site wave function. However, competing factors can also be involved, in particular the orthorhombic anisotropy, which has been the major factor in controlling the impurity mode enhancement in CoF_2 + Mn crystals, according to the scheme in [361]. In FeF_2 + Mn the orthorhombic anisotropy is also of considerable magnitude (2.68 cm^{-1} according to the measurements [363] of the electron paramagnetic resonance of Fe^{2+} atoms in ZnF_2). Hence, we should expect that it has an important (if not dominating) influence on the increasing impurity mode coefficient. Similar considerations are inferred from the above-mentioned paper [356] presenting a calculation of the local (band) mode width in FeF_2 + Mn. These authors have shown that the orthorhombic anisotropy strongly mixes the impurity mode and the magnon mode.

Further, Rezende predicts an increase in the optical impurity mode $\omega_{s_0} =$ 95 cm^{-1} in MnF_2 + Fe as it approaches the optically inactive upper edge of the magnon band, viz., $\omega_m = 52$ cm^{-1} (the numerical values are for $H = 0$). According to Rezende, the increase in intensity should be from 1.3 to 1.5 over the frequency range 95 to 105 cm^{-1}, while the experiment of Blewitt and Weber [364] has shown a constant intensity through a much wider range, 70 to 103 cm^{-1}. Measurements for a FeF_2 + Co crystal have also been made by Dürr and Uwira [331]. By controlling the applied magnetic field, they brought the impurity mode substantially closer to the upper edge of the magnon band and observed a decrease in the intensity of the impurity mode.

Before starting the presentation of these results we note that a phenomenological interpretation of the increase of the intensity of the impurity mode has been suggested in [355]. This paper also contains many interesting results concerning the impurity mode and the polariton effect in $FeF_2 + c$ Mn. For example, in $FeF_2 + (0.5 - 4) \times 10^{-2}$ Co crystals a magnetic dipole absorption band of width 0.8 cm^{-1} was revealed at 85.5 cm^{-1}. The band lies above the magnon band of FeF_2 (52.6 to 79 cm^{-1}) and has been identified as an optical impurity mode. Its behavior in an applied magnetic field is illustrated in Fig. 5.15. The Zeeman component of the optical mode that moves towards higher frequencies is characterized by $g = 2.2 \pm 0.1$, which is very close to the g-factor of the host magnons. The mode did not change its intensity as the magnetic field was varied. The impurity mode that moved towards lower frequencies as the magnetic field was increased attenuated rapidly and was "repelled" from the upper edge of the magnon band. The linewidths of the modes were practically unchanged. Such behavior of the optical mode resembles that of an impurity line near the top of the exciton band in molecular crystals [334]. A quantitative description of this experiment [331] has been suggested using the idea of the impurity delocalization [332, 323].

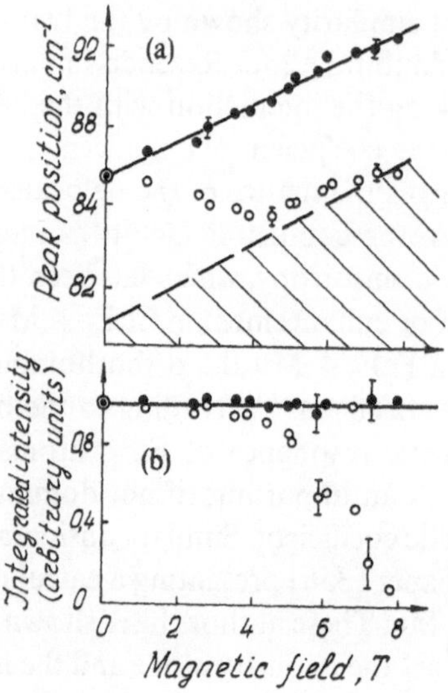

Fig. 5.15. The magnetic field dependence of the optical mode in $FeF_2 + 10^{-2}$ Co: (a) position of the peak; (b) integrated absorption. The parameters are $g = 2.2 \pm 0.1$, $H \parallel C_4$, $T = 2$ K, and $T_N = 78$ K [331].

5.2. Localization and Delocalization States in the Anderson and Lifshitz Models

In the preceding section we have considered impurity mode effects that could be interpreted to a sufficient accuracy within the single-impurity approximation. The presence of other impurities could be taken into account by introducing the concentration as a mere factor (save in the analysis of the linewidth versus concentration dependence). The corresponding excitation involved just the impurity atom itself and/or its nearest environment consisting solely of host matrix ions. Hence, we have so far not been able to consider the behavior of impurity line frequencies, intensities, widths, and shapes near the region where the interaction between the impurity and intrinsic excitations is essential.

The presence of impurity atoms in the crystal violates the translation symmetry. However, with the assumption of randomly distributed impurities the crystal can be described approximately in terms of the perfect crystal, as if the translation symmetry were restored. Indeed, the physically observable quantities relating to an impurity-containing crystal with randomly distributed atoms can be regarded as some mean quantities, averaged over all impurity configurations. With low concentrations the probability that any lattice site is occupied by any impurity atom is independent of the site number, and equal to the concentration c; hence the configurational averaging corresponds to a sort of spreading the impurity through the crystal, so that the translational symmetry is restored on average. As a result, the energy levels of excitations in the system can again be characterized by the quasi momentum, as was the case with the perfect crystal. However, the corresponding states are no longer stationary, being rather characterized by a finite lifetime.

Naturally, this is only true for those states of the crystal which are not "too" different from states of the perfect crystal, i.e., the states belonging to the quasi-continuous spectrum of excitations. In the case of a single impurity atom in the crystal they are represented as a distorted plane wave, where the distortion is only essential near the impurity center, with the size of the distorted region being the "radius" of the quasi-localized resonant state. Such states are localized in energy space too, occupying some range Γ (resonant level width) near the resonant levels of the system.

In a crystal with a finite concentration of impurities the states belonging to the quasi-continuous spectrum can be described upon averaging over the impurity configurations, in terms of plane waves. The state energies are markedly different from the corresponding excitation energies of the perfect crystal only near the resonant levels of the single-impurity system. Thus, the energy-momentum relation of a doped crystal should reflect the features of the quasi-continuous spectrum shown by the crystal with a single impurity atom.

The excitations of the imperfect crystal lying outside the quasi-continuous spectral range should be considered in a quite different way. Such excitations do not exist in the perfect crystal, while in the single-impurity system they correspond to local states characterized by spatially localized wave functions. If the crystal contains many impurities, these levels should split owing to the impurity–impurity interactions. In contrast to the principal band, the impurity state bands thus produced generally cannot be described in terms of energy-momentum relations, as the corresponding states have nothing in common with plane waves. For the impurity bands to be describable in terms of an energy-momentum relation, the impurity centers should be subject to a sufficiently strong coupling.

5.2.1. The Anderson Model

Let the impurities be arranged in a regular sublattice with a period $\ell \gg a$ (where a is the lattice period of the host crystal). However, we will not assume the impurities to make a perfect lattice. Let the electron energy level be different for all the impurity sites. Thus, we are considering a periodic set of potential wells of different depths. The site energies are uncorrelated random values (i.e., the energy at some particular site is independent of the energies at the other sites). Let there be just one electron. The basic question to be analyzed in the Anderson [316] model is as follows. Are the electron wave functions localized near some site or are they extended over the entire system?

Near the impurity, the wave function resembles that of the site representation in both cases, in view of the small amount of overlap. What is to be understood is whether a coherent state is formed, which is a superposition of an infinite number of "site" functions taken with roughly equal weights, that is, extended over macroscopic distances. The alternative case is when the site functions participate in the superposition with a weight decreasing exponentially with separation from a certain site. This corresponds to a state localized in the vicinity of that site. Should all the states be localized, the zero-temperature conductivity of the system would be zero.

Anderson formulated the localization condition in the following way. As is known, the squared modulus of the wave function $|\Psi_i(t)|^2$ is equal to the probability of detecting the excitation at site i. Let the impurity electron be at that site at the initial time $t = 0$, hence let its wave function at that moment coincide with the "site" wave function. With nonlocalized states we would have $\lim_{t\to\infty}|\Psi_i(t)|^2 = 0$. On the contrary, for localized states $\lim_{t\to\infty}|\Psi_i(t)| \neq 0$, i.e., the wave function does not spread with time, just acquiring exponentially low "tails" at neighboring sites. It remains concentrated in roughly the same spatial area as at the initial moment.

The Anderson model involves one dimensionless parameter, namely W/I, where W is the range over which the potential well energies are distributed uniformly and I is the magnitude of the overlap integral for neighboring sites. Anderson's result is as follows: With large enough W/I all the states are localized. There exists some critical value W_c/I at which delocalized states can appear. As W/I is further decreased, the area of delocalized states grows to include almost the entire crystal volume (see Fig. 5.16).

Consider an interpretation of the Anderson model [366]. We will analyze a band of energies $-\frac{1}{2}\Delta < \varepsilon < \frac{1}{2}\Delta$, with the bandwidth Δ almost equal to the overlap integral I. The sites whose energies lie within this band will be called resonant, while those with energies outside the band are called nonresonant. The essence of this definition becomes clear from the fact that an electron state is collectivized by two resonant sites if the latter are nearest neighbors. Two resonant sites will be called coupled sites if they are nearest neighbors. A set of many coupled resonant sites is called a cluster. The electron states corresponding to a cluster are characterized by wave functions whose squared modulus is of the same order at all the sites of the cluster and drops to low values away from the cluster.

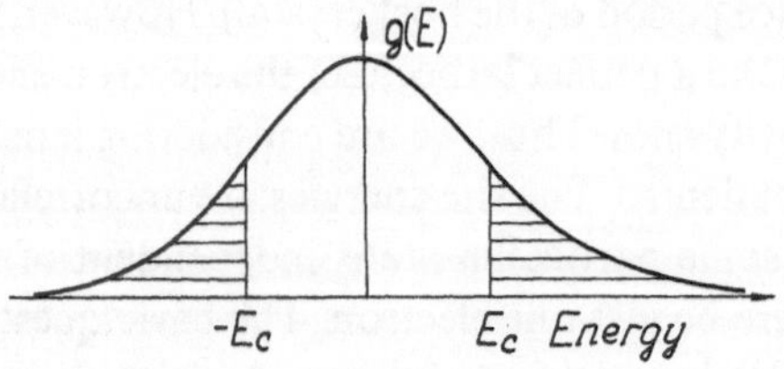

Fig. 5.16. The density of states in the Anderson model. Localized states have been marked by hatching. The energies E_c and $-E_c$, separating the localized from the delocalized state regions, are mobility thresholds [365].

The next step consists in neglecting nonresonant sites totally. This procedure seems doubtful as Anderson's Hamiltonian couples neighboring sites only. Nevertheless, two reasonant sites can have an electron state in common if they have a nonresonant site in between. The effective overlap integral for this case is of the order of I^2/W rather than I. If I/W is small, interaction via the nonresonant site would result in state collectivization over a much narrower band than the Δ considered; hence the effect may be neglected. Thus, disregarding the nonresonant sites we can conclude that the spatial scale of the wave function is controlled by the size of clusters of coupled resonant sites. Within the Anderson model, the distribution of electron energies for the ith site is uniform over the interval W. Therefore, the fraction of resonant sites should be of the order of I/W. If this parameter is small, the resonant sites are few and mainly solitary. However, at a certain critical value I/W_c an infinite cluster of coupled resonant sites appears, i.e., paths leading to infinity may be formed. It is along these paths that wave functions of the electron states are extended. Essentially, this is the Anderson transition.

To estimate the critical value W_c/I, it is necessary to establish the relation between the resonant bandwidth Δ and the overlap integral I. This is the most complicated and delicate part of the analysis. The point is that the resonant site cannot be defined uniquely, since collectivization of the electron states of two sites with decreasing $|\varepsilon_i - \varepsilon_j|$ occurs continuously rather than abruptly. Yet the Anderson transition should occur at a single, precisely defined vlaue W_c/I, since the appearance of a delocalized state with decreasing W/I in an infinite system is a distinctly critical effect, necessarily associated with a threshold value W_c/I.

The Anderson transition should be understood as the appearance of a band of delocalized states. However, the term is sometimes interpreted in a different sense. Consider a situation where the band of interest already contains a distinct boundary separating localized and delocalized states. If the electron occupation of the band were somehow changed, the Fermi level would be changed accordingly. In case the Fermi level can cross the boundary of the localization area, the electronic properties of the system can be altered essentially. Before all, this may be true for disordered systems, e.g., amorphous semiconductors. According to the current theory, the band structure of

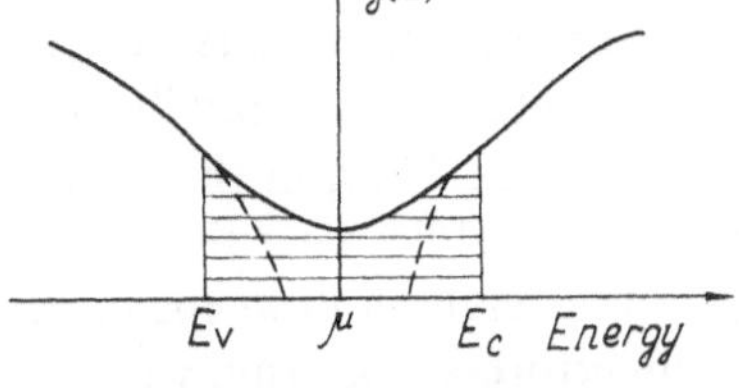

Fig. 5.17. The density of states of an amorphous semiconductor. Localized states have been hatched.

an amorphous semiconductor looks as in Fig. 5.17. The short-range ordering that exists in such materials results in the persistence of allowed and forbidden bands (the dashed line in the figure relates to a perfect semiconductor with a forbidden gap). However, the numerous structural imperfections produce "tails" of the density of states extending into the forbidden band, i.e., the boundaries of the band (in the sense of zero densities of states) become fuzzy, with a quasi gap arising in lieu of the gap. The states in the forbidden band are localized, the localization thresholds playing the role of a sort of band boundaries. Should the Fermi level cross the mobility threshold separating localized from current states (as a result of a changed electron concentration or due to a high applied pressure), the semiconductor would transform to a metal or vice versa.

5.2.2. The Lifshitz Model

In the Anderson model, potential wells of different depths were located at the sites of a regular sublattice. Now we will consider a different case, namely identical potential wells scattered randomly in space. This model is often referred to as the "structurally disordered model." The potentials of the wells will be assumed to be short-ranged, and the mean separation between the wells to be large compared with both the range of the potential and the "radius" of the wave function of a single well. This is the model first considered by Lifshitz [317]. The question of major interest is whether the electron states will be localized near individual wells or be extended over the entire structure.

The Lifshitz model is essentially different from the Anderson model, which suggested that the scatter in electron energies and the overlap were physically different, so that their quantitative relation could be chosen arbitrarily. The electron states were localized if the scatter could be regarded as much larger than the overlap, and delocalized in the opposite case. In the Lifshitz model the overlap and the scatter have the same physical origin and, on average, are of the same magnitude. The only small parameter of the problem is the ratio of the "radius" of the wave function to the mean well separation. Lifshitz developed a special mathematical technique allowing the use of that parameter.

We will not dwell on any details of the technique but rather present the principal results of Lifshitz. The electron states can either be localized near individual potential wells or pertain to pairs of neighboring wells. In both cases, the wave functions contain exponentially small admixtures of other states (of "two's," "three's," etc.). Thus, in the Lifshitz model the states are localized. When an impurity level lies near the edge of the quasi-continuous spectrum the density of crystal states follows the pattern of Fig. 5.18. Here E_g^* is the boundary of the unperturbed spectrum without impurities, i.e., the initial band boundary of the perfect crystal; E_g is the true boundary for the density of states in the doped crystal and E_c^* the renormalized boundary of the band of intrinsic states; finally, E_0 is the energy level of the isolated impurity

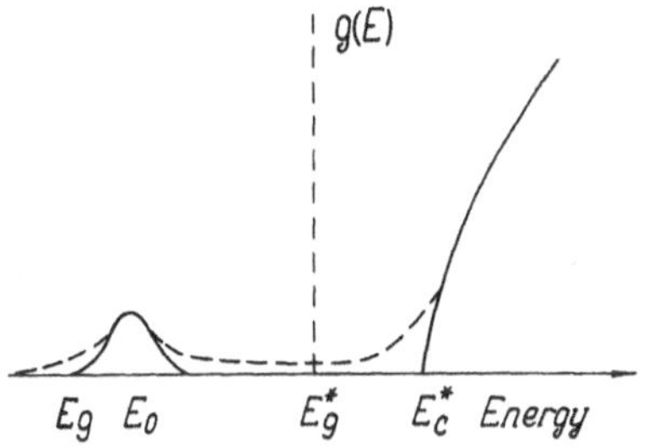

Fig. 5.18. The density of states near an unperturbed boundary E_g^* [317].

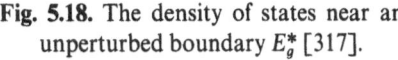

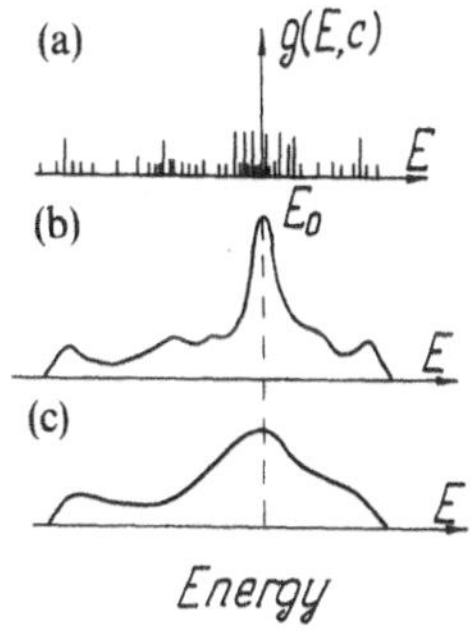

Fig. 5.19. Variation of the impurity band structure with concentration (schematic) [317].

center. The dashed curve denotes the fluctuation levels that appear because of fluctuations in the locations of the impurities in the crystal. The interval between E_c^* and E_g mainly contains contributions from relatively small ensembles of impurities. The density of states near the boundary E_c^* is controlled by the net probability of fluctuations (i.e., in practice, by the fluctuation with the highest probability). Thus, the local level with energy E_0 is replaced by an impurity band characterized by a finite density of states (showing, however, no dependence on the quasi momentum).

The structure of this impurity band is shown separately in Fig. 5.19. With $c \to 0$, the point E_0 corresponds to the levels of isolated impurity centers and is characterized by an "intensity" c. It is associated with impurities separated by an average distance $\bar{r} \propto c^{-1/3}$, which is the most probable of all possible separations of impurity centers. The points $\varepsilon_{1,2}(\mathbf{r}) = E_0 - E_{1,2}(\mathbf{r})$ on the E scale (where $\mathbf{r}$ is the separation between two impurity centers) apparently represent pairs of impurities. They correspond to discrete levels of intensity c^2, whose limiting accumulation points is E_0. In their turn, each of the $\varepsilon_{1,2}(\mathbf{r})$ levels is a limiting point for the levels $\varepsilon(\mathbf{r}_1, \mathbf{r}_2)$ corresponding to impurity trials, of intensity c^3, etc. However, interaction with remote impurities leads to a concentration-dependent broadening of each discrete level, hence smears the "fine structure" described. Obviously, the split-level fine structure can remain only for such energies where the separation between neighboring discrete levels of "pairwise" origin is $\delta\varepsilon \gg \Delta$, with Δ denoting the concentration dependent width of the level. As the concentration is increased the maxima are "washed out" and the fine structure disappears (Figs. 5.19(b) and (c)).

Widely separated impurities correspond to the immediate vicinity of the main impurity level E_0, whereas closely spaced impurities are associated with relatively large separations from that level. Naturally, all these considerations are invalid if some two impurity atoms are nearest neighbors in the crystal lattice. The corresponding impurity level lies outside the impurity band and is characterized by a very low value of the energy, owing to the low probability of the event.

5.3. Coherent and Incoherent Impurity Excitations. Impuritons

The prediction following from the Lifshitz model, namely that electron states are all localized, results from the assumption that the mean separation between the impurities is large compared to the "radius" of the wave function. If the two values were of similar magnitude, then apparently delocalized states might appear in this model too. The lengths could become equal, e.g., if the mean impurity–impurity separation were reduced owing to an increased concentration of impurities. A corresponding computer experiment has been performed for a two-dimensional crystal, and the localized-to-delocalized state transition examined [367, 368]. The appearance at high impurity concentrations of a full-fledged impurity magnon band is also well known. The effect was detected for $Mn_{0.3}Co_{0.7}F_2$ and $KMn_{0.29}Co_{0.71}F_3$ crystals [369, 370].

However, a different, nontrivial approach also seems possible, namely increasing by some other means the radius of the impurity wave function with the aim of obtaining delocalization at low impurity concentrations. This implies the necessity to analyze the interaction of widely separated impurities and to suggest a method for resonantly increasing the magnitude of the interaction. This is the key point in understanding the results obtained by ILP [318–327] which we will present below.

The model system chosen by ILP for their analysis was the hybrid model of Anderson [371] representing the interaction of the states of the impurity band with the electron states of the continuous spectrum. Within this framework, the electronic spectrum near the electron band edge has been analyzed for different relative magnitudes of the impurity parameters. The nature of the Anderson transitions within the band is discussed, as well as the question of the minimal metallic electrical conductivity in the impurity band.

5.3.1. Electron Excitations in the Presence of Large-Radius Impurity States. Theory

Let an impurity level lie near the edge of the electron band and let the corresponding wave function be essentially localized at the impurity site with the probabilities of transfer from the impurity to neighboring sites decreasing exponentially with distance. According to Anderson [371], the Hamiltonian of such a system can be written as

$$\mathscr{H} = \sum_{\mathbf{k}} \varepsilon_{\mathbf{k}} a_{\mathbf{k}}^{+} a_{\mathbf{k}} + \varepsilon_0 \sum a_p^{+} a_p + N^{-1/2} \sum_{\mathbf{k},p} (\gamma_{\mathbf{k}} e^{ikR_p} a_{\mathbf{k}}^{+} a_p + \text{h.c.}), \quad (5.38)$$

where $a_{\mathbf{k}}^{+}$ and $a_{\mathbf{k}}$ are the Fermi operators of the band electrons characterized by the wave vector $\mathbf{k}$ and the unperturbed energy $\varepsilon_{\mathbf{k}}$ (the spin indices have been suppressed); ε_0 is the unperturbed energy of the impurity state; a_p^{+} and a_p are the creation and annihilation operators for electrons at the impurity sites p; and N denotes the number of sites in the lattice. The parameter $\gamma_{\mathbf{k}}$ describes "hybridization" of the band and impurity states. Its values are controlled by the transfer integrals from the impurity to neighboring sites. These integrals rapidly decrease with distance, so that with $c \ll 1$ direct jumps between impurity sites can be neglected.

The density of states in the system described by the Hamiltonian of (5.38) can be expressed in terms of advanced double-time Green functions, viz.,

$$\rho(E) = \rho_1(E) + \rho_2(E), \tag{5.39}$$

with

$$\rho_1(E) = \frac{2}{\pi N} \operatorname{Im} \lim_{\delta \to 0} \sum_{\mathbf{k}} \langle\langle a_{\mathbf{k}} | a_{\mathbf{k}}^+ \rangle\rangle^{E - i\delta},$$
$$\rho_2(E) = \frac{2}{\pi} \operatorname{Im} \lim_{\delta \to 0} G_{\text{imp}}(E - i\delta), \tag{5.40}$$

$$G_{\text{imp}} = \frac{1}{N} \sum_{p} \langle\langle a_p | a_p^+ \rangle\rangle^{E}, \tag{5.41}$$

where the factor 2 is to take account of the spin degeneracy. (In what follows, the superscript $E - i\delta$ will be suppressed.) The Green functions $\langle\langle a_{\mathbf{k}} | a_{\mathbf{k}}^+ \rangle\rangle$ and G_{imp} are diagonal.

Constructing the set of coupled equations of motion and restricting the expansion to groups of all possible impurity pairs (i.e., neglecting triads and higher-order groups) we can arrive at the following representation for the diagonal Green functions averaged over impurity configurations:

$$\langle\langle a_{\mathbf{k}} | a_{\mathbf{k}}^+ \rangle\rangle = (E - \varepsilon_{\mathbf{k}} - R_{\mathbf{k}})^{-1}, \tag{5.42}$$

where

$$R_{\mathbf{k}} = \frac{c|\gamma_{\mathbf{k}}|^2}{D(E)}(1 + cB_{\mathbf{k}} + \cdots), \tag{5.43}$$

$$D(E) = E - \varepsilon_0 - \frac{1}{N} \sum_{\mathbf{k}' \neq \mathbf{k}} \frac{|\gamma_{\mathbf{k}}|^2}{E - \varepsilon_{\mathbf{k}} - R_{\mathbf{k}'}}, \tag{5.44}$$

$$B_{\mathbf{k}} = \sum_{l \neq 0} \frac{A_{0l} e^{-i\mathbf{k}\mathbf{R}_e} + A_{0l} A_{l0}}{1 - A_{0l} A_{l0}},$$
$$A_{0l} = \frac{1}{D(E) N} \sum_{\mathbf{k}' \neq \mathbf{k}} \frac{|\gamma_{\mathbf{k}}|^2 e^{i\mathbf{k}'\mathbf{R}_e}}{E - \varepsilon_{\mathbf{k}'} - R_{\mathbf{k}'}}, \tag{5.45}$$
$$G_{\text{imp}} = \frac{c}{D(E)}(1 + cB + \cdots),$$
$$B = \sum_{l \neq 0} \frac{A_{0l} A_{l0}}{1 - A_{0l} A_{l0}}.$$

The index l in (5.44) and (5.45) runs over all lattice sites except the zeroth. In deriving (5.43)–(5.45) we have taken into account the "self-averaging" of the diagonal Green functions $\langle\langle a_{\mathbf{k}} | a_{\mathbf{k}}^+ \rangle\rangle$ and G_{imp}, which allows us to perform the averaging over the random distribution of impurities only in the polarization operator $R_{\mathbf{k}}$ (both in (5.42) and in $D(E)$ and A_{0l}). The Green function representations obtained (i.e., (5.43)–(5.45)) are completely renormalized. These are self-consistent expressions as the polarization operator $R_{\mathbf{k}}$ is involved in $D(E)$ and A_{0l}, with $R_{\mathbf{k}}$ and G_{imp} containing all terms quadratic in the impurity concentration. The quantities $B_{\mathbf{k}}$ and B describe interaction effects for pairs of impurities separated by arbitrary distances. If these wre neglected, the formulas would be similar to those of the single-site approximation in the coherent potential method (with $c \ll 1$). The terms that have actually been omitted in (5.43)–(5.45) correspond to groups of three and more impurities.

Along with the completely renormalized representation, ILP have also constructed a nonrenormalized representation of the diagonal Green functions that might prove convenient outside the region where the states are described in terms of the wave vector, e.g., in the vicinity of the impurity level. The authors have examined the convergence of various group representations and found the electron spectra for various energies and impurity concentrations.

Near the band edge, i.e., where the energy-wave number relation takes the form $\varepsilon_{\mathbf{k}} = k^2/2m$, the resonant denominator $D^{(0)}(E)$ becomes

$$D^{(0)}(E) = E - \tilde{\varepsilon}_0 - \frac{\gamma^2}{E_1^{3/2}}(-E)^{1/2}, \tag{5.46}$$

with

$$\tilde{\varepsilon}_0 = \varepsilon_0 - \frac{1}{N}\,\text{Re}\sum_{\mathbf{k}} \frac{|\gamma_{\mathbf{k}}|^2}{\varepsilon_{\mathbf{k}}}, \qquad E_1 = \frac{(4\pi)^{2/3}}{2mv^{2/3}}, \qquad \gamma = |\gamma_{\mathbf{k}=0}|^2.$$

Higher-order terms in E/E_1 (where E_1 is of the order of the electron bandwidth) have been omitted.

The energy ε_{loc} of the localized impurity state can be found from the equation

$$D^{(0)}(\varepsilon_{\text{loc}}) = 0. \tag{5.47}$$

From (5.46), the solution is

$$\varepsilon_{\text{loc}} = \tilde{\varepsilon}_0\left(1 - \frac{1}{2\delta} + \frac{1}{2\delta}(1 - 4\delta)^{1/2}\right), \qquad \delta = \tilde{\varepsilon}_0 E_1^3/\gamma^4. \tag{5.48}$$

As can be seen, the position of the impurity level is controlled by two parameters, namely ε_0 (position of the impurity level without interaction with the electron band, or the "primer" impurity level) and δ (or γ), which parameter is a feature of the model discussed. The solution given by (5.48) is physically significant with either $\delta < 0$ or $\delta > \frac{1}{4}$. Accordingly, $\tilde{\varepsilon}_0 > 0$ corresponds to a local impurity level (with a real value of ε_{loc}), whereas $\tilde{\varepsilon}_0 < 0$ corresponds to a virtual level.

First consider $\tilde{\varepsilon}_0 < 0$. The wave function of the impurity state of energy ε_{loc} existing in a crystal with an isolated impurity atom at site $p = 0$ can be represented as

$$|\psi\rangle = \varphi_0|a_0\rangle + \sum_{l\neq 0}\varphi_l|a_l\rangle, \qquad a_l = N^{-1/2}\sum_{\mathbf{k}} e^{-i\mathbf{k}\mathbf{R}_l}a_{\mathbf{k}}, \qquad \varphi_l = \varphi_0 G'_{0l},$$

$$G'_{0l} = \frac{1}{N}\sum \frac{\gamma_{\mathbf{k}} e^{ikR_l}}{\varepsilon_{\text{loc}} - \varepsilon_{\mathbf{k}}} = \frac{\gamma v^{1/3}\exp(-R_l/r_{\text{imp}})}{(4\pi)^{1/3}E_1 R_1}, \tag{5.49}$$

$$r_{\text{imp}} = (v/4\pi)^{1/3}|E_1/\varepsilon_{\text{loc}}|^{1/2} = (2m|\varepsilon_{\text{loc}}|)^{-1/2}.$$

Hence the probability of finding an electron in this state at site $p = 0$ is

$$p_0 = |\varphi_0|^2 = \left(1 + \frac{\gamma^2}{2|\varepsilon_{\text{loc}}|^{1/2}E_1^{1/2}}\right)^{-1} \approx \begin{cases} 1 - \frac{1}{2}|\delta|^{1/2}, & |\delta| \gg 1, \\ 2\delta, & |\delta| \ll 1. \end{cases} \tag{5.50}$$

As can be seen from (5.49) and (5.50), with $|\delta| \gg 1$, the wave function of the impurity state is localized essentially at the impurity site, with just a small fraction (about $2|\delta|)^{-1/2}$) of the function extended over a volume of radius r_{imp}, the impurity state radius. On the other hand, with $|\delta| \ll 1$, which corresponds to strong hybridization, practically all of the wave function is extended over the total volume. Hence, the characteristic impurity concentration, at which a substantial rearrangement of the spectrum may be expected, equals

$$c_0 = v/4\pi r_{\text{imp}}^3 = |\varepsilon_{\text{loc}}/E_1|^{3/2}. \tag{5.51}$$

In accordance with the different behavior of the wave function for $|\delta| \ll 1$ and $|\delta| \gg 1$, the rearrangement of the energy spectrum of a doped crystal with $c \gg c_0$ should be qualitatively different in the two cases. At low impurity concentrations, $c \ll c_0$, the average impurity–impurity separations $\bar{r} \propto (v/c)^{1/3}$ is much larger than the impurity state radius r_{imp}; hence a cooperative rearrangement of the spectrum cannot occur. A detailed analysis of the rearrangement of the energy spectrum has been given in [321] for crystals with $c \ll c_0$ and $c \gg c_0$, $|\delta| \ll 1$ and $|\delta| \gg 1$.

We shall concentrate on the case $c \gg c_0$, $|\delta| \gg 1$ with $\tilde{\varepsilon}_0 < 0$, as can be easily achieved in practice. Besides, important new results have been obtained for this case, which corresponds to an impurity level lying outside the unperturbed band. The wave function of the level is mainly localized at the impurity site. With $c \ll c_0$, the width of the area near ε_{loc}, where the nonrenormalized Green function representation diverges, is of the order of $c\gamma^2/|\varepsilon_{\text{loc}}|$. Meanwhile, the expansion parameter for the renormalized representation is $c(\gamma^2/E_1|E - \varepsilon_{\text{loc}}|)^3$; hence this representation diverges in a narrower area of width

$$|E - \varepsilon_{\text{loc}}| \lesssim \Delta_2 \simeq c^{1/3}\gamma^2/E_1. \tag{5.52}$$

Therefore, with $c \gg c_0$ the renormalized representation is employed, for which the concentration dependent broadening is given by (5.52). Within the convergence region of the renormalized representation, electron states can be either localized or delocalized. The eigenenergies of delocalized (i.e., "band") states characterized by the wave vector can be found from the secular equation

$$E - \varepsilon_{\mathbf{k}} - \mathrm{Re}\,\frac{c|\gamma_{\mathbf{k}}|^2}{D^{(0)}(E)} = 0. \tag{5.53}$$

They are

$$\varepsilon_{1,2}(\mathbf{k}) = \tfrac{1}{2}(\varepsilon_{\mathrm{loc}} + \varepsilon_{\mathbf{k}}) \pm [\tfrac{1}{4}(\varepsilon_{\mathrm{loc}} - \varepsilon_{\mathbf{k}})^2 + c\gamma^2]^{1/2}. \tag{5.54}$$

The decay rate of these states due to scattering on the impurities considered is

$$\Gamma_{1,2}(\mathbf{k}) = \left|\frac{\varepsilon_{1,2}(\mathbf{k}) - \varepsilon_{\mathrm{loc}}}{\varepsilon_1(\mathbf{k}) - \varepsilon_2(\mathbf{k})}\right| \mathrm{Im}\, R_{\mathbf{k}}(\varepsilon_{1,2}(\mathbf{k}))$$

$$= \frac{c\gamma^4 \varepsilon_{\mathbf{k}}^{1/2}}{E_1^{3/2}|[\varepsilon_{1,2}(\mathbf{k}) - \varepsilon_{\mathrm{loc}}][\varepsilon_1(\mathbf{k}) - \varepsilon_2(\mathbf{k})]|}. \tag{5.55}$$

The states with the energy $\varepsilon_2(\mathbf{k})$ make up an impurity band where states can be described in terms of a wave vector and which is isolated from the principal band. The bandwidth Δ_0 is much larger here than the concentration dependent width Δ_2 of the impurity level. If the impurity concentration c satisfies the condition $c_0 \ll c \ll c_1 = c_0|\delta_p|^{1/2}$, then $\varepsilon_2(\mathbf{k})$ takes the form

$$\varepsilon_2(\mathbf{k}) = \varepsilon_{\mathrm{loc}} + \frac{c\gamma^2}{\varepsilon_{\mathrm{loc}} - \varepsilon_{\mathbf{k}}}. \tag{5.56}$$

The energy-momentum relation for this case is shown graphically in Fig. 5.20. The width Δ_0 of the impurity band becomes

$$\Delta_0 = c\gamma^2/|\varepsilon_{\mathrm{loc}}| \ll |\varepsilon_{\mathrm{loc}}|. \tag{5.57}$$

The limiting values assumed by the wave vector in this band can be estimated from the condition that the decay of the state per one wavelength be small, i.e.,

$$\mathbf{k}|\partial\varepsilon_2(\mathbf{k})/\partial\mathbf{k}| \gg \Gamma_2(\mathbf{k}); \tag{5.58}$$

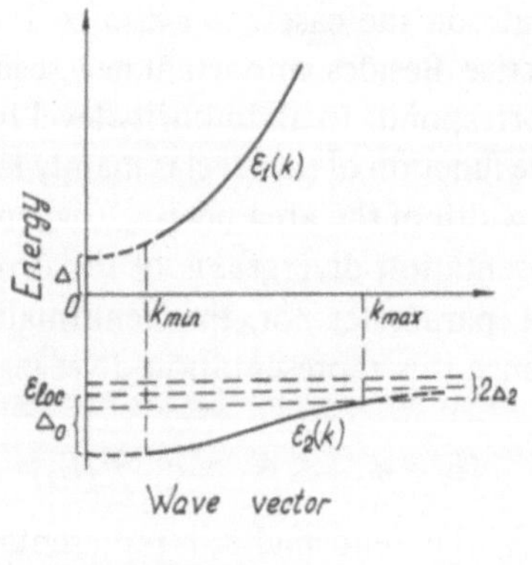

Fig. 5.20. The states in the host and the impurity band for $c_0 \ll c \ll c_0|\delta|^{1/2}$, $\tilde{\varepsilon}_0 < 0$, and $|\delta| \gg 1$ [321].

hence we have

$$k_{\min} \simeq (c_0^{4/3}/c)v^{-1/3}, \qquad k_{\max} \simeq c^{1/3}v^{-1/3}. \tag{5.59}$$

With $k \to k_{\max}$ the energies of the band states approach the magnitude of the concentration-dependent impurity level width. The states below the impurity band are fluctuational. The width Δ_1' of the transition region is controlled here by the magnitude of $\varepsilon_2(k_{\min})$, viz.

$$\Delta_1' = \varepsilon_2(k_{\min}) - \varepsilon_2(0) \simeq \Delta_0(c_0/c)^2. \tag{5.60}$$

The magnitudes of $k_{\min}$ and $k_{\max}$ as given by (5.59) determine the range of energies corresponding to delocalized impurity states. The energies pertaining to these values of the wave vector are of the same order of magnitude as the mobility thresholds in the band. This suggests that, as in the simple diagonally disordered Anderson model [316], a critical concentration c_{cr} might exist at which the first delocalized states appear in the impurity band. The magnitude of c_{cr} differs from c_0 by a numerical factor $\approx 10^{-1}$, whose precise evaluation needs different methods of analysis. It seems noteworthy that, in contrast to the common Anderson transition, the mobility thresholds $k_{\min}$ and $k_{\max}$ in this nondiagonally disordered model both lie on the same side of the maximum of the density of states in the impurity band.

At this point we are able to formulate the cooperative rearrangement (CR) criterion. For cooperative rearrangement to occur and the impurity band to be formed it is necessary that the scale of the rearrangement region be larger than that of the concentration dependent broadening. In other words, the energy of indirect impurity–impurity interaction at average distances, $E_{\text{ave}} \propto c^{1/3}\gamma^2/E_1$, should be less than the energy separation from the impurity level to the renormalized band edge, i.e.,

$$|E_{\text{ave}}| \ll |\varepsilon_1(\mathbf{k} = 0) - \varepsilon_{\text{loc}}|. \tag{5.61}$$

This condition is always met with $c \gg c_0$, provided the hybridization energy is not too high, i.e., $|\delta| \gg 1$. If, however, $|\delta| \ll 1$, then the condition of (5.61) is met for $c \gg c_2 = c_0/|\delta|^3$. Accordingly, the CR criterion can be written as

$$c \gg c_0 \max\{1, |\delta|^{-3}\}.$$

Thus, the spectral area ranging from the nearest vicinity of the local level to that of the boundary of the renormalized impurity band has been analyzed throughout by the technique of expanding diagonal elements of the average Green function in powers of the impurity concentration. The use of this expansion has revealed the actual small parameter $c_0 \approx 3v/4\pi r_{\text{imp}}^3$, allowing one to analyze a number of singular points in the energy spectrum and to discover the cooperative spectrum rearrangement. While not denying the possibility of such a rearrangement, in [328] doubts are expressed as to whether the expansion in powers of the concentration can provide a rigorous description of the transition from localized states to plane waves.

5.3.2. Rearrangement of the Spectrum of Spin Excitations in an Antiferromagnet with Magnetic Impurities in an External Magnetic Field. Magnetic Impuritons

As stated above, the spectrum of elementary excitations in a crystal with large-radius impurity states can be rearranged collectively by increasing either the impurity concentration or the impurity state radius r_{imp}. Antiferromagnetic crystals are highly suitable objects for the implementation of the latter method. Indeed, the frequency difference between the impurity level ω_0 and the magnon band edge $\omega(0)$ can be easily varied in such crystals by varying the external field **H**. With an increase

$$r_{\text{imp}} = a\left(\frac{\Omega}{\omega_0 - \omega(0)}\right)^{1/2},$$

where Ω is the coefficient of the quadratic term of $\omega(\mathbf{k})$, viz., $\omega(\mathbf{k}) = \omega(0) + (4\pi)^{-2/3}\Omega(a\mathbf{k})^2$, assuming values of the order of the magnon bandwidth,

$$c_0(H) \simeq (4\pi)^{-1}(a/r_{\text{imp}})^3$$

can drop below the true concentration c. The result would be a cooperative spectrum rearrangement and the appearance of impuritons.

Below, we will outline briefly a theory for the cooperative rearrangement of the spectrum in antiferromagnetic $CoF_2 + Mn^{2+}$ in an external magnetic field $\mathbf{H} \parallel C_4$, following the papers of ILP [324, 325]. The spin Hamiltonian of CoF_2 has the form [63]

$$\mathscr{H}_{\text{Co}} = \frac{1}{2z}\sum_{\mathbf{n}\alpha\boldsymbol{\rho}} [I_{\parallel} S^z_{\mathbf{n}\alpha} S^z_{\mathbf{n}\alpha+\boldsymbol{\rho}} + I_{\perp}(S^x_{\mathbf{n}\alpha} S^x_{\mathbf{n}\alpha+\boldsymbol{\rho}} + S^y_{\mathbf{n}\alpha} S^y_{\mathbf{n}\alpha+\boldsymbol{\rho}})]$$
$$- \sum_{\mathbf{n}\alpha} \{A(S^z_{\mathbf{n}\alpha})^2 - B_\alpha[(S^x_{\mathbf{n}\alpha})^2 - (S^y_{\mathbf{n}\alpha})^2] + \mu g_{\text{Co}} H S^z_{\mathbf{n}\alpha}\}, \tag{5.62}$$

where $\mathbf{n}\alpha$ is the nth site of the αth magnetic sublattice ($\alpha = 1, 2$); $I_{\parallel}$ and $I_{\perp}$ are the longitudinal and the transverse exchange constants; $\boldsymbol{\rho}$ is the vector connecting nearest neighbors from different sublattices; $A > 0$ and $B_\alpha = (-1)^\alpha B$, with $B > 0$, are signle-ion anisotropy parameters; finally, g_{Co} is the Co^{2+} g-factor in the crystal field of CoF_2. The exchange interaction between nearest neighbors within the same sublattice has been neglected, being much weaker than the intersublattice exchange; z is the number of next-nearest neighbors ($z = 8$).

Since for $H < H^{\parallel}_{\text{cr}}$ (where $H^{\parallel}_{\text{cr}} \approx 22$ T) the collinear orientation of the sublattice spins along $0z$ remains, so that $\langle S^{x,y}_{\mathbf{n}\alpha}\rangle = 0$ and $\langle S^z_{\mathbf{n}\alpha}\rangle \neq 0$, and the Hamiltonian of (5.62) can be brought into the form

$$\mathscr{H}_{\text{Co}} = \sum_{\mathbf{k}j=1,2} \omega_j(\mathbf{k})\beta_j^+(\mathbf{k})\beta_j(\mathbf{k}), \tag{5.63}$$

where $\beta_j^+(\mathbf{k})$ and $\beta_j(\mathbf{k})$ are the creation and annihilation operators for the jth magnon branch

$$\omega_j(\mathbf{k}) = \omega(\mathbf{k}) + (-1)^{j-1}\mu g H,$$

with

$$\omega(\mathbf{k}) = \sqrt{\varepsilon^2 - \zeta\gamma_{\mathbf{k}}^2}, \qquad \gamma_{\mathbf{k}} = z^{-1}\sum_{1}^{z} e^{i\mathbf{k}\boldsymbol{\rho}},$$

$$\varepsilon = I_{\parallel}S_0 + \sqrt{(I_{\parallel}S_0 + A)^2 + 3B^2} - \sqrt{(I_{\parallel}S_0 - A)^2 + 3B^2},$$

$$\zeta = I_{\perp}(x^2 + y^2)/2, \qquad x = \sqrt{3}\cos(\varphi - \chi), \qquad y = 2\sin\varphi\cos\chi,$$

$$\tan 2\varphi = \sqrt{3}\,B/(I_{\parallel}S_0 + A), \qquad \tan 2\chi = \sqrt{3}\,B/(I_{\parallel}S_0 - A),$$

$$g = g_{\mathrm{Co}}(1 - \xi)(1 + \cos\varphi - \cos 2\chi),$$
$$\xi = I_{\parallel}\sin^3 2\varphi/(\sqrt{3}\,B + I_{\parallel}\sin^3 2\varphi). \tag{5.64}$$

The mean value S_0 assumed by the spin at the site (with $H \to 0$) is given by the equation

$$S_0 = \tfrac{1}{2} + \cos 2\varphi. \tag{5.65}$$

Let us assume that the crystal sites $\mathbf{p}_\alpha$ are occupied by impurity Mn^{2+} ions (with the probability c equal for the two sublattices, $\alpha = 1$ or 2). Allowing for isotropy of Mn^{2+} we may assume $g_{\mathrm{Mn}} = 2$ and consider its exchange interaction with the neighboring Co^{2+} ions as isotropic. Then the Hamiltonian of the doped crystal takes the form

$$\mathscr{H} = \mathscr{H}_{\mathrm{Co}} + \mathscr{H}',$$

$$\mathscr{H}' = \frac{1}{z}\sum_{\mathbf{p}\alpha,\boldsymbol{\rho}} \{I'\sigma_{\mathbf{p}\alpha}S_{\mathbf{p}\alpha+\boldsymbol{\rho}} - I_{\parallel}S_{\mathbf{p}\alpha}^z S_{\mathbf{p}\alpha+\boldsymbol{\rho}} - I_{\perp}(S_{\mathbf{p}\alpha}^x S_{\mathbf{p}\alpha+\boldsymbol{\rho}}^x + S_{\mathbf{p}\alpha}^y S_{\mathbf{p}\alpha+\boldsymbol{\rho}}^y)$$
$$+ A(S_{\mathbf{p}\alpha}^z)^2 - B_\alpha[(S_{\mathbf{p}\alpha}^x)^2 - (S_{\mathbf{p}\alpha}^y)^2] + \mu(g_{\mathrm{Co}}S_{\mathbf{p}\alpha}^z - g_{\mathrm{Mn}}\sigma_{\mathbf{p}\alpha}^z)\mathbf{H}\}. \tag{5.66}$$

Using the Loktev–Ostrovski [117] expansion for the operators $S_{\mathbf{n}\alpha}$, viz.,

$$S_{\mathbf{n}1}^+ = xb_{\mathbf{n}1} + yb_{\mathbf{n}1}, \qquad S_{\mathbf{n}2}^+ = xb_{\mathbf{n}2} - yb_{\mathbf{n}2},$$

$$S_{\mathbf{n}\alpha}^z = (-1)^\alpha[S_0 - (1 + \cos 2\varphi - \cos 2\chi)b_{\mathbf{n}\alpha}^+ b_{\mathbf{n}\alpha}],$$

and the Holstein–Primakoff expansion for $\sigma_{\mathbf{p}\alpha}$,

$$\sigma_{\mathbf{p}1}^+ = \sqrt{2\sigma}\,a_{\mathbf{p}1}, \qquad \sigma_{\mathbf{p}2}^+ = \sqrt{2\sigma}\,a_{\mathbf{p}2}^+,$$

$$\sigma_{\mathbf{p}\alpha}^z = (-1)^{\alpha-1}(\sigma - a_{\mathbf{p}\alpha}^+ a_{\mathbf{p}\alpha}),$$

we bring (5.66) into the form

$$\mathscr{H}' = \frac{1}{z}\sum \{\omega_\alpha a_{\mathbf{p}\alpha}^+ a_{\mathbf{p}\alpha} - \varepsilon_\alpha b_{\mathbf{p}\alpha}^+ b_{\mathbf{p}\alpha}$$
$$+ [a_{\mathbf{p}\alpha}(m_{1\alpha}b_{\mathbf{p}\alpha+\boldsymbol{\rho}}^+ + m_{2\alpha}b_{\mathbf{p}\alpha+\boldsymbol{\rho}}) - \zeta b_{\mathbf{p}\alpha}b_{\mathbf{p}\alpha+\boldsymbol{\rho}} + \text{h.c.}]\},$$

$$\omega_\alpha = \omega_0 + (-1)^{\alpha-1}\mu\tilde{g}H, \qquad \omega_0 = I'S, \tag{5.67}$$

$$\tilde{g} = g_{\mathrm{Mn}} - \xi g_{\mathrm{Co}}I'/I_{\parallel}, \qquad \varepsilon_\alpha = \varepsilon + (-1)^\alpha\mu gH,$$

$$m_{1\alpha} = (-1)^{\alpha-1}I'\sqrt{\sigma/2}\,y, \qquad m_{2\alpha} = I'\sqrt{\sigma/2}\,x,$$

$$\Delta = (I'\sigma - I_{\parallel}S_0)(1 + \cos 2\varphi\cos 2\chi).$$

The first term under the summation sign in (5.67) corresponds to two excitation frequencies of the impurity spins in the average field of the host matrix, split by the external field. The other terms are responsible for the dynamic interaction of the host and impurity excitations (the corresponding constants are $m_{j\alpha}$) and the magnon scattering by the impurity center (terms diagonal in $b_{p\alpha}$).

The characteristic frequency spectrum of the impure crystal and, accordingly, its absorption spectrum, are described by diagonal Green functions that make up the matrix $G(\mathbf{k}, \mathbf{k}')$ in the fourspace L_4, corresponding to two types of excitations in each sublattice, viz.,

$$G(\mathbf{k}, \mathbf{k}'; \omega) = \langle\langle |B(\mathbf{k})\rangle_4 | _4\langle B^+(\mathbf{k}')| \rangle\rangle_\omega, \tag{5.68}$$

The line vector from the space L_4 is

$$_4\langle B^+(\mathbf{k})| = \{b_1^+(\mathbf{k}), b_2(-\mathbf{k}), b_2^+(\mathbf{k}), b_1(-\mathbf{k})\},$$
$$b_\alpha(\mathbf{k}) = N^{-1/2} \sum_n b_{\mathbf{n}\alpha} e^{i\mathbf{k}\mathbf{n}_\alpha}.$$

Its approximate form linear in the concentration is

$$\hat{G}(\mathbf{k}, \mathbf{k}) = [(\hat{G}^{(0)})^{-1}(\mathbf{k}) - \hat{R}_\mathbf{k}]^{-1}, \tag{5.69}$$

where the matrix of unperturbed Green functions has the form

$$\hat{G}^{(0)}(\mathbf{k}) = \begin{pmatrix} \hat{g}_1(\mathbf{k}) & 0 \\ 0 & \hat{g}_2(\mathbf{k}) \end{pmatrix}, \tag{5.70}$$

with

$$\hat{g}_{1,2}(\mathbf{k}) = \Delta_{1,2}^{-1}(\mathbf{k}) \begin{pmatrix} \varepsilon_{2,1} + \omega & -\zeta\gamma_\mathbf{k} \\ -\zeta\gamma_\mathbf{k} & \varepsilon_{1,2} - \omega \end{pmatrix},$$

$$\Delta_j(\mathbf{k}) = [\omega + (-1)^j \mu g H]^2 - \omega^2(\mathbf{k}), \qquad \gamma_\mathbf{k} = \frac{1}{z} \sum_\rho e^{i\mathbf{k}\rho},$$

and the polarization operator is

$$\hat{R}_\mathbf{k} = c \sum_\alpha {}_2\langle \psi_\mathbf{k}^+ | \hat{V}_\alpha (\hat{1} - \hat{G}_0 \hat{V}_\alpha)^{-1} | \psi_\mathbf{k} \rangle_2. \tag{5.71}$$

The poles, yielding the spin excitation spectrum of the doped crystal, are solutions of the equation

$$D_\mathbf{k} = \det[(\hat{G}^{(0)})^{-1}(\mathbf{k}) - \hat{R}(\mathbf{k})] = \Delta_1(\mathbf{k})[\Delta_2(\mathbf{k}) - R^1(\mathbf{k}) - R^2(\mathbf{k})] = 0. \tag{5.72}$$

The quantities $R^{(\alpha)}(\mathbf{k})$ correspond to resonances between the upper (with $\alpha = 1$) or the lower ($\alpha = 2$) impurity sublevels in the field $\mathbf{H}$ and the $\omega_2(\mathbf{k})$ branch of the host excitations. They are

$$R^{(\alpha)}(\mathbf{k}) = c\gamma_\mathbf{k}^2 V^\alpha [D^\alpha(\omega, H)]^{-1}, \tag{5.73}$$

with

$$D^{\alpha}(\omega, H) = 1 - V^{\alpha} f_2(\omega, H),$$

$$V^{\alpha} = \zeta^2 + (-1)^{\alpha-1}[\varepsilon_\alpha + (-1)^{\alpha}\omega] V^{\alpha}_{\alpha'} + (V^{\alpha}_3)^2(\varepsilon^2_\alpha - \omega^2) f_r(\omega, H)/W^{\alpha}, \qquad \alpha' \neq \alpha,$$

$$V^{\alpha}_{\alpha'} = \Delta - \sum_{j=1,2} (-1)^{\alpha'+j} m^2_{j\alpha}[\omega + (-1)^{\alpha'+j}\omega_\alpha]^{-1}, \tag{5.74}$$

$$V^{\alpha}_3 = 2m_{1\alpha} m_{2\alpha}\omega_\alpha(\omega^2 - \omega^2_\alpha)^{-1},$$

$$W^{\alpha} = 1 - [\zeta^2 + V^{\alpha}_{\alpha'}(\varepsilon_\alpha - \omega)] f_1(\omega, H),$$

$$f_j(\omega, H) = N^{-1} \sum_{\mathbf{k}} \gamma^2_{\mathbf{k}} \Delta^{-1}_j(\mathbf{k}).$$

The frequencies $\omega^{\alpha}_0(H)$ of the two impurity levels are given in the single-impurity approximation by the roots of the equation

$$D^{\alpha}(\omega^{\alpha}_0(H), H) = 0. \tag{5.75}$$

Various group expansions of the polarization operator $\hat{R}_{\mathbf{k}}$ can be derived as before. The expansion terms of higher order than linear in c correspond to indirect impurity–impurity interactions, $A_{\mathbf{pp}'}$. They are responsible for the nature of the excitations in the range of the cooperative rearrangement of the spectrum. Various representations are convergent in various frequency ranges, whereas the range where none of the representations is convergent is identified as the concentration dependent broadening $\Delta^{\alpha}(H)$ of the impurity levels $\omega^{\alpha}_0(H)$, viz.,

$$|\omega - \omega^{\alpha}_0(H)| \lesssim \Delta^{\alpha}(H) \approx |A_{\mathbf{pp}'} D^{\alpha}(\omega, H)| \left|\frac{\partial D^{\alpha}(\omega, H)}{\partial \omega}\right|_{\omega^{\alpha}_0 H}, \tag{5.76}$$

where $|\mathbf{p} - \mathbf{p}'| \approx ac^{-1/3}$. The expression for $A_{\mathbf{pp}'}$ can be found in [322].

The frequency of the doubly degenerate impurity level $\omega^{1,2}_0(0) \equiv \omega_0(0)$ in $CoF_2 + c$ Mn lies, for $H = 0$, somewhat lower than the AFMR frequency $\omega(0)$. In a field $\mathbf{H} \parallel C_4$ the degeneracy is removed and with sufficiently small $\mathbf{H}$ we can write (retaining only terms linear in H and $\omega - \omega(0)$ in the left-hand side of (5.75))

$$\omega^{\alpha}_0(H) = \omega_0(0) + (-1)^{\alpha-1}\mu g_{\mathrm{eff}} H, \tag{5.77}$$

where

$$g_{\mathrm{eff}} = \frac{(\partial V^{\alpha}/\partial H)_0 f(0)\mu^{-1} + g(V^{\alpha}\partial f_2(\omega, 0)/\partial\omega)_{\omega_0(0)}}{[\partial(V^{\alpha} f_2(\omega, 0))/\partial\omega]_{\omega_0(0)}},$$

$$f(0) = f_j(\omega_0(0), 0).$$

As the field is increased, the dependence $\omega^{\alpha}_0(H)$ becomes essentially nonlinear (which has been confirmed by numerical analysis of (5.75)), and for some resonant field strength H^{α}_r the excitation frequency of an isolated impurity in the αth sublattice approaches the band boundary, viz.,

$$\omega_0^\alpha(H_r^\alpha) = \omega(0) - \mu g H_r^\alpha \equiv \omega_r^\alpha. \tag{5.78}$$

Over a sufficiently wide range about the resonance $(\omega_r^\alpha, H_r^\alpha)$, this quantities V^α in (5.74) can be expanded in a Taylor series to first-order terms in $(\omega - \omega_r^\alpha)$ and $(H - H_r^\alpha)$, using the expansion

$$f_2(\omega, H) = f_0 + \sqrt{\mu g(H_r^\alpha - H) - \omega + \omega_r^\alpha}/[2\omega(0)\Omega^{3/2}] \tag{5.79}$$

for $f_0(\omega, H) \simeq -0.39\zeta^2$, where ζ is determined by (5.64). Then the impurity level frequency can be represented analytically as

$$\omega_0^\alpha(H) = \omega_r^\alpha + \mu g_r^\alpha(H - H_r) - \tfrac{1}{2}\delta_\alpha + [\tfrac{1}{4}\delta_\alpha^2 + \mu(g + g_r)(H - H_r)\delta]^{1/2}, \tag{5.80}$$

with

$$\begin{aligned} \delta_\alpha &= m_\alpha^4/\Omega^3, \\ m_\alpha^2 &= -\{V^\alpha[2\omega(0) f_0\, \partial V^\alpha/\partial\omega]^{-1}\}_{\omega_r^\alpha, H_r^\alpha}, \\ g_r^\alpha &= -[(\partial V^\alpha/\partial H)/(\mu\, \partial V^\alpha/\partial\omega)^{-1}]_{\omega_r^\alpha H_r^\alpha}. \end{aligned} \tag{5.81}$$

If $\mu(g + g_r^\alpha)(H_r^\alpha - H) \ll \delta_\alpha$, then the impurity level moves close to the edge of the band according to the square law

$$d_\alpha(H) \equiv \omega_2(0) - \omega_0^\alpha(H) \approx [\mu(g + g_r^\alpha)(H - H_r^\alpha)]^2/\delta_\alpha. \tag{5.82}$$

In the magnetic system described the resonant areas for the two impurity levels are sufficiently widely separated. Therefore, the spectrum of band states described in terms of a quasi momentum can be sought separately for each of the areas, using the equation

$$\Delta_2(\mathbf{k}) - \operatorname{Re} R^\alpha(\mathbf{k}) = 0. \tag{5.83}$$

In the region of validity of the approximate (5.79) this equation takes the form

$$\omega - \omega_r^\alpha + \mu g(H - H_r^\alpha) - \omega(\mathbf{k}) + \omega(0) - \frac{c m_\alpha^2 \gamma_\mathbf{k}^2}{\omega - \omega_r^\alpha - \mu g_r^\alpha(H - H_r^\alpha) - \{[\mu g(H_r^\alpha - H) - \omega + \omega_r^\alpha]\delta_\alpha\}^{1/2}} = 0. \tag{5.84}$$

The solutions are physically significant if they are in the range of convergence of the renormalized group representation for $R^\alpha(\mathbf{k})$. Far from the resonance (i.e., for $c \ll c_0^\alpha(H) \simeq 4\pi d_\alpha(H)/\Omega^{3/2}$) the physically meaningful solutions are those representing weakly perturbed states from the principal band, whose frequencies are

$$\tilde{\omega}_j(\mathbf{k}) = \omega_j(\mathbf{k}) + \operatorname{Re}\sum_\alpha R^\alpha(\mathbf{k}, \tilde{\omega}_j(\mathbf{k})). \tag{5.85}$$

The corresponding concentration-dependent broadening of the impurity level is exponentially small, viz.,

$$\Delta^\alpha(H) \propto \frac{c^{1/3} m_\alpha^2}{\Omega(1 + \frac{1}{2}\sqrt{\delta_\alpha/d_\alpha(H)})} \exp[-v(c_0^\alpha(H)/c)^{1/3}], \qquad v \simeq 1. \tag{5.86}$$

This estimate is determined by the convergence radius of the nonrenormalized representation, while the region where the renormalized representation diverges is much wider, i.e.,

$$|\omega - \omega_0^\alpha(H)| - \Delta'_\alpha(H) \simeq c^{1/2} m_\alpha^2 \Omega^{-3/4} d_\alpha^{-1/4}(H).$$

For this reason, although (5.84) has some solutions within a range of width $cm_\alpha^2/d_\alpha(H)$ lying below $\omega_0^\alpha(H)$, off $\Delta^\alpha(H)$, they do not correspond to real band states as they fall within $\Delta'_\alpha(H)$. Therefore, for $c \ll c_0^\alpha(H)$ all the states near $\omega_0^\alpha(H)$ are localized.

Now we will analyze the immediate vicinity of the resonant field H_r^α, where $|d_\alpha(H)| \lesssim c^{2/3}\Omega$. In this region, the impurity state radius is larger than the average separation $\bar{r}$ between impurities, i.e., $r_{\text{imp}}(H) \gtrsim \bar{r}$ and $c_0^\alpha(H) \leq c$. As a result, the indirect interaction of impurities becomes essential, rendering the single-impurity approximation inapplicable. The broadening $\Delta^\alpha(H)$ increases from the exponentially small value of (5.86) to

$$\Delta^\alpha \simeq \begin{cases} c^{2/3}\Omega, & c \ll c_{\text{cr}}^\alpha = \eta(m_\alpha/\Omega)^6, \\ c^{1/3} m_\alpha^3/\Omega, & c \gg c_{\text{cr}}^\alpha. \end{cases} \tag{5.87}$$

We wish to emphasize particularly the existence of a critical concentration $c_0 \ll c_{\text{cr}} \ll 1$ that divides the cooperative rearrangement area into two regions, that of coherent (CR) and that of incoherent rearrangement (ICR). For $c \ll c_{\text{cr}}^\alpha$ the broadening Δ^α is of the same order of magnitude as the separation $d_\alpha(H)$ between the impurity level and the magnon band. Hence the impurity line merges with the AFMR line. The vicinity of the boundary of the band, of width $\simeq c^{2/3}\Omega$, is occupied by fluctuational levels. The notion of band states is not meaningful here, nor is that of local or quasi-local levels. This structure of the spectrum corresponds to ICR. Above the ICR region, states of the principal band remain, which are characterized by the $\tilde{\omega}_j(\mathbf{k})$ law of (5.85) with $ak \gtrsim c^{1/3}$. As the field is further increased, $c_0^\alpha(H)$ becomes larger than c again and the ICR region disappears. With $\mu(g + g_r^\alpha)(H - H_r^\alpha) \gg \delta_\alpha$ a quasi-local level appears in the magnon spectrum at the frequency

$$\omega_0^\alpha(H) \simeq \omega_r^\alpha + \mu g_r^\alpha(H - H_r^\alpha).$$

For $c > c_{\text{cr}}^\alpha$ coherent rearrangement can occur near the crossover point, where $c_0^\alpha(H) < c$ (or $\mu(g + g_r^\alpha)(H - H_r^\alpha) \lesssim c^{2/3}\Omega$). Both solutions of (5.84) are physically signficant here, namely

$$\begin{aligned} \Omega_j^\alpha(\mathbf{k}, H) = \omega_r^\alpha &+ \tfrac{1}{2}[\omega(\mathbf{k}) - \omega(0)] - \tfrac{1}{2}\mu(g - g_r^\alpha)(H - H_r^\alpha) \\ &- (-1)^j\{[\tfrac{1}{2}\omega(\mathbf{k}) - \omega(0)] \\ &- \tfrac{1}{2}\mu(g + g_r)(H - H_r^\alpha)]^2 + cm_\alpha^2\gamma_{\mathbf{k}}^2\}^{1/2}, \qquad j = 1, 2. \end{aligned} \tag{5.88}$$

As we see, in the CR case there are two band state regions instead of one. The range of permissive wave vector values in the impurity band can be estimated from the condition that the decay per wavelength should be small, i.e., $ak \leq c^{1/3}$. The width $|\Omega_2^\alpha(0, H) - \omega_0^\alpha(H)|$ of the impurity band (which is about $c^{1/2}m_\alpha$

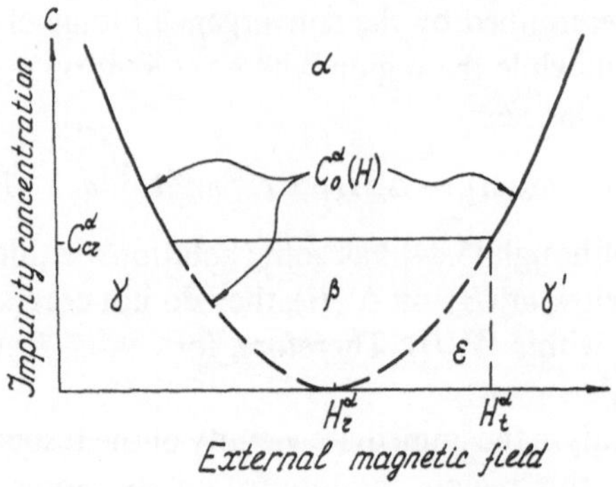

Fig. 5.21. Phase diagram of states in a doped antiferromagnet [322–324].

for $\mu(g + g_r^\alpha)(H - H_r^\alpha) \lesssim c^{1/2}m_\alpha$) is greater than the magnitude of the broadening Δ^α. The smallest value of the separation $|\Omega_1^\alpha(0, H) - \Omega_2^\alpha(0, H)|$ between the two limiting frequencies of the absorption lines, that is assumed for $H \simeq H_r^\alpha$ equals $2c^{1/2}m_\alpha > \Delta^\alpha$. Hence, in the case of CR, two lines in the adsorption spectrum can be resolved through the entire range of field strengths. With H larger than H_r^α the impurity band is separated from the principal band by a narrow quasi gap, whose width becomes equal to $\Delta^\alpha(H)$ for $\mu(g + g_r^\alpha)(H - H_r^\alpha) \propto c^{2/3}$. After this, the quasi gap rapidly closes, leaving in the spectrum only the range of band states and a quasi-local level similar to the one discussed in connection with ICR.

The phenomena concerning the cooperative rearrangement of the spectrum in an impure antiferromagnet can be generally represented in the form of a phase diagram as in Fig. 5.21. The parabolic curve represents the concentration $c_0^\alpha(H)$ at which the cooperative rearrangement occurs. For $H = H_r^\alpha$ the impurity line and the AFMR frequencies coincide, $c_0^\alpha(H) = 0$. When $|H - H_r^\alpha|$ is increased, $c_0(H)$ increases too, to become roughly equal to c_{cr} at $|\omega_1 - \omega_0| \simeq \delta_\alpha$. The field dependence of c_0^α can be represented as

$$c_0^\alpha(H) \propto |H - H_r^\alpha|^3, \qquad c_0^\alpha < c_{cr}^\alpha,$$

$$c_0^\alpha(H) \propto |H - H_r|^{3/2}, \qquad c_0^\alpha > c_{cr}.$$

The point H_t^α corresponds to the field strength at which

$$\mu(g + g_r^\alpha)(H_t' - H_r^\alpha) \simeq \delta^\alpha, \qquad c_0(H_t^\alpha) \simeq c_{cr}.$$

The ranges of parameters where the Anderson transition may be expected (i.e., appearance of collectivized delocalized states on the background of localized states, and the reverse transition) are shown in the figure by the solid lines. The region labeled γ, where $c \ll c_0^\alpha(H)$ corresponds to the presence of a local impurity level. Dominant among the impurity states here are the states localized near noninteracting impurities, i.e., in a sense the system can be compared with a weakly nonideal gas. In the region α (where $c > c_0^\alpha(H)$ and

$c > c_{cr}^{\alpha}$), CR has occurred and two ranges of band states are present. The coherent impurity states possess long-range ordering, and hence in this respect can be compared with the perfect crystal. The range β ($c > c_0^{\alpha}(H)$, yet $c < c_{cr}^{\alpha}$) is characterized by ICR. Most of the impurity states are localized at many impurity sites at a time. This range may be compared to a liquid phase. In the ranges α and β magnetic impuritons can exist. The region γ', where $c < c_0^{\alpha}(H)$ and $H > H_r^{\alpha}$, corresponds to a "gas" of quasi-local excitations with a finite lifetime. Finally, in the region ε ($c_0^{\alpha}(H) \gg c$) the decay is strong, and hence impurity excitations are virtual and cannot produce resonances in the spin wave (magnon) band.

Now we set out to calculate the magnetic dipole absorption of the electromagnetic radiation polarized within the plane perpendicular to C_4. The absorption coefficient in the region where the downward-going edge of the magnon band approaches (with increasing H) the upward-going impurity level can be expressed in the conventional way in terms of retarded Green functions, viz.,

$$\sigma(\omega, H) = \frac{K}{\pi} \lim_{\mathbf{k}\to 0} \Bigg(\langle\beta_2(\mathbf{k})|\beta_2^+(\mathbf{k})\rangle_\omega + \frac{1}{N}\sum_{\mathbf{p}} [e^{ikp^+}\langle\langle a_{\mathbf{p}}|\beta_2^+(\mathbf{k})\rangle\rangle_\omega + e^{-i\mathbf{kp}}\langle\langle\beta_2(\mathbf{k})|a_{\mathbf{p}}^+\rangle\rangle_\omega] + \frac{1}{N}\sum_{\mathbf{p},\mathbf{p}'} e^{i\mathbf{k}(\mathbf{p}-\mathbf{p}')}\langle\langle a_{\mathbf{p}}|a_{\mathbf{p}'}^+\rangle\rangle_\omega \Bigg). \tag{5.89}$$

Here K is a factor that weakly depends on frequency. Away from the collective rearrangement area, (5.89) describes two peaks, one of which represents the AFMR line at $\omega_2(H)$, while the other corresponds to the absorption at the impurity-induced frequency $\omega_0^1(H)$. In the region of validity of the single-impurity approximation the integrated absorption intensity (i.e., that integrated over ω with $H = \text{const.}$) of the electromagnetic radiation at $\omega_0^1(H)$ can be written as

$$\mathscr{I}_{\text{imp}}(H) = cK\Lambda(H)P(H), \tag{5.90}$$

with

$$P(H) = |\psi_0|^2 = [1 + \tfrac{1}{2}\sqrt{\delta_1 d_1^{-1}(H)}]^{-1},$$
$$\Lambda(H) = [1 + m_1 d_1^{-1}(H)]^{-1}, \tag{5.91}$$
$$\delta_1 \simeq m_1^4/\Omega^3, \qquad d_1(H) = \omega_2(H) - \omega_0^1(H).$$

The quantity $P(H)$ is the squared modulus of the impurity state wave function ψ taken at the impurity site. In other words, it gives the field dependence of the oscillator strength. Sufficiently far from the rearrangement area ψ is almost completely localized at the impurity site, hence $P(H) = 1$. Closer to the rearrangement area the state described by ψ gets delocalized to become

$$|\psi_0^2| = \mu(g + g_r^1)(H - H_r^1)/\delta_1,$$

when $d_1(H) \ll \delta_1$. Only a small fraction of this function is localized near the impurity site, hence $P(H)$ decreases. However, (5.90) and (5.91) contain another factor, $\Lambda(H)$, that grows faster than $P(H)$ decreases. The increase in $\Lambda(H)$ is associated with the fact that the absorption at the impurity mode frequency is contributed to not by the impurity excitation alone but by the host excitation with $\mathbf{k} = 0$ coupled to the former via the hybridization constant m_1. This coupling corresponds to an effective increase in r_0. The resonant increase in $\Lambda(H)$ results in an enhancement of the impurity line in the vicinity of the AFMR line, at the expense of the latter.

It should be emphasized that if the impurity line approached an optically inactive (in this case, the upper) edge of the band, $\Lambda(H)$ would not experience a resonant increase, while ψ would be delocalized as before. As a result, the impurity line would become weaker rather than be enhanced. We will discuss this case in some detail below, when analyzing the impurity line intensity in $FeF_2 + Co$.

In the proximity of the cooperative rearrangement region of the spectrum (i.e., $d_1(\hat{n}) \to c^{2/3}\Omega$) the intensity of the impurity line is close to the AFMR line intensity, both lines merging for $c < c^1_{\mathrm{cr}}$. With $c > c^1_{\mathrm{cr}}$ the two absorption lines remain in the CR region and the intensity is transferred from one line to another. The ratio of the intensities of the two absorption lines at $\Omega_1(H)$ and $\Omega_2(H)$ in the CR range can be written as

$$\frac{\mathscr{I}_1}{\mathscr{I}_2} = \frac{\Omega_1(H) - \omega_0^1(H)}{\omega_0^1(H) - \Omega_2(H)} \tag{5.92}$$

The dependence of $\mathscr{I}_1$ and $\mathscr{I}_2$ on H is shown in Fig. 5.22.

Thus, the interaction of the impurity and host magnetic excitations causes delocalization of the impurity state. The interaction strength (and hence, the amount of delocalization) is higher, the less the energy separation between the impurity level and the bottom of the magnon band. In antiferromagnets this separation can easily be varied with the aid of an applied magnetic field. Delocalization of the impurity wave function and the corresponding enhance-

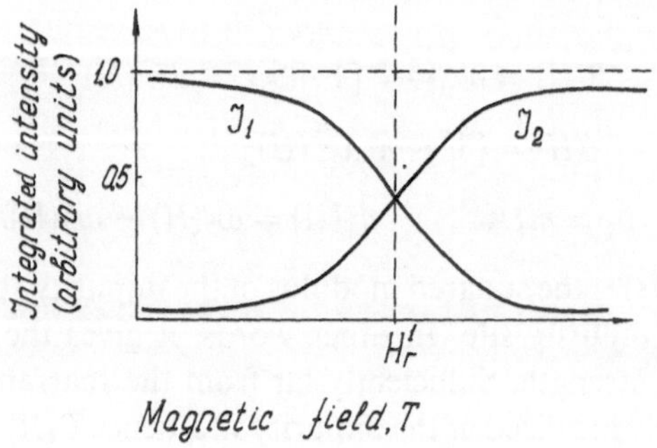

Fig. 5.22. Integrated intensities of the AFMR (J_1) and the impurity (J_2) lines as functions of H in the case of a coherently rearranged spectrum [322–324].

ment of the impurity state radius make the host matrix and the impurity excitations mutually dependent. This leads to an increased intensity of impurity absorption. When r_{imp} is ultimately increased to the extent that it becomes comparable with the average impurity separation, the impurity–impurity interaction via the host ions becomes of sustantial importance; the spectrum undergoes a cooperative rearrangement and the formerly localized impurity states become impuritons. The spectrum rearrangement is coherent if $c > c_{cr} = \eta(m/\Omega)^6$, which condition is necessary to overcome the scatter in impurity level energies. With CR, the impurity excitations form a full-fledged impurity–magnon band where they can be described in terms of a wave vector. The absorption spectrum demonstrates split lines with a transfer of intensity between them. If $c < c_{cr}$, the rearrangement of the spectrum is incoherent and the vicinity of the spin wave band edge is filled with fluctuational levels. The impurity states do not form a magnon band. The notion of an isolated local or quasi-local impurity level cannot be applied. The absorption lines of the AFMR and of the impurity excitation merge.

5.3.3. Experimental Studies of Magnetic Impuritons (CoF_2 + c Mn)

As follows from the above analysis, the appearance of magnetic impuritons is accompanied either by splitting (in the CR case) or merging (with ICR) the impurity and the intrinsic lines, with a considerable enhancement of the impurity line (ICR) or a transfer of intensity between the impurity and the intrinsic line (CR), which effects occur in either the absorption or Raman scattering spectra. The behavior of absorption lines in CoF_2 + 0.01 Mn and $CoCO_3$ + c Fe described in the Introduction suggested the existence of magnetic impuritons. Purposeful experiments were designed to establish general features of the behavior of magnetic impuritons, evaluating c_{cr} and η, and to find some experimental signature of CR and ICR, i.e., of the existence of coherent and incoherent impuritons. They also examined the role of the crystal symmetry and the physical nature of the impurity ion. Suitable systems turned out to be CoF_2 + c Mn and $CoCO_3$ + c_1 Fe + c_2 Mn + c_3 Ni.

With regard to the first of these crystals, observation of a gap impurity mode was reported in [372–374, 375]. The mode was detected near 28 cm^{-1} (for $H = 0$), with the AFMR line in the vicinity of 36 cm^{-1}. In [200] an increased intensity of the impurity line was reported, observed when the AFMR line was approached with increasing **H**.

The theory of magnetic impuritons in this crystal was developed by ILP [324, 325] and has been presented above. It was tested experimentally in [358, 376]. To that end, single crystals with impurity weight concentration between 2×10^{-5} and 10^{-2} were used. The samples were parallel-plane or slightly tapered plates of 3×3 mm^2. Particular attention was given to the choice of an optimal thickness, since the area of efficient interaction between the intrinsic and the impurity excitations cannot be properly examined if the absorption lines are saturated. On the other hand, too thin samples might have restricted the region where impurity lines could be observed. The sample thickness was varied from 0.06 to 0.01 mm, with the corresponding absorption intensity in the interaction region reaching 50 to 70%, which was sufficient for resonably accurate measurements of the line intensities and widths. The measurements were performed in the regions where the low-frequency AFMR mode approached up- or down-going impurity levels, with the field **H** oriented either along or perpendicular to the crystal axis C_4. The corresponding interaction regions have been marked as 1, 2, and 3.

High-frequency measurements (22 to 42 cm^{-1}) were carried out with a diffractional FIR spectrometer [377, 378]. A magnetic field of up to 6 T was provided by a superconducting solenoid. The spectrum could be scanned either by varying the frequency at a fixed magnetic field or by varying the field strength for a fixed frequency. To improve the accuracy of the line shape and linewidth measurements, both types of scanning were employed. The magnetic field could be measured to a high accuracy (about 0.2%) by measuring the current flowing through the solenoid.

For frequencies of 12 to 30 cm^{-1} a pulsed, transmission-type submillimeter spectrometer was used [379]. The radiation sources were backward wave tubes, the detector was an n-InSb crystal cooled to 4.2 K. The wavelength was measured to 0.2% accuracy with an interferometer. The wavelength stability was provided mainly by the constant (to 10^{-4}) voltage at the slowing structure of the tube. With a calibration curve available, the wavelength could be determined to a high accuracy by measuring the potential of the slowing structure with a digital voltmeter.

The sample was placed in the center of a conventional solenoid (or a Helmholtz solenoid) and held between two conical teflon or quartz lenses. The use of conical lenses and spring-loaded mounts provided a broad band of the spectrometer and low transmission losses of the radiation fed to the sample. The major advantage of this arrangement is the possibility of adjusting the pressure at the sample such that the resonance lines would not be distorted even at the transition field (about 22 T) for the CoF_2 crystal, which is characterized by a high magnetostriction constant.

The magnetic field strength was evaluated by integrating the response of a probe coil placed in the bore hole of the pulsed solenoid, right near the sample. To calibrate the magnetic field, either the EPR line of DPPH or the AFMR line of $RbMnF_3$ were measured. A polyethylene pellet with a small amount of one of the materials was placed near the sample. By using calibration marks, taking into account the nonlinearity shown by the field recording channel, and monitoring the zero field level, the magnetic field could be measured to an accuracy of 1 to 2%.

An accurate determination of the resonance and impurity line intensities was a matter of prime importance for this work. While satisfactory techniques have been developed for conventional (i.e., optical) spectrometers, the problem remains difficult for submillimeter spectrometers operating with pulsed

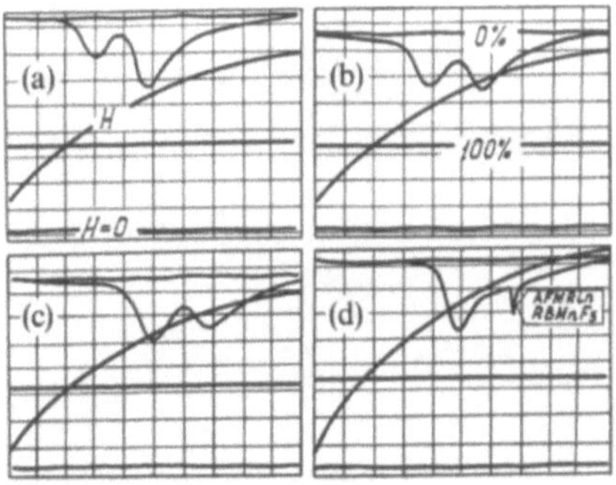

Fig. 5.23. Oscillograms for $CoF_2 + 3 \times 10^{-2}$ Mn in the interaction region 2 (coherent rearrangement of the spectrum). The magnetic field orientation is $\mathbf{H} \parallel C_4$, the sample thickness 0.07 mm. (a) Enhancement of the impurity absorption line approaching the AFMR line, $\nu = 22.4$ cm^{-1}; (b) absorption lines near the cross-over point, $\nu = 19.6$ cm^{-1}; (c) absorption lines past the cross-over point, $\nu = 19.6$ cm^{-1}; (d) the impurity-dependent absorption manifests itself through distortions in the right wing of the AFMR line, $\nu = 15.6$ cm^{-1}, and $T = 4.2$ K, $T_N = 37.7$ K [358].

magnetic fields and radiation sources of low intensity, which are characterized by highly nonuniform amplitude characteristics. Many of the difficulties could be avoided when the 100% and zero absorption levels were memorized on the screen of a memory oscilloscope prior to each magnetic pulse. The spectrogram photographed from the screen would contain records of the field H and the zero magnetic field level ($H = 0$), as well as the field-scanned spectrum of the sample and the 0% and 100% levels of the absorption for $H = 0$ (see Fig. 5.23). By analyzing all of these levels, absorption line parameters can be evaluated to a high accuracy.

In calculating absolute integrated intensities, corrections for the spectrometer apparatus function and line truncation were introduced. Normally, the sample linewidths were several times larger than the width of the spectrometer apparatus function. Hence, the major correction was that for the line truncation. Taking into account all the necessary corrections and the deterioration of the signal-to-noise ratio that occurred at some of the frequencies, the net accuracy of the absolute integrated intensities measured can be estimated as 30%.

The field dependences of the intrinsic and impurity absorption line frequencies of $CoF_2 + 4 \times 10^{-3}$ Mn^{2+} have been investigated with $\mathbf{H} \parallel C_4$ through the range of 12 to 42 cm^{-1} using d.c. magnetic fields up to 6 T and pulsed fields up to 17 T. For $H = 0$ the widths of the impurity absorption line and the AFMR line were about 1.5 and 2.2 cm^{-1}, respectively, showing no substantial changes with variation of H. The lines were of a slightly asymmetrical shape, with each half somewhere between Lorentzian and Gaussian. The general absorption spectrum of the $CoF_2 + 4 \times 10^{-3}$ Mn crystal is shown in

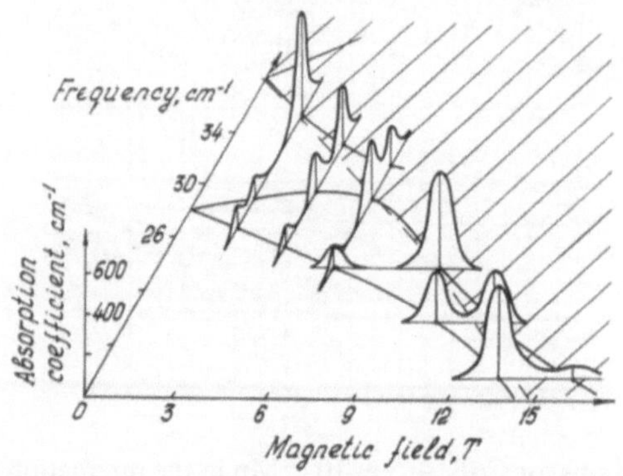

Fig. 5.24. General appearance of the absorption spectrum of $CoF_2 + 4 \times 10^{-3}$ Mn with $\mathbf{H} \parallel C_4$. Contours of the absorption lines are shown for a frequency-scanned spectrum in the interaction region 1 and for a field-scanned spectrum in region 2. The hatched region is the magnon band of a pure crystal. The solid lines connect experimental points.

Fig. 5.24. One can see how the lines transform as the magnetic field H is varied. Measured and calculated data for the same crystal are compared in Fig. 5.25. The theoretical curves have been calculated for $A = 24$ cm^{-1}, $B = 43$ cm^{-1}, and $g_0 = 2.6$ [380]. The values of $I_{\parallel} = 35$ cm^{-1} and $I_{\perp} = 42$ cm^{-1} have been selected to provide the best fit of the theoretical and the measured data. The good agreement provides support for the theoretical model chosen for $CoF_2 + c\ Mn^{2+}$.

The interaction region 1 was investigated for CoF_2 samples with weight concentrations of the Mn^{2+} impurity equal to $c = 10^{-3}$, 4×10^{-3}, and 10^{-2}.

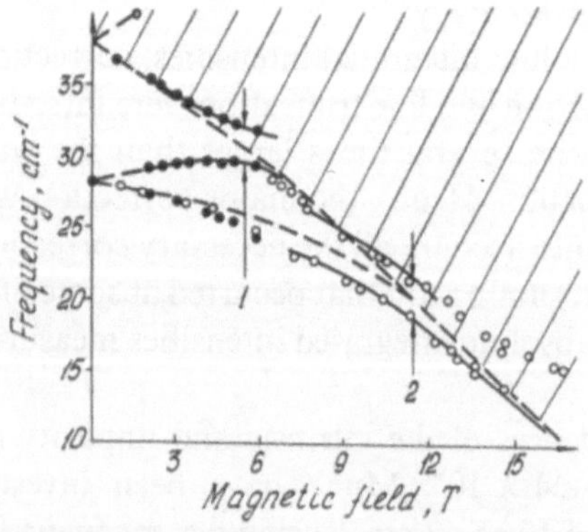

Fig. 5.25. The frequency-field dependence of absorption lines in $CoF_2 + 4 \times 10^{-3}$ Mn ($\mathbf{H} \parallel C_4$). The dashed lines correspond to the single-impurity approximation. The solid lines are a theoretical prediction for a frequency scanned spectrum. The solid dots are measured data obtained with frequency scanning, the open dots correspond to field scanning. The hatched region is the magnon band; (1) is the interaction region 1; (2) is the interaction region 2 [358].

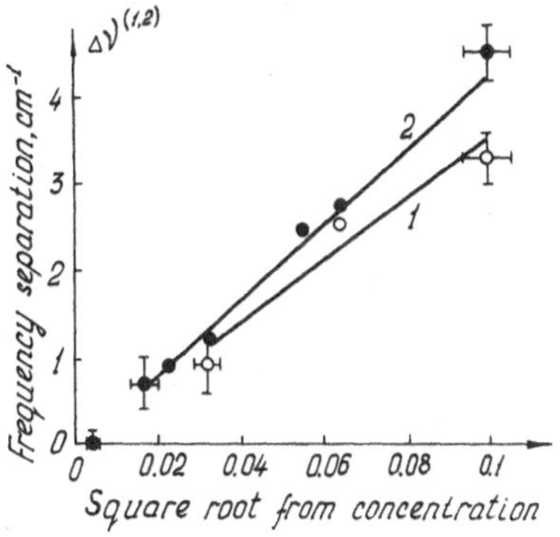

Fig. 5.26. Impurity concentration dependence of the frequency separation $\Delta\nu^{(1,2)}$ between the absorption lines at the cross-over points 1 and 2 [358].

With all three values the spectrum was rearranged coherently, i.e., the impurity line was enhanced in the vicinity of the AFMR line and the absorption lines showed splitting with transfer of intensity from one split component to the other. If within the magnon band, the impurity line decreased in intensity and became wider. The concentration dependence of $\Delta\nu^{(1,2)}$, i.e., the splitting of the absorption line at the cross-over point (the point where the integrated intensities of the two lines are equal) with the coordinates ν_{cr}^{α} and H_{cr}^{α}, such that $H_{cr}^{\alpha} \approx H_r^{\alpha}$, is shown in Fig. 5.26. (To avoid confusion, superscripts will be put in parentheses in some cases, so as not to be mistaken for power indices, e.g., $\Delta\nu^{\alpha}$ with $\alpha = 2$ will be written as $\Delta\nu^{(2)}$.) As can be seen in the figure, the $\Delta\nu^{(1,2)}$ versus $\sqrt{c}$ curve is linear to within the experimental error. This is in agreement with the theory.

Figure 5.27 shows the ratio of the integrated intensities of the impurity absorption and AFMR lines for the interaction regions 1 and 2, with the

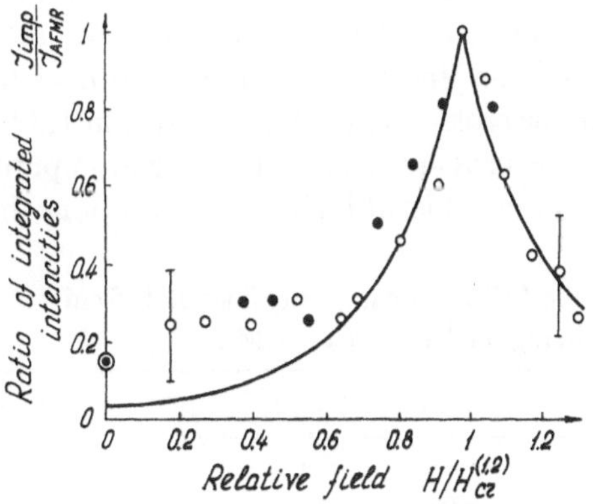

Fig. 5.27. The ratio of integrated absorption line intensities in $CoF_2 + 4 \times 10^{-3}$ Mn as a function of the magnetic field. The solid dots are for the interaction region 1, the open dots for region 2. The solid line is a theoretical curve, $H_{cr}^{(1,2)} \simeq H_r^{(1,2)}$ are the fields at the cross-over points for regions 1 and 2 [358].

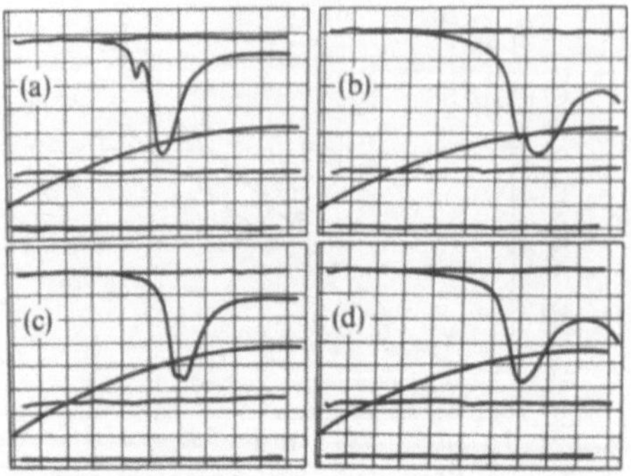

Fig. 5.28. Oscillograms illustrating ICR in the interaction region 2 of a $CoF_2 + 2 \times 10^{-5}$ Mn sample of thickness 0.1 mm: (a) $\nu = 21.2\ cm^{-1}$; (b) $\nu = 19.0\ cm^{-1}$; (c) $\nu = 18.1\ cm^{-1}$; (d) $\nu = 16.7\ cm^{-1}$.

impurity concentration $c = 4 \times 10^{-3}$. The theoretical curve has been calculated from (5.92). The discrepancy between this curve and the measured data observed for $H/H_{cr}^{(1,2)} < 0.5$ indicates once again that the equation is inapplicable away from the cross-over point. It might also be evidence for the necessity to take into account all three interacting oscillation modes in this range of magnetic fields.

The interaction region 2 was investigated on CoF_2 samples with weight concentrations of Mn^{2+} equal to $c = 2 \times 10^{-5}$, 3×10^{-4}, 10^{-3}, 3×10^{-3}, 4×10^{-3}, and 10^{-2}. The spectrum of the sample containing $2 \times 10^{-5}\ Mn^{2+}$ was rearranged incoherently. In contrast to CR, the impurity absorption line merged with the AFMR line, showing no sizeable variation in width (in case of CR, the two line widths tended to becoming equal). After the lines had merged, the impurity line did not manifest itself in any way in the spin wave band of the intrinsic excitations of the crystal. The ICR is illustrated by the oscillograms of Fig. 5.28. The rest of the samples showed a coherent rearrangement of the excitation spectrum, similar to that described for region 1. The impurity absorption linewidths and the width of the AFMR line at 22.5 cm^{-1} (far from the cross-over points) are listed in Table 5.2, together with the coordinates of the cross-over point for several impurity concentrations. As can be seen from this table, at higher concentrations the cross-over point

Table 5.2. Widths of the AFMR and impurity lines at 22.5 cm^{-1} and coordinates of the cross-over point in region 2 for $CoF_2 + c$ Mn.

C	2×10^{-5}	3×10^{-4}	5×10^{-4}	10^{-3}	3×10^{-3}	4×10^{-3}	10^{-2}
Δ_{AFMR} (cm^{-1})	1.0	1.3	1.2	1.2	1.6	2.2	2.4
Δ_{imp} (cm^{-1})	0.2	0.5	0.5	1.0	1.6	1.5	3.2
$\nu_{cr}^{(2)}$ (cm^{-1})	17 (merging)	18.3	18.1	19	20	20.5	21.7
$H_{cr}^{(2)}$ (T)	13.5 (merging)	12.9	13.2	12.6	11.8	11.1	10.2

Table 5.3. Experimental and theoretical values of the cooperative interaction constants for $CoF_2 + c$ Mn.

Constant	$m_1\,(\mathrm{cm}^{-1})$	$m_2\,(\mathrm{cm}^{-1})$	$c_{cr}^{(1)}$	$c_{cr}^{(2)}$	η	$\Omega\,(\mathrm{cm}^{-1})$
Experiment	17.5	20.5	7×10^{-5}	1.5×10^{-4}	0.02	45
Theory (ILP 1981)	22.3	27.3	6×10^{-4}	2×10^{-3}	0.05	46.5

shifts towards higher values of ν ($\omega = 2\pi c\nu$), i.e., lower H, because the spectrum rearrangement begins at lower magnetic field strengths (or smaller r_{imp}) owing to the smaller $\bar{r}$.

With a transverse orientation of the magnetic field, $\mathbf{H} \perp C_4$, the interaction of the low-frequency AFMR mode with the impurity excitation (interaction region 3) was investigated for impurity concentrations $c = 3 \times 10^{-4}$, 10^{-3}, 3×10^{-3}, and 10^{-2}. The spectrum was rearranged in a coherent way at all the concentrations studied. Although the transverse case $\mathbf{H} \perp C_4$ showed some peculiarities, we shall not discuss them in any detail as the basic ideas concerning delocalization of impurity excitations are valid here, too.

Let us estimate some of the parameters involved in the theory. Since in interaction region 2 an impurity concentration $c = 2 \times 10^{-5}$ brings about an incoherent rearrangement of the spectrum, whereas $c = 3 \times 10^{-4}$ results in a coherent rearrangement, the critical concentration c_{cr} lies between two values (i.e., $2 \times 10^{-5} < c_{cr}^{(2)} < 3 \times 10^{-4}$). The most probable value is $c_{cr}^{(2)} = (1.5 \pm 1) \times 10^{-4}$. The slope of the straight lines in Fig. 5.26 corresponds to $m_1 = (17.5 \pm 2)\ \mathrm{cm}^{-1}$ and $m_2 = (20.5 \pm 1.5)\ \mathrm{cm}^{-1}$, following from the equation $\Delta\nu^{(1,2)} = 2m_{1,2}\sqrt{c}$. Using the value $\Omega = (45 \pm 5)\ \mathrm{cm}^{-1}$ calculated from neutron scattering data on the dispersion of spin waves in CoF_2 [375], we can evaluate the magnitude of η from $c_{cr}^{(1,2)} = \eta(m_{1,2}/\Omega)^6$, proceeding exclusively from measured data. From the numerical values of $c_{cr}^{(2)}$ and m_2 for region 2 we obtain $\eta = 0.02$, which is in good agreement with the theoretical value $\eta = 0.05$ (taking experimental errors into account, the allowable value is 0.08). Now taking in region 1, $\eta = 0.02$, $\Omega = 45\ \mathrm{cm}^{-1}$, and $m_1 = 17.5\ \mathrm{cm}^{-1}$, we obtain $c_{cr}^{(2)} = 7 \times 10^{-5}$.

Table 5.3 presents the physical constants, as obtained in this paper, and the values predicted theoretically [325]. As can be seen from the comparison, the experimental and the predicted values of m_1 and m_2, the resonance interaction constants, and of η are in fair agreement. The discrepancy shown by $c_{cr}^{(1,2)}$ is due to the high power with which the constants $m_{1,2}$ participate in the formula for $c_{cr}^{(1,2)}$.

5.3.4. Orthogonal- and Collinear-Type Impurities in $CoCO_3$ + Fe + Mn + Ni

Borovik-Romanov and Meshcheryakov [314] were the first to observe split frequency-field dependences of the low-frequency AFMR mode and the impurity mode in an "easy plane" antiferromagnet (namely, $CoCO_3$ + Fe). The unusually low frequency (about 1.5 cm^{-1}) of the impurity excitation and its low g-factor are surprising. The common explanation in terms of a weak exchange between Fe and Co ions is not satisfactory as numerous experiments have shown the strength of the exchange interaction to be roughly the same for all transition metals. A realistic explanation has been given in [326], where it was suggested that the impurity spin might not be oriented perpendicular to the trigonal axis as the host spins are, but rather might be just slightly declined from that axis, so that the impurity spin is nearly orthogonal to the host spin direction. This was quite a reasonable assumption, particularly because the spins in $CoCO_3$ and $FeCO_3$ are oriented, respectively, perpendicular and parallel to the trigonal axis, thus manifesting the crystal symmetry and the nature of the ion. If the impurity spin is orthogonal to the direction of the host spins, then the impurity excitation frequency is low and depends quadratically on the exchange interaction. To confirm this, consider the behavior of a Fe^{2+} ion in $CoCO_3$.

The ground state ^5D of a free Fe^{2+} ion in the $CoCO_3$ crystal field (symmetry D_{3d}^6) is split, owing to the cubic field component, into the orbital triplet $^5T_{2g}$ and the doublet ^5E lying some 10^5 cm^{-1} higher. The effective Hamiltonian, all-owing for splitting of the $^5T_{2g}$ state can be represented as owing for splitting of the $^5T_{2g}$ state, can be represented as

$$\mathscr{H}_{LS} = -\zeta L_z^2 + \lambda \mathbf{L}\cdot\mathbf{S}, \tag{5.93}$$

where $\mathbf{L}(L = 1)$ and $\mathbf{S}(S = 2)$ are, respectively, the orbital and the spin moment operators; the z-axis is along the C_3-axis of the crystal; the trigonal field constant ζ and the spin-orbit coupling parameters λ are, respectively, [381], $\zeta = 1500\ cm^{-1}$ and $\lambda \approx 100\ cm^{-1}$. The ground state described by the Hamiltonian of (5.93) is doubly degenerate. Neglecting corrections of the order of λ/ζ one can express this states as a linear combination of $|M_S = 2, M_L = -1\rangle \equiv |2, -1\rangle$, and $|-2, 1\rangle$, i.e., states characterized by maximum projections of the spin and the orbital moments on the z-axis. Hence, it is feasible to say that the spin of Fe^{2+} in $CoCO_3$ is oriented along the C_3-axis.

Using the molecular field approximation, the effective Hamiltonian of Fe^{2+} can be written as

$$\mathscr{H}_0 = \mathscr{H}_{LS} - IS_x, \qquad I = (I_1 - I_2)S, \tag{5.94}$$

where x is the antiferromagnetic axis, S is the mean spin value in the host matrix, and I_1 and I_2 are integrals of the exchange interaction between the impurity spin and the magnetic sublattices of the matrix. Making use of the Hamiltonian of (5.94), the energies and wave functions of the two lowest states can be calculated [326], and which can be used to find the frequency of the lowest excitation for an orthogonal impurity, viz.,

$$\omega_{ort} = 12I^2/\zeta. \tag{5.95}$$

It seems noteworthy that, owing to the exchange field of the host, the impurity spin is deflected by a small angle from the trigonal axis, hence its average ground state projection on the x-axis is not zero, $\langle S_x \rangle_{gr} \neq 0$, although it is small. The amount of deflection is determined mainly by the competition of the exchange interaction and the spin-orbit coupling as $\langle S_x \rangle_{gr} = 2I/\lambda + 14I/\zeta$.

The dependence of ω_{ort} on the applied magnetic field has the form

$$\omega_{ort}^2(\mathbf{H}) \simeq \omega_{ort}^2 + 2\omega_{ort}\mu g_{ort}^{\perp} H_y + (\mu g_{ort}^{\parallel} H_z)^2$$

with

$$g_{ort}^{\perp} = 24\frac{\gamma}{\zeta}, \qquad \gamma = \frac{2\chi_{\perp} H_D}{\mu g_{\perp}}\left(\frac{2\chi_{\perp} I_1 I_2}{\mu^2 g_{\perp}} - I_1 - I_2\right),$$
$$g_{ort}^{\parallel} = 2\left(5 - \frac{I_1 + I_2}{\mu^2 g_{\parallel}}\chi_{\parallel}\right). \tag{5.96}$$

Here H_D is the Dzyaloshinsky field; $\chi_{\parallel}$ and $\chi_{\perp}$ are, respectively, the longitudinal and the transverse host susceptibilites; and $g_{\parallel}$ and $g_{\perp}$ are the longitudinal and the transverse g-factors. The temperature dependence of $\omega_{ort}(H)$ is controlled essentially by a quantity proportional to the square of the susceptibility.

In impurity ions where the ground state is not orbitally degenerate in the crystal field (like Mn^{2+} or Ni^{2+} in $CoCO_3$) the spins are oriented along the exchange field of the host, with the spin-orbit coupling resulting in a weak single-ion anisotropy. Such impurities are known as collinear. For the case $\mathbf{H} \parallel y$ (x being the antiferromagnetic axis) ILP [326] have obtained the following field dependence of the impurity excitation frequency

$$\omega_{col}^2(H_y) = \omega_{col}^2 + 2\omega_{col}\mu g_{col}^{\perp} H_y + \alpha(\mu H_y^2), \tag{5.97}$$

with

$$\omega_{col} = I\left[1 + \left(\frac{I_1 + I_2}{I_1 - I_2}\right)^2 \left(\frac{\chi_{\perp} H_D}{\mu g_{\perp} S}\right)^2\right]^{1/2}, \qquad g_{col}^{\perp} = \gamma/\omega_{col}, \quad \alpha \approx 1.$$

Comparing (5.96) and (5.97) we can see that ω_{col} is higher than ω_{ort} for the same values of I_1 and I_2, by a factor of $\varkappa = \zeta/12\omega_{col}$. For sufficiently low fields, $\mu H_y < \gamma$, the dependence $\omega_{col}(H_y)$ is linear, with $g_{col}^{\perp} = g_{ort}^{\perp}\varkappa/2$. Note that the cases $g_{col}^{\perp} > 0$ and $g_{col}^{\perp} < 0$ are equally possible. The temperature dependence of ω_{col} is determined by the first power of the susceptibility, unlike ω_{ort}.

Thus, orthogonal and collinear impurities can be distinguished by the frequencies of the appropriate lowest lying impurity excitations and the nature of their temperature and field dependences. First consider an Fe^{2+} impurity in $CoCO_3$. Taking the frequency-field dependence as measured in [315] and the values $\chi_{\perp} = 53 \times 10^{-3}$ CGSM/mole, $\chi_{\parallel} = 35 \times 10^{-3}$ CGSM/mole, $H_D = 2.7$ T, $g_{\perp} = 2.5$, $g_{\parallel} = 3.1$, and $S = 1.15$, we can use (5.95) to obtain

$$I_1 \simeq 26\ \text{cm}^{-1}, \qquad I_2 \simeq 14\ \text{cm}^{-1},$$

i.e., both integrals happen to be positive (antiferromagnetic coupling). Then we have $g_{ort}^{\perp} \simeq 0.06$, which explains the lack of a noticeable dependence of ω_{ort} on H_y observed in the experiment. The data on the temperature shift of $\omega_{ort}(T)$ [315] do not contradict this determination of the Fe^{2+} parameters either.

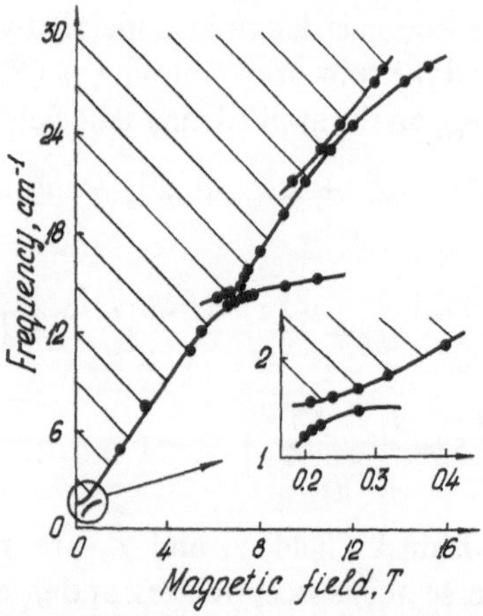

Fig. 5.29. Frequency-field dependences of the intrinsic and the impurity-dependent absorption in $CoCO_3 + 10^{-3}$ Fe + 10^{-4} Mn + 10^{-4} Ni. The hatched region is the magnon band of the crystal. $T = 4.2$ K ($T_N = 17.1$ K), $H \perp C_3$ [353].

Along with the Fe^{2+} impurity excitation near 1.5 cm^{-1}, two more excitations were discovered [353] in a $CoCO_3 + 10^{-3}$ Fe + 10^{-4} Mn + 10^{-4} Ni crystal near the frequencies 14.3 cm^{-1} and 22 cm^{-1} (Fig. 5.29). The 14.3 cm^{-1} excitation is due to Mn ions and that at 22 cm^{-1} to Ni. These frequencies are in agreement with the ILP theoretical predictions if the excitations are regarded as collinear, with the same values of I_1 and I_2 for the first frequency as in the previous case and with $(-I_1)$, $(-I_2)$ for the second frequency (ferromagnetic interaction with the host matrix). The respective g-factors are 0.25 and 1.5, which is in good agreement with the experimental data presented in Fig. 5.29.

The type of spectrum rearrangement in doped crystals of D_{3d}^6 symmetry was analyzed in [327]. In its major features, the rearrangement occurs as in the $CoF_2 + c$ Mn crystal we have just discussed. Naturally, the numerical values of the parameters are different, e.g., the critical impurity concentration is much lower, namely $c_{cr} \simeq 10^{-9}$. Accordingly, the rearrangement pattern shown in Fig. 5.29 is coherent, as has been obtained for concentrations about 10^{-4} or 10^{-3}. In the vicinity of the cross-over points (which were 1.5 cm^{-1} and 0.24 T for Fe, 14.3 cm^{-1} and 7 T for Mn, and 22 cm^{-1} and 10.2 T for Ni) the absorption lines (i.e., that of impurity absorption and AFMR) exchanged their integrated intensities. Upon entering the magnon band, the impurity absorption ceased appearing as a resonance.

5.3.5. Magnetic Impuritons Near the Top of a Magnon Band (FeF_2 + Co)

In Sections 5.3.2, 5.3.3, and 5.3.4 we have given a detailed discussion of the rearrangement occurring in the energy spectrum of an antiferromagnet as a local impurity level approaches the bottom of the magnon band. The appearance of magnetic impuritons and a gigantic enhancement of the impurity line intensities have been noticed. Now the question is what would happen should a local impurity level approach an optically inactive top of the magnon band. A partial answer was given in Section 5.3.2, when we discussed the single-impurity approximation, i.e., the impurity excitation would be delocalized and the impurity absorption line intensity would decrease.

Consider an optical impurity mode approaching the top of the magnon band in more detail [322, 323]. Let the crystal be FeF_2 + Co and $\mathbf{H} \parallel C_4$. The Hamiltonian of the doped crystal has the form

$$\mathscr{H} = \sum_{\mathbf{k}} [\omega(\mathbf{k}) + \mu g H] a_{\mathbf{k}}^{+} a_{\mathbf{k}} \sum_{\mathbf{p}} (\omega_0 - \mu g H) a_{\mathbf{p}}^{+} a_{\mathbf{p}}$$

$$+ \frac{1}{\sqrt{N}} \sum_{\mathbf{k},\mathbf{p}} (m e^{-i\mathbf{k}\mathbf{p}} a_{\mathbf{k}}^{+} a_{\mathbf{p}} + \text{h.c.}), \qquad (5.98)$$

where $a_{\mathbf{k}}^{+}$ and $a_{\mathbf{k}}$ are the creation and annihilation operators for magnons, $\mathbf{p}$ enumerates the sites where the impurity spin is oriented opposite to the field, m is the resonant interaction parameter for the intrinsic and the impurity excitation, and N is the number of magnetic cells in the crystal. The dispersion law for the magnons near the top of the band is

$$\omega(\mathbf{k}) = \omega_m - (4\pi)^{-2/3} \Omega a^2 (\mathbf{k} - \mathbf{k}_m)^2.$$

According to the measurements in [331] the g-factor of the impurity and the host have been assumed equal ($g \approx 2.2$).

The absorption coefficient of radiation polarized prpendicular to the C_4-axis can be expressed, as is commonly done, in terms of retarded Green functions, viz.,

$$\sigma(\omega) = A \lim_{\mathbf{k}=0} \Bigg[\langle\langle a_{\mathbf{k}} | a_{\mathbf{k}}^{+} \rangle\rangle^{\omega} + \frac{1}{\sqrt{N}} \sum_{\mathbf{p}\mathbf{p}'} (e^{i\mathbf{k}\mathbf{p}} \langle\langle a_{\mathbf{k}} | a_{\mathbf{p}}^{+} \rangle\rangle + e^{-i\mathbf{k}\mathbf{p}} \langle\langle a_{\mathbf{p}} | a_{\mathbf{k}}^{+} \rangle\rangle) + \frac{1}{N} \sum_{\mathbf{p}\mathbf{p}'} e^{i\mathbf{k}(\mathbf{p}-\mathbf{p}')} \langle\langle a_{\mathbf{p}} | a_{\mathbf{p}}^{+} \rangle\rangle^{\omega} \Bigg], \qquad (5.99)$$

where A is a parameter depending only weakly upon frequency. Near the top $\omega_m(H)$ of the magnon band, where $|\omega - \omega_m - \mu g H| \ll \omega - \omega(0) - \mu g H| \approx \Omega$.

$$\sigma(\omega) = cA \left(1 + \frac{m}{\omega - \omega(0) - \mu g H}\right)^2$$

$$\times \frac{\Gamma_0/2 + \operatorname{Im} P(\omega, H)}{[\omega - \omega_0 + \mu g H - \operatorname{Re} P(\omega, H)]^2 + [\Gamma_0/2 + \operatorname{Im} P(\omega, H)]^2}. \qquad (5.100)$$

Here Γ_0 is the intrinsic width of the impurity line and $P(\omega, H)$ is the polarization operator, whose specific form depends on the choice of the representation (i.e., renormalized or not). Let $c < c_{cr}$, so that the rearrangement of the spectrum is incoherent. Sufficiently far from the region of rearrangement, i.e., where the single-impurity approximation is valid, $P(\omega, H)$ has the form

$$P(\omega, H) = \frac{m^2}{N} \sum_{\mathbf{k}} [\omega - \omega(\mathbf{k}) - \mu g H]^{-1}. \tag{5.101}$$

If $\Gamma_0 > c^{1/3} m^2/\Omega$, then the indirect interaction of impurities via magnons can be neglected even in the range of collective rearrangement and (5.100) and (5.101) describe the impurity absorption through the entire frequency range. Right near the top fo the band, the frequency and the field dependences of the polarization operator can be written explicitly, viz.,

$$P(\omega, H) = P_0 - \sqrt{(\omega - \omega_m - \mu g H)\delta}, \qquad \delta = m^4/\Omega^3. \tag{5.102}$$

Higher-order terms in $(\omega - \omega_m - \mu g H)/\Omega \ll 1$ have been neglected. The parameter δ controls the width of the threshold range, being a scale size for such phenomena as the nonlinear field dependence of the frequency and suppression of the impurity line intensity near the top of the band.

The impurity level frequency $\omega_0(H)$ is given by the equation

$$\omega - \omega_0 + \mu g H - \operatorname{Re} P(\omega, H) = 0. \tag{5.103}$$

Bearing in mind that the second term in (5.102) is only real outside the range $\omega > \omega_m + \mu g H$, we have

$$\begin{aligned} \omega_0(H) &= \omega_m + \mu g H + 2\mu g (H_r - H) + \delta/2 \\ &\quad - \sqrt{\delta^2/4 + 2\mu g (H_r - H)\delta}, \qquad H < H_r, \\ 2\mu g H_r &= \omega_0 + P_0 - \omega_m. \end{aligned} \tag{5.104}$$

As can be seen, if H is increased, $\omega_0(H)$ first decreases to $\omega_{\min} = \omega_m + \mu g H_r - \delta/8$ for $2\mu g(H_r - H) = \frac{3}{4}\delta$, and then increases, merging with the upper edge of the band at $H \approx H_r$. This is in agreement with the experimental data [331] under the assumption that $\delta = 15\ \text{cm}^{-1}$ and $2\,\mu g H_r = 15.6\ \text{cm}^{-1}$. The curve given by (5.104) is represented by the solid line of Fig. 5.30(a). The discrepancy noticeable at low field values can be reduced by using a more accurate expression for $P(\omega, H)$ away from the top of the band. The effective g-factor,

$$g_{\text{eff}} = \frac{1}{\mu} \frac{d\omega_0(H)}{dH} = -g\left(1 - \frac{\delta}{\sqrt{2\mu g (H_r - H)\delta + \delta^2/4}}\right), \tag{5.105}$$

changes quite strongly near the upper edge of the band.

With $H < H_r$ the absorption coefficient $\sigma(\omega)$ is characterized by a sharp peak in its frequency dependence. At a fixed H, the peak width Γ and inte-

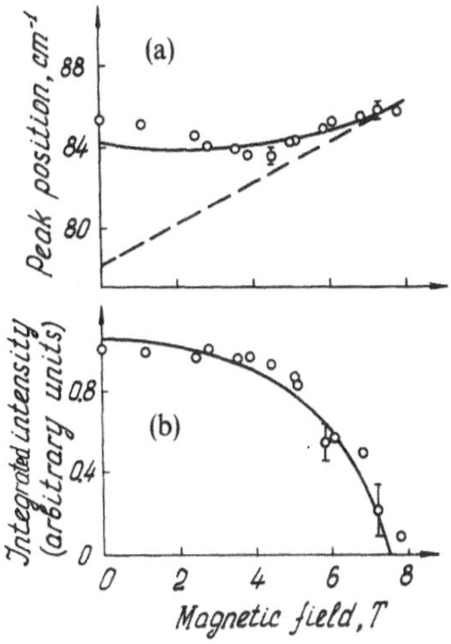

Fig. 5.30. Field dependences of (a) the peak position and (b) the integrated intensity of an impurity absorption line approaching the top of the magnon band in $FeF_2 + 10^{-3}$ Co. The dots represent experimental data of [331] and the solid lines are theoretical curves [323].

grated (over ω) intensity $\mathscr{I}_1$ are, respectively,

$$\Gamma = \Gamma_0 k(H),$$

$$\begin{aligned} \mathscr{I}_1(H) &= cA\left(1 + \frac{m}{\omega_0(H) - \omega(0) - \mu g H}\right)^2 k(H), \\ k(H) &= \left(1 - \left|\frac{\partial P(\omega, H)}{\partial \omega}\right|_{\omega=\omega_0(H)}\right)^{-1} \\ &= \left(1 + \frac{1}{2}\sqrt{\frac{\delta}{\omega_0(H) - \omega_m - \mu g H}}\right)^{-1} \\ &= \frac{\sqrt{2\mu g(H_r - H) + \delta/4} - \sqrt{\delta}/2}{\sqrt{2\mu g(H_r - H) + \delta/4}}\Bigg) < 1, \\ k(H) &\approx 4\mu g(H_r - H)/\delta \qquad \text{for} \quad 2\mu g(H_r - H) \ll \delta/4. \end{aligned} \tag{5.106}$$

Hence, the integrated intensity $\mathscr{I}_1(H)$ of the impurity line decreases with increasing H and the depndence of $\mathscr{I}_1$ on H is linear near H_r. In Fig. 5.30(b) these predictions are compared with experimental data. The parameter m has been taken $m = 22\ \text{cm}^{-1}$, though it is understood that $\mathscr{I}_1$ is only weakly sensitive to its value. As follows from (5.106), for $H \to H_r$ the impurity line should become narrower while retaining the height of its peak.

Along with the narrow peak at $\omega_0(H)$, the theory predicts a broad asymmetrical absorption maximum inside the magnon band ($\omega < \omega_0 + \mu g H$),

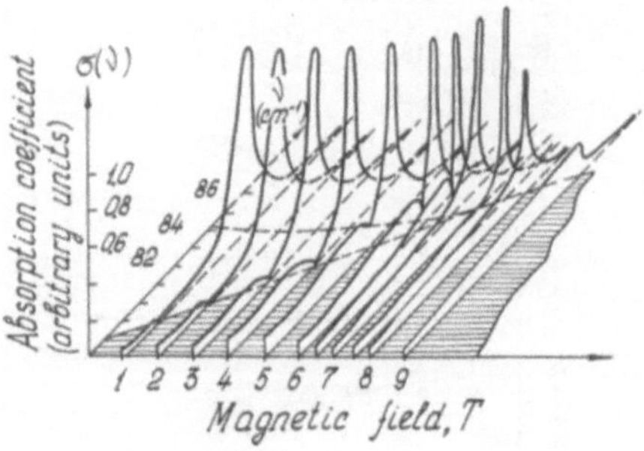

Fig. 5.31. Theoretical frequency dependences of the far IR absorption $\sigma(\nu)$ in FeF_2 + 10^{-3} Co for different values of the applied field [323].

whose intensity decreases as $(\omega_m + \mu g H - \omega)^{-3/2}$. Its highest point is at

$$\tilde{\omega}(H) = \omega_m + \mu g H - [2\mu g(H_r - H)]^2 \qquad \text{for} \quad 2\mu g(H_r - H) \ll \delta.$$

The total integrated intensity of the two peaks is constant. Since broad lines are hard to register in the experiment, the experimentlist normally follows the narrow line only. The pattern described is shown in Fig. 5.31. Perhaps it should be noted that the theory predicts a reduction in the width of the principal line, whereas an experiment [331] has revealed a constant width.

Consider now the case of coherent rearrangement of the spectrum, $c > c_{cr}$. To obtain an expression for $P(\omega, H)$, it is now necessary to use the renormalized representation, taking into account, among other things, the concentration-dependent shift of the top of the band. The CR is characterized by a new impurity branch of the spectrum, with the same polarizaton and wave vectors of the excitation as at the nearby top of the band. The additional band has a width of $m\sqrt{c}$ (at $H \approx H_r$), being separated from the main band by a quasi gap of the same width. Let $m\sqrt{c} > \Gamma_0 > c^{1/3}m^2/\Omega > \delta$. Since the tops of both the principal and the impurity band are not optically active, radiation will be absorbed at some frequency between the tops, at localized levels corresponding to the "unperturbed" impurity frequency $\omega_0(H) \approx \omega_0 - \mu g H$. For $c > c_{cr}$ the concentration-dependent width, $\approx c^{1/3}m^2/\Omega$, of the impurity line is larger than the near-threshold region δ. Hence manifestations of the nonlinear dependence $\omega_0(H)$ become insignificant. For $H > H_r$ the impurity line, of roughly constant intensity, penetrates deep in the band, to become quasi local for $2\mu g(H - H_r) > c^{2/3}\Omega$. The quasi-local linewidth is $\Gamma_0 + [2\mu g(H - H_r)\delta]^{1/2}$, i.e., it increases slowly with $H - H_r$.

Estimates for the FeF_2 + Co crystal yield $c_{cr} = \eta(m/\Omega)^6 \approx 2.5\%$ and the behavior of the absorption lines observed in the experiment [331] corresponds to ICR.

Antiferromagnetic crystals containing small amounts of impurities have proved to be very suitable objects for the study of impurity states delocalized at the approach of the intrinsic excitation band of the crystal, and of magnetic impuritons. Up to the present time, only initial steps in this direction have been made. This relates equally to the interaction of impurity excitations with magnons near the bottom and top of the magnon band. Among the problems of primary importance stand out measurements of the energy-momentum relation for the magnetic impuritons existing at the coherent rearrangement of the spectrum. The most complete data could be obtained by means of neutron spectroscopy. However, the technique is likely to face considerable difficulties in providing the necessary experimental accuracy. Indeed, with a 1% concentration of the impurity the theoretical prediction for the bandwidth is only 1 to 2 cm^{-1}. Variations of the shape and width of the intrinsic and impurity lines, resulting from the variation of the impurity state radius, have not been analyzed either, theoretically or experimentally. The effect of temperature on the cooperative rearrangement of the spectrum has been studied in just a single experiment. All of these problems await detailed investigations.

References

1. Borovik-Romanov, A.S.: *Results of Science. Antiferromagnetism.* Moscow: Akademizdat, vol. 4, 1976, 215 pp.
2. Turov, E.A., Shavrov, V.G.: On Some Galvano- and Thermomagnetic Effects in Antiferromagnets. *Zh. Èksper. Teoret. Fiz.* (1962), **43**, No. 6, pp. 1582–1589.
3. Brown, W.F. Jr., Shtrikman, S., Treves, D.: Possibility of Visual Observation of AFM Domains. *J. Appl. Phys.* (1963), **34**, No. 4, pp. 1233–1234.
4. Fuchs, R.: Wave Propagation in a Magneto-Electric Medium. *Phil. Mag.* (1965), **11**, No. 3, pp. 647–651.
5. Birss, R.R., Shrubsall, R.G.: The Propagation of Electromagnetic Waves in Magnetoelectric Crystals. *Phil. Mag.* (1967), **15**, No. 136, pp. 687–700.
6. Hornreich, R.M., Shtrikman, S.: Theory of Gyrotropic Birefringence. *Phys. Rev.* (1968), **171**, No. 3, pp. 1065–1074.
7. Lyubimov, V.N.: Crystalo-optics with an Account of Magnetoelectric Effect. *Dokl. Akad. Nauk. SSSR* (1968), **181**, No. 4, pp. 858–861.
8. Lyubimov, V.N.: Magnetoelectric Effect and Irreversibility of Light Propagation in Crystals. *Crystallography* (1969), **14**, No. 2, pp. 213–217.
9. Portigal, D.L., Burstein, E.: Magneto-Spatial Dispersion Effects on the Propagation of Electro-Magnetic Radiation in Crystals. *J. Phys. Chem. Solids* (1971), **32**, No. 3, pp. 602–608.
10. Markelov, V.A., Novikov, M.A., Turkin, A.N.: Experimental Observation of New Unreciprocal Magneto-Optical Effect. *Pis'ma v JETP* (1977), **25**, No. 9, pp. 404–407.
11. Novikov, M.A.: Unreciprocal Optical Effects in an External Magnetic Field. *Crystallography* (1979), **24**, No. 4, pp. 666–671.
12. Krinchik, G.S., Chetkin M.V.: Transparent Ferromagnets. *Uspekhi Fiz. Nauk* (1969), **98**, No. 1, pp. 3–26.
13. Krinchik, G.S.: *Magneto-optical Phenomena in Ferromagnets. Magnetism Problems.* Moscow: Nauka, 1972, 133 pp.
14. Dillon, J.F., Jr.: *Magneto-optics of Magnetic Materials. Magnetic Properties of Materials.* New York: McGraw-Hill, 1971, Part 5, p. 149.
15. Kharchenko, N.F., Eremenko, V.V.: Optical and Magneto-Optical Studies of Magnetically Ordered Dielectrics and Semiconductors (Ferrum Oxides). *Fiz. Kondensir. Sostoyaniya*'(1969), No. 4, pp. 144–202.
16. Kharchenko, N.F., Eremenko, V.V.: Optical and Magneto-Optical Studies of Magnetically Ordered Dielectrics and Semiconductors (Ferro- and Antiferro-

magnetic Compounds of Transient and Rare-Earth Ions). *Fiz. Kondensir. Sostoyaniya* (1971), No. 13, pp. 3–29.
17. Pisarev, R.V.: *Magnetic Ordering and Optical Phenomena in Crystals. Fiz. Magn. Dielectrics.* Leningrad: Nauka, 1974, 454 pp.
18. Smolenskii, G.A., Pisarev, R.V., Sinii, I.G.: Birefringence of Magnetically Ordered Crystals. *Uspekhi Fiz. Nauk* (1975), **116**, No. 2, pp. 231–270.
19. Portigal, D.L.: Intrinsic Magneto-Optical Effects in Magnetic Crystals. *Phys. Rev.* (1978), **18B**, No. 7, pp. 3637–3644.
20. Dillon, J.F., Jr.: Magneto-Optics and its Uses *J. Magn. and Magn. Mater.* (1983), **31/34**, pp. 1–9.
21. Ferre, J., Gehring, G.A.: Linear Optical Birefringence of Magnetic Crystals. *Rep. Progr. Phys.* (1984), **47**, No. 5, pp. 513–611.
22. Prokhorov, A.M., Smolenskii, G.A., Ageev, A.N.: Optical Phenomena in Thin Film Magnetic Waveguides and Their Technical Application. *Uspekhi Fiz. Nauk* (1984), **143**, No. 1, pp. 33–72.
23. Landau, L.D., Lifshitz, E.M.: *Electrodynamics of Continuous Media.* Moscow: Gostekhizdat, 1959, 525 pp.
24. Ginzburg, M.A.: Gyrotropic Waveguide. *Dokl. Akad. Nauk SSSR* (1954), **95**, No. 3, pp. 489–492.
25. Ginzburg, M.A.: On Propagation of Electromagnetic Waves in Gyrotropic Layer. *Dokl. Akad. Nauk SSSR* (1954), **95**, No. 4, pp. 753–756.
26. Krinchik, G.S.: On Magneto-Optical Phenomena in Ferromagnetics. *Vestnik Moskov. Univ.* (1955), **12**, pp. 61–68.
27. Krinchik, G.S., Chetkin, M.V.: To the Problem on Determination of Tensors of Dielectric and Magnetic Medium Permitivity. *Zh. Èksper. Teoret. Fiz.* (1959), **36**, No. 6, pp. 1924–1925.
28. Pershan, P.S.: Magneto-Optical Effects. *J. Appl. Phys.* (1967), **38**, No. 3, pp. 1482–1490.
29. Ignatov, A.M., Rukhadze, A.A.: On Nonunique Determination of Magnetic Permittivity of Material Media. *Uspekhi Fiz. Nauk* (1981), **135**, No. 1, pp. 171–174.
30. Eritsyan, O.S.: Optical Problems in Electrodynamics of Gyrotropic Media. *Uspekhi Fiz. Nauk* (1982), **138**, No. 4, pp. 645–674.
31. Arguers, P.N.: Theory of the Faraday and Kerr Effects in Ferromagnets. *Phys. Rev.* (1985), **97**, No. 2, pp. 334–345.
32. Shen, Y.R.: Faraday Rotation of Rare-Earth Ions, I. Theory. *Phys. Rev.* (1964), **133**, No. 2A, pp. 511–515.
33. Benett, H.S., Stern, E.A.: Faraday Effect in Solids. *Phys. Rev.* (1965), **1377**, No. 2A, pp. 448 461.
34. Murano, T., Ebina, A.: Faraday Effect for Localized Electrons in Insulators. *J. Phys. Soc. Japan* (1965), **20**, No. 6, pp. 997–1008.
35. Krinchik, G.S., Chetkin, M.V.: Exchange Interaction and Magneto-Optical Effects in Ferrites–Garnets. *Zh. Ekssper. Teoret. Fiz.* (1961), **41**, No. 3, pp. 673–680.
36. Hougen, J.T.: Anomalous Faraday Dispersion of O_2. *J. Chem. Phys.* (1960), **32**, No. 4, pp. 1122–1125.
37. Finkel, J.: Faraday Rotation in a Spin System. *J. Opt. Soc. Amer.* (1963), **53**, No. 9. pp. 1115–1116.
38. Kharchenko, N.F., Eremenko, V.V.: Magneto-Resonance Faraday Effect in Antiferromagnets MnF_2 and $RbMnF_3$. *Soviet Phys. Solid State* (1967), **9**, No. 6, pp. 1302–1305.

39. Chetkin, M.V., Shalygin, A.I.: Exchange Interaction and Temperature Dependence of Faraday Effect in Ferrimagnets. *Zh. Èksper. Teoret. Fiz.* **52**, No. 4, pp. 882–884; (1967), *J. Appl. Phys.* (1968), **39**, No. 2, pp. 561–563.
40. Johnson, B., Tebble, R.: The Infra-Red Faraday Effect and g-Values in Rare-Earth Garnets. *Proc. Phys. Soc.* (1966), **87**, No. 4, pp. 935–944.
41. Fedorov, F.I.: *Optics of Anisotropic Media.* Minsk: Izdat. Acad. Sci. Belorussian SSR, 1958, 348 pp.
42. Fedorov, F.I.: *Gyrotropy Theory.* Minsk: Izdat. Acad. Sci. Belorussian SSR, 1976, 346 pp.
43. Kaganov, M.I., Yankelevich, R.P.: Propagation of Electromagnetic Wave in Gyroanisotropic Medium. *Fiz. Tverd. Tela.* (1968), **10**, No. 11, pp. 2771–2773.
44. Malakhovskii, A.V.: Magneto-Optical Phenomena in Bigyrotropic Media. *Fiz. Tverd. Tela.* (1974), **16**, No. 2, pp. 632–634.
45. Barkovskii, L.M.: Electromagnetic Waves in Bigyrotropic Media with Noncommutating Tensors ε and μ. *Opt. i Spektrosk.* (1975), **38**, No. 1, pp. 115–119.
46. Akhmediev, N.I., Zvezdin, A.K.: Spatial Dispersion and New Magneto-Optical Effects in Magnetically Ordered Crystals *Pis'ma v JETP* (1983), **38**, No. 4, pp. 167–169.
47. Onsager, L.: Reciprocal Relations in Irreversible Process, 1, 2. *Phys. Rev.* (1931), **37**, No. 4, pp. 405–426; **38**, No. 12. pp. 2265–2279.
48. Callen, H., Barasch, M.L., Jackson, J.J.: Statistical Mechanics of Irreversibility. *Phys. Rev.* (1952), **88**, No. 6, pp. 1382–1386.
49. Ginzburg, V.M. Agronovich, V.L.: *Crystalo-Optics with an Account of Spatial Dispersion and Exciton Theory.* Moscow: Nauka, 1965, 374 pp.
50. Sirotin, Yu. I., Shaskolskaya, M.P.: *Fundamentals of Crystalo-physics.* Moscow: Nauka, 1979, 617 pp.
51. Birss, R.R.: Macroscopic Symmetry in Space–Time. *Rep. Progr. Phys.* (1963), **26**, No. 3–4, pp. 307–360.
52. Kleiner, W.H.: Space–Time Symmetry of Transport Coefficients, I. *Phys. Rev.* (1966), **142**, No. 2, pp. 318–326.
53. Kleiner, W.H.: Space–Time Symmetry of Transport Coefficients, II. *Phys. Rev.* (1976), **153**, No. 3, pp. 726–727.
54. Kleiner, W.H.: Soace–Time Symmetry Restrictions on Transport Coefficients, III. Thermogalvanomagnetic Coefficients. *Phys. Rev.* (1969), **182**, No. 3, pp. 705–709.
55. Birss, R.R., Fletcher, D.S.: Space–Time Symmetry Restrictions on Transport Coefficients. *J. Magn. and. Magn. Mater.* (1980), **15/18**, pt. 2, pp. 915–916.
56. Bukinhem, A., Stefens, P.: *Magnetic Optical Activity Dispersion of Optical Rotation and Circular Dichroism in Organic Chemistry.* Moscow: Mir, 1970, pp. 399–428.
57. Kizel, V.A., Burkov, V.I.: *Gyrotropy of Crystals.* Moscow: Nauka, 1980, 340 pp.
58. Kharchenko, N.F.: Linear Magneto-Optic Effect in Magnetically Ordered Crystals. *All-Union Conference on Physics of Magnetic Phenomena* (*Kharkov*, 1979): *Abstract.* Kharkov, 1979, p. 180.
59. Kharchenko, N.F.: Magneto-Optic Studies of Magnetically Ordered Crystal Structure. Author's Essay of Dissertation of Doctor of Physics and Mathematics. Kharkov, 1983.
60. Eremenko, V.V., Kharchenko, N.F.: The linear Magneto-Optic Effect in Magnetically Ordered Crystals. *Soviet Sci. Rev. A* (1984), **5**, pp. 1–97.

61. Bhagavantan, S.: Birefringence and Optical Activity in Magnetic Crystals. *Mat. Res. Bull.* (1969), **4**, pp. 477–488.
62. Kharchenko, N.F., Bibik, A.V., Eremenko V.V.: Quadratic Magnetic Rotation of Light Polarization Plane in Antiferromagnet CoF_2. *JETP Lett.* (1985), **42**, No. 11, pp. 553–556.
63. Eremenko, V.V., Kharchenko, N.F., Bibik, A.V., Milner, A.A.: Quadratic Magnetic Rotation of the Polarization Plane of Light in Antiferromagnets. *Abstr. Intern. Sympos. on Magneto-Optics* (*Kyoto*, 1987). Kyoto, 1987, pp. 27–28.
64. Girgal, S.S.: Faraday Effect of Even Parity in Magnetically Ordered Crystals. *Dokl. Akad. Nauk BSSR* (1982), **26**, No. 10. pp. 890–891.
65. Loktev, V.M., Ostrovskii, V.S.: On the Nature of Even Parity Effect in CoF_2. Kiev, 1986, 7 pp. (Preprint/Acad. Sci. Ukrain. SSR Inst. Teoret. Fiz., pp. 86–87.)
66. Belyi, V.N., Serdyukov, A.N.: On Linear Effect of Magnetic Field on Optical Activity. *Crystallography* (1974), **19**, No. 16, pp. 1279–1280.
67. Vloch, O.G.: *Spatial Dispersion Phenomena in Parametric Crystalo-Optics*. L'vov: L'vov University, 1984, 155 pp.
68. Zheludev, I.S.: *Physics of Crystals and Symmetry*. Moscow: Nauka, 1987, 188 pp.
69. Nye, J.: *Physical Properties of Crystals*. Moscow: Izdat. Inostran. Literat, 1960, 315 pp.
70. Tavger, B.A.: Symmetry of Antiferromagnet Piezomagnetism. *Crystallography* (1958), **3**, No. 3. pp. 342–345.
71. Le Corre, Y.: Les Groupes Cristallographiques Magnetiques et Leurs Proprietes. *J. Phys. Radium* (1958), **19**, No. 10, pp. 750–764.
72. Dzyaloshinskii, I.E.: Thermodynamic Theory of "Weak" Ferromagnetism of Antiferromagnets. *Zh. Èksper. Teoret. Fiz.* (1957), **32**, No. 6, pp. 1547–1562.
73. Turov, E.A.: Physical Properties of Magnetically Ordered Crystals. Moscow: Izdat. Acad. Sci. SSSR, 1963, 130 pp.
74. Turov, E.A., Shavrov, V.G.: Galvano- and Thermomagnetic Effects in Antiferromagnets and Ferromagnets. *Izv. Akad. Sci. SSSR* (1963), **27**, No. 12. pp. 1487–1495.
75. Pisarev, R.V.: Optical Gyrotropy and Birefringence in Magnetically Ordered Crystals. *Zh. Èksper. Teoret. Fiz.* (1970), **58**, No. 4. pp. 1421–1427.
76. Vlasov, K.B., Rozenberg, E.A., Turov, E.A., Shavrov, V.G.: *Kinetic and Optical Effects of* AFM *Ordering. Dynamic and Kinetic Properties of Magnets*. Moscow: Nauka, 1986, 248 pp.
77. Mitsek, A.I., Shavrov, V.G.: Piezomagnetism in Antiferromagnets. *Solid State Phys.* (1964), **6**, No. 1. pp. 210–218.
78. Loktev, V.M., Ostrovskii, V.S.: Microscopic Theory of Anisotropic Antiferrodielectrics with a Spin 3/2. Kiev, 1977, 49 pp. (Preprint/Acad. Sci. Ukrain. SSR Inst. Teoret. Fiz.; ITF-77-105 pp.)
79. Loktev, V. M., Ostrovskii, V.S.: To the Theory of Carbonate Magnetic Properties. *Ukrain. Fiz. Zh.* (1980), **25**, No. 6, pp. 1043–1045.
80. Loktev, V.M.: On Theory of Optical Properties of Anisotropic Magnetic Crystals in External Magnetic Fields. Kiev, 1984, 10 pp. (Preprint/Acad. Sci. Ukrain. SSR Inst. Teoret Fiz.; ITP-84-33E.) *Solid State Commun.* (1984), **50**, No. 10, pp. 933–936.
81. Loktev, V.M.: On Theory of Odd Parity Linear Dichroism of Exciton–Magnon Transitions in Antiferromagnets. *Dokl. Akad. Sci. SSSR* (1986), **286**, No. 6, pp. 1382–1386.

82. Bokut', B.V., Fedorov, F.I.: On Theory of Optical Activity of Crystalls, III. *Opt. i Spektrosk.* (1959), **6** No. 4, pp. 537–541.
83. Konstantinova, A.F., Ivanov, N.R., Grechushnikov, B.I.: Optical Activity of Crystals in Directions Different from Optical Axis, I, II. *Crystallography* (1969), **14** No. 2, pp. 283–292; (1970); **15**, No. 3, pp. 490–499.
84. Tron'ko, V.D.: Magneto-Optical Effects in Birefringent Media. *Opt. i Spektrosk.* (1971), **30**, No. 4, pp. 739–744.
85. Born, M.: *Optics.* Berlin: Springer-Verlag, 1933, 785 pp.
86. Ramachandaran, G., Ramaseshan, S.: Magneto-Optical Rotation in Birefringent Media: Application of the Poincaré Sphere. *J. Opt. Soc. Amer.* (1952), **42**, No. 1, pp. 49–59.
87. Jerrard, H.G.: Transmission of Light Through Birefringent and Optically Active Media: The Poincaré Sphere. *J. Opt. Soc. Amer.* (1954), **44**, No. 8, pp. 634–640.
88. Shurkliff, U.: *Polarized Light.* Moscow: Mir, 1965, 264 pp.
89. Born, M. Wolf, E.: *Fundamentals of Optics.* Moscow: Nauka, 1973, 720 pp.
90. Kharchenko, N.F., Tutakina, O.P., Belyi, L.I.: Birefringence of Linearly Polarized Light in AFM $CoCO_3$ in High Magnetic Fields. 19 *All-Union Conference on Low Temperature Physics (Minsk, 1976): Abstracts.* Minsk, 1976, pp. 650–651.
91. Kharchenko, N.F., Eremenko, V.V., Titakina, O.P.: Bilinear in Ferromagnetic and Antiferromagnetic Vectors Birefringence in Cobalt Carbonate. *JETP Lett.* (1978), **27**, No. 7, pp. 437–440.
92. Kharchenko, N.F., Eremenko, V.V., Belyi, L.I.: Longitudinal Magnetic Field Induced Lowering of Optical Class of AFM Crystal. *JETP Lett.* (1979), **28**?, No. 6, pp. 325–327.
93. Kharchenko, N.F., Gnatchenko, S.L.: Linear Magneto-Optic Effect and Visualization of AFM Domains in Orthorhombic Crystal $DyFeO_3$. *Fiz. Nizk. Temp.* (1981), **7**, No. 4, pp. 475–493.
94. Borovik-Romanov, A.S.: Piezomagnetism in Antiferromagnetic Cobalt Fluoride and Mangenese Fluoride. *ZhETF* (*Soviet Phys. JETP*) (1960), **38**, No. 4, pp. 1088–1097.
95. Bibik, A.V., Kharchenko, N.F., Petrov, S.V.: Monodomainization and Remagnetization of AFM Cobalt Fluoride. *XVI All-Union Conference on Physics of Magnetic Phenomena (Tula,* 1983): *Abstracts.* Tula, 1983, Pt. 3, pp. 95–96.
96. Shtrikman, S., Treves, D.: Non-Linearity of the Susceptibility in Weak Ferromagnets and Antiferromagnets. *International Conference on Magnetism (Nottingham,* 1965). London, 1965, pp. 484–487.
97. Gorodetsky, G., Sharon, B., Shtrikman, S.: Linear Effect of the Magnetic Field on the Magnetic Susceptibility in Antiferromagnetic $DyFeO_3$. *Solid Stat. Commun.* (1967), **5**, No. 9, pp. 739–741.
98. Bazhan, A.N., Bazan, Ch.: Weak Ferromagnetism of CoF_2 and NiF_2. *ZhETF* (*Soviet Phys. JETP*) (1975), **69**, No. 5, pp. 1768–1781.
99. Bibik, A.V., Kharchenko, N.F., Lebedev, P.P.: Visual Studies of Magnetic Field-Induced $\mathbf{H} \parallel [11h]$ Remagnetization of AFM CoF_2. *XVIII All-Union Conference on Physics of Magnetic Phenomena (Donetsk,* 1985): *Abstracts.* Donetsk, 1985, Fs-34. pp. 71–72.
100. Dillon, J.F., Jr., Tallay, L.D., Chen, E. Yi.: Optical Birefringence in DyAlG in Fields Along [111]. *AIP Conf. Proc.* (1976), **34**, pp. 388–390.
101. Le Gall, H., Leycuras, C., Minell, D., Rudashewsky, E.G., Merculov, V.S.: Anomalous Evolution of the Magnetic and Magneto-Optical Properties of Hematite at

Temperature Near and Lower Than the Morin Phase Transition. *Physica* (1977), **86–88B**, pp. 1223–1226.
102. Borovik-Romanov, A.S., Ozhogin, V.I.: Weak Ferromagnetism in Antiferromagnetic Single Crystal $CoCO_3$. *ZhETF* (*Soviet Phys. JETP*) (1960), **39**, No. 1, pp. 27–36.
103. Kacer, J.: Gexagonal Anisotropy and Magnetization Curves of Antiferromagnet $CoCO_3$. *ZhETF* (*Soviet Phys. JETP*) (1962), **43**, No. 6, pp. 2042–2049.
104. Bazhan, A.N., Kreines, N.M.: Spontaneous Magnetic Moment Directed Along the Trigonal Axis in $CoCO_3$. *Pis'ma v ZhETF* (*JETP Lett.*) (1972), **15**, No. 9, pp. 533–537.
105. Bazhan, A.N.: Weak Ferromagnetism Along Trigonal Axis in Antiferromagnetic $CoCO_3$ and $NiCO_3$. *ZhETF* (*Soviet Phys. JETP*) (1974), **67**, No. 4, pp. 1520–1526.
106. Alikhanov, R.A.: Antiferromagnetism of $CoCO_3$. *ZhETF* (*Soviet Phys. JETP*) (1960), **39**, No. 5, pp. 1481–1483.
107. Brown, P.J., Welford, P.J., Forsyth, J.B.: Magnetization Density and the Magnetic Structure of Cobalt Carbonate. *J. Phys. C* (1973), **6**, No. 8, pp. 1405–1421.
108. Kharchenko, N.F., Eremenko, V.V., Tutakina, O.P.: Magnetic Birefringence and Domain Structure of Antiferromagnetic Cobalt Carbonate. *Soviet Phys. JETP* (1973), **37**, No. 4, pp. 672–677.
109. Borovik-Romanov, A.S., Kreines, N.M., Pankov, A.A., Talalaev, M.A.: Magnetic Birefringence in AFM $MnCO_3$, $CoCO_3$, $CsMnF_3$. *ZhETF* (*Soviet Phys. JETP*) (1974), **66**, No. 2, pp. 782–791.
110. Eremenko, V.V., Kharchenko, N.F., Belyi, L.I., Tutakina, O.P.: Birefringence of the AFM Crystals Linear in a Magnetic Field. *J. Magn. and Magn. Mater.* (1980), **15–18**, pp. 791–792.
111. Tutakina, O.P., Kharchenko, N.F.: Anisotropy of Linear Birefringence Induced by a Longitudinal Magnetic Field in $CoCO_3$. *XV ALL-Union Conference on Physics of Magnetic Phenomena* (*Perm', 1984*): *Abstracts*. Sverdlovsk: UNC Acad. Sci. SSSR, 1981, Pt. 3, pp. 86–87.
112. Kucab, M.: Magnetic Space Groups of Magnetic Structures Determined by Neutron Diffraction. Report No. 25/ps. Institute of Nuclear Techniques, Cracow, April 1972, 35 pp.
113. Pisarev, R.V., Krichevtsov, B.B.: Linear and Quadratic in Magnetic Field Birefringence in CoF_2. *Pis'ma v ZhETF* (*JETP Lett.*) (1979), **5**, No. 5, pp. 312–316.
114. Prokhorov, A.S., Rudashevskii, E.G.: Magnetostriction of AFM CoF_2. *Pis'ma v ZhETF* (*JETP Lett.*) (1969), **10**, No. 4, pp. 175–179.
115. Popkov, Yu. A., Fomin, V.I.: Mandelstam–Brillouin Scattering of Light in MnF_2, CoF_2, $KMnF_3$, and $RbMnF_3$ Crystals. *Proceedings of the Second International Conference on Light Scattering in Solids*. Paris: Flammarion, 1971, pp. 502–507.
116. Lines, M.E.: Magnetic Properties of CoF_2. *Phys. Rev.* (1965), **135**, No. 3A, pp. 982–993.
117. Ostrovskii, V.S., Loktev, V.M.: On New Magneto-Optic Effect in Antiferromagnetic Fluorides of Transition Metals in Longitudinal Magnetic Fields. *Pis'ma v ZhETF* (*JETP Lett.*) (1977), **26**, No. 3, pp. 139–141.
118. Eremenko, V.V., Kharchenko, N.F., Mil'ner, A.A., Bibik, A.V.: Linear Dichroism of Odd Parity in *H* in Magnetically Ordered Crystals. *All-Union Congress on Spectroscopy* (*Tomsk*, 1983): *Abstracts*. Tomsk, 1983, pp. 86–89.

119. Van der Ziel, J.P., Guggenheim, H.J.: Optical Spectrum of CoF_2. *Phys. Rev.* (1968), **166**, No. 2, pp. 479–487.
120. Gorodetsky, G., Sharon, B., Shtrikman, S.: Magnetic Properties of an Antiferromagnetic Orthoferrite, *J. Appl. Phys.* (1968), **39**, No. 2, pp. 1371–1372.
121. Cocure, P., Challeton, D.: Properties Magneto-Optiques des Orthoferrites. *Solid State Commun.* (1970), **8**, No. 17, pp. 1345–1348.
122. Tabor, W.J., Anderson, A.W., Van Vitert, L.G.: Visible and Infra-Red Faraday Rotation and Birefringence of Rare-Earth Orthoferrites *J. Appl. Phys.* (1970), **41**, No. 7, pp. 3018–3021.
123. Belov, K.P., Zvezdin, A.K., Kadomtseva, A.M., Levitin, R.Z.: *Orientational Transitions in Rare-Earth Magnets.* Moscow: Nauka, 1979, 317 pp.
124. Merkulov, V.S., Rudashevskii, E.G., Le Gall, H., Leycuras, C.: Birefringence Studies in Hematite Under Uniaxial Mechanical Stress in Magnetic Field. *ZhETF* (*Soviet Phys. JETP*) (1978), **75**, No. 2(8), pp, 628–640.
125. Merkulov, V.S., Rudashevskii, E.G., Le Gall, H., Leycuras, C.: Magnetic Linear Birefringence in Hematite near Morin Temperature. *ZhETF* (*Soviet Phys. JETP*) (1981), **80**, No. 1, pp. 161–170.
126. Kharchenko, N.F., Eremenko, V.V., Gnatchenko, S.L., Mil'near, A.A., Sofroneev, S.V.: Magneto-Optical Determination of Magnetic Point Group of $Ca_3Mn_2Ge_3O_{12}$. *Soviet J. Low Temp. Phys.* (1985), **11**, No. 2, pp. 116–117.
127. Eremenko, V.V., Gnatchenko, S.L., Kharchenko, N.F., Sofroneev, S.V., Desvignes, J.M., Feldmann, P., Le Gall, H.: Linear Magneto-Optic Effect in Tetragonal Antiferromagnetic Garnet $Ca_3Mn_2Ge_3O_{12}$. *Acta Phys. Polon.* (1985), **A68**, No. 3, pp. 419–422.
128. Kharchenko, N.F., Eremenko, V.V., Mil'ner, A.A., Bibik, A.V.: Magnetic Gyrational Properties of AFM Fluorides of Transition Metals. *24th All-Union Conference on Low Temperature Physics* (*Tbilisi*, 1986): *Abstracts.* Tbilisi, 1986, **1**, Pt. 3, pp. 3–4.
129. Pershan, P.S.: Absence of Antiferromagnetic Domain Walls in MnF_2. *Phys. Rev. Lett.* (1961), **7**, No. 7, pp. 280–281.
130. Shulman, R.G.: Nuclear Magnetic Resonance in NiF_2 Domain Walls. *J. Appl. Phys.* (1961), **32**, No. 3, pp. 126–128.
131. Alperin, H.A., Brown, P.J., Nathans, R., Pickart, S.J.: Polarized Neutron Study of Antiferromagnetic Domains in MnF_2. *Phys. Rev. Lett.* (1962), **8**, No. 6, pp. 237–239.
132. Birss, R.R., Anderson, J.C.: Linear Magnetostriction in Antiferromagnets. *Proc. Phys. Soc.* (1963), **81**, No. 524, pp. 1139–1140.
133. Anderson, J., Birss, R., Scott, R.: Linear Magnetostriction in Hematite. *Proceedings of the International Conference on Magnetism* (*Nottingham, 1964*). London: Inst. Phys. and Phys. Soc., s.a. 1964, pp. 597–599.
134. Voskanyan, R.A., Levitin, R.Z., Shurov, V.A.: Magnetostriction of Single Crystal of Hematite in Fields up to 150 kOe. *ZhETF* (*Soviet Phys. JETP*) (1968), **54**, No. 3, pp. 790–795.
135. Guy, C.N., Strom-Olsen, J.O., Cochrane, R.W.: Direct Macroscopic Observation of Antiferromagnetic Order in Zero Applied Field. *Phys. Rev. Lett.* (1979), **25**, No. 4, pp. 257–260.
136. Schlenker, M., Baruchel, J.: Neutron Techniques for the Observation of Ferro- and Antiferromagnetic Domains. *J. Appl. Phys.* (1978), **49**, No. 3, pp. 1996–2001.

137. Baruchel, J., Schlenker, M., Barbara, B.: Antiferromagnetic Domains in MnF_2 by Neutron Topography. *J. Magn. and Magn. Mater.* (1980), **15–18**, pp. 1510–1512.
138. Dillon, J.F., Chen, E.Yu., Giordane, N., Wolf, W.P.: Time-Reversed Antiferromagnetic State in Dysprosium Aluminium Garnet. *Phys. Rev. Lett.* (1974), **33**, No. 2, pp. 98–101.
139. Dillon, J.F., Jr., Chen, E.Yu., Guggenheim, H.J.: Magneto-Optical Studies of Metamagnets. *AIP Conf. Proc.* (1975), **24**, pp. 200–206.
140. Kharchenko, N.F., Eremenko, V.V., Belyi, L.I.: Visualization of 180° Antiferromagnetic Domains. *JETP Lett.* (1979), **29**, No. 7, pp. 392.
141. Eremenko, V.V., Kharchenko, N.F., Belyi, L.I.: Visualization of the 180-degree AFM Domains by Means of the New Magneto-Optical Effect. *J. Appl. Phys.* (1979), **50**, No. 11, pp. 7751–7753.
142. Kharchenko, N.F., Belyi, L.I.: Visualization of 180° AFM Domains in a Magnetic Field, I. Tetragonal Collinear Antiferromagnetic. *Izv. Akad. Nauk SSSR, Ser. Fiz.* (1980), **44**, No. 7, pp. 1451–1459.
143. Gnatchenko, S.L., Eremenko, V.V., Kharchenko, N.F.: Antiferromagnetic Wall the Nucleation Line of WFM Phase at the Phase Transition AFM ↔ WFM in $DyFeO_3$. *Soviet J. Low Temp. Phys.* (1981), **7**, No. 12, pp. 742–746.
144. Sapriel, J.: Domain-Wall Orientations in Ferroelastics. *Phys. Rev. B* (1975), **12**, No. 11, pp. 5128–5140.
145. Lifshitz, I.M.: On Kinetics of Ordering at the Second-Order Phase Transitions. *ZhETF* (*Soviet Phys. JETP*) (1962), **42**, No. 5, pp. 1354–1359.
146. Hubert, A.: *Theory of Domain Walls in Ordered Media.* Moscow: Mir, 1977, 306 pp.
147. Li, Y.Y.: Domain Walls in Antiferromagnets and the Weak Ferromagnetism of α-Fe_2O_3. *Phys. Rev.* (1956), **101**, No. 5, pp. 1450–1454.
148. Farztdinov, M.M.: Structure of Antiferromagnets. *Uspekhi Fiz. Nauk* (1964), **84**, No. 4, pp. 611–649.
149. Kharchenko, N.F., Eremenko, V.V., Belyi, L.I.: Magneto-Optical Studies of the Noncollinear State of Antiferromagnetic Cobalt Fluoride Induced by Longitudinal Magnetic Field. *Soviet Phys. JETP* (1982), **55**, No. 3, pp. 490–498.
150. Ozhogin, V.I.: Antiferromagnets $CoCO_3$, CoF_2, $FeCO_3$ in High Fields. *ZhETF* (*Soviet Phys. JETP*) (1963), **45**, No. 5, pp. 1687–1690.
151. Kocharyan, K.N., Rudashevskii, E.G.: Antiferromagnetic Resonance in Cobalt Fluoride. *Izv. Akad. Nauk SSSR, Ser. Fiz.* (1972), **36**, No. 7, pp. 1556–1558.
152. Shapiro, V.G., Ozhogin, V.I., Gurtovoi, K.T.: Orientational Dependence of Antiferromagnetic Resonance in CoF_2. *Izv. Akad. Nauk SSSR, Ser. Fiz.* (1972), **36**, No. 7, pp. 1559–1561.
153. Kuleshov, V.S., Popov, V.A.: Spin-Flop of Magnetic Sublattices in a Tetragonal Weak Ferromagnet. *Fiz. Nizk. Temper.* (1972), **20**, pp. 119–125.
154. Kuleshov, V.S., Popov, V.A.: Magnetic Phase Transition in Tetragonal Antiferromagnet with Superweak Ferromagnetism. *Fiz. Tverd. Tela.* (1973), **15**, No. 3, pp. 937–940.
155. Litvinenko, Yu.G., Shapiro, V.V.: Spectral Observation of Phase Transition in Iron Fluoride Induced by a High Magnetic Field. *Fiz. Nizk. Temper.* (1976), **2**, No. 2, pp. 233–235.
156. Eremenko, V.V., Kharchenko, N.F.: Field-Induced Spin-Orientational Phase Transitions in Neel Ferrimagnets, I. *Phase Transitions* (1980), **1**, pp. 207–268.

157. Borovik-Romanov, A.S., Kreines, N.M., Tankov, A.A., Talalaev, M.A.: Magnetic Birefringence in Antiferromagnetic Fluorides of Transition Metals. *ZhETF* (*Soviet Phys. JETP*) (1973), **64**, No. 5, pp. 1762–1775.
158. Gurtovoi, K.T.: Phase Transition in CoF_2 in a Transverse Field. *Fiz. Tverd. Tela.* (1978), **20**, No. 9, pp. 2666–2671.
159. Kazei, Z.A., Mill', B.V., Sokolov, V.I.: Cooperative Jahn–Teller Effect in Garnet $Ca_3Mn_2Ge_3O_{12}$. *Pis'ma v ZhETF* (*Soviet Phys. JETP Lett.*) (1976), **24**, No. 4, 229–232.
160. Belov, K.P., Sokolov, V.I.: Antiferromagnetic Garnets. *Uspckhi Fiz. Nauk* (1977), **121**, No. 2, pp. 285–317.
161. Kazei, Z.A., Novak, P., Sokolov, V.I.: Cooperative Jahn–Teller Effect in Garnets. *ZhETF* (*Soviet Phys. JETP*) (1982), **83**, No. 4(10), pp. 1483–1499.
162. Gnatchenko, S.L., Eremenko, V.V., Sofroneev, S.V., Kharchenko, N.F., Desvignes, J.M., Feldmann, P., Le Gall, H.: Spontaneous Phase Transitions and Optical Anisotropy in Manganese–Germanium Garnet $Ca_3Mn_2Ge_3O_{12}$ *Soviet Phys. JETP* (1986), **63**, No. 1, pp. 102–109.
163. Plumier, R.: Determination par Diffraction des Neutrons de la Structure Antiferromagnetique du Granat $Ca_3Mn_2Ge_3O_{12}$. *Solid State Commun.* (1971), **9**, No. 20, pp. 1723–1725.
164. Plumier, R., Esteve, D.: Reinvestigation of the Magnetic Structure of $Ca_3Mn_2Ge_3O_{12}$. *Solid State Commun.* (1979), **31**, No. 12, pp. 921–925.
165. Plumier, R., Esteve, D., Lecomte, M., Songi, M.: Neutron Diffraction Investigation of a Highly Anisotropic Garnet: $Ca_3Mn_2Ge_3O_{12}$. *J. Appl. Phys.* (1978), **49**, No. 3, Pt. II, pp. 1525.
166. Eremenko, V.V., Kharchenko, N.F., Gnatchenko, S.L., Milner, A.A., Sofroneev, S.V., Le Gall, H., Desvignes, J.M., Feldmann, P.: Magneto-Optical Determination of the MnGeG Magnetic Point Group. *J. Magn. and Magn. Mater.* (1986), **54–57**, pp. 1397–1398.
167. Vainstein, B.K., Fridkin, V.M., Indenbom, V.L.: *Modern Crystallography*. Vol. 2. *Crystal Structure*. Moscow: Nauka, 1979, 359 pp.
168. Indenbom, V.L.: Phase Transitions without a Change of Atom Numbers in a Crystal Unit Cell. *Crystallography* (1960), **5**, No. 1, pp. 115–125.
169. Gnatchenko, S.L., Eremenko, V.V., Kharchenko, N.F., Sofroneev, S.V., Bedarev, V.A., Desvignes, J.M., Feldmann, P., Le Gall, H.: Phase H–T Diagrams of AFM MnGeG near Neel Temperature. 17*th All-Union Conference on Physics of Magnetic Phenomena* (*Donetsk*, 1985): *Abstracts*. Donetsk, 1985, Fc-83, pp. 169–170.
170. Aizu, K.: Possible Species of Ferromagnetic, Ferroelectric and Ferroelastic Crystals. *Phys. Rev. B* (1970), **2**, No. 3, pp. 754–772.
171. Rightburd, A.L.: Theory of Heterophase Structure Formation at Phase Transformations in Solid State. *Uspekhi Fiz. Nauk* (1974), **113**, No. 1, pp. 69–104.
172. White, R., Jebell, T.: *Long-Range Order in Solids*. Moscow: Mir, 1982, 447 pp.
173. Huebener, R.P.: *Magnetic Flux Structures in Superconductors*. Berlin: Springer-Verlag, 1979, 378 pp.
174. Baryakhtar, V.G., Borovik, A.E., Popov, V.A.: Theory of Intermediate State of Antiferromagnets. *ZhETF* (*Soviet Phys. JETP*) (1972), **62**, No. 6, pp. 2233–2242.
175. Dudko, K.L., Eremenko, V.V., Fridman, V.M.: Studies of the Transition from an AFM to a Paramagnetic State in $FeCO_3$ under Magnetic Field. *Soviet Phys. JETP* (1975), **41**, No. 2, pp. 326–332.
176. Eremenko, V.V., Kharchenko, N.F., Belyi, L.I., Guillot, M., Marchand, A.,

Feldmann, P.: Magnetic Phase Diagram of $FeCO_3$ Ising AFM with AFM Inter- and Intrasublattice Exchange Interactions. *Soviet Phys. JETP* (1986), **62**, No. 5, pp. 988–999.
177. King, A., Paquette, D.: Spin-Flop Domains in MnF_2. *Phys. Rev. Lett.* (1973), **30**, No. 14, pp. 662–666.
178. Dudko, K.L., Eremenko, V.V., Fridman, V.M.: Magnetic Stratification upon Sublattice Flop in AFM MnF_2. *Soviet Phys. JETP* (1972), **34**, No. 2, pp. 362–367.
179. Dillon, J.F., Jr., Chen, Y.E., Guggenheim, H.J.: Microscope Studies of the Metamagnetic Transition in $FeCl_2$. *Solid State Commun.* (1975), **16**, No. 4, pp. 371–375.
180. Kharchenko, N.F., Eremenko, V.V., Gnatchenko, S.L.: Studies of Orientational Transitions and the Coexistence of Magnetic Phases in a Cubic Ferrimagnet GdIG. *ZhETF* (*Soviet Phys. JETP*) (1975), **69**, No. 5 (11), pp. 1697–1709.
181. Kharchenko, N.F., Szymczak, G., Eremenko, V.V., Gnatchenko, S.L., Szymaczak, R.: Magnetic Intermediate State in Dysprosium Orthoferrite. *JETP Lett.* (1977), **25**, No. 5, pp. 237–240.
182. Kharchenko, N.F., Eremenko, V.V., Gnatchenko, S.L., Szymczak, R., Szymczak, H.: Magnetic Field-Induced Intermediate State in Dysprosium Orthoferrite, *Solid State Commun.* (1977), **22**, No. 7, pp. 463–465.
183. Gnatchenko, S.L., Kharchenko, N.F., Szymczak, R.G.: Visual and Magneto-Optical Studies of Magnetic Phase Coexistence in near Morin Temperature in Dysprosium Orthoferrite. *Izv. Akad. Nauk SSSR* (1980), **44**, No. 7, pp. 1460–1472.
184. Belov, K.P., Zvezdin, A.K., Kadomtseva, A.M., Levitin, R.Z.: Transitions of Spin Reorientation in Rare-Earth Magnets. *Uspekhi Fiz. Nauk* (1976), **119**, No. 2, pp. 447–486.
185. Turov, E.A., Nite, V.E.: On the Theory of Weak Ferromagnetism in Rare-Earth Orthoferrites. *Fiz. Metal. i Metallovedenie* (1960), **9**, No. 1, pp. 10–18.
186. Yamaguchi, T.: Theory of Spin Reorientation in Rare-Earth Orthochromites and Orthoferrites. *J. Phys. Chem. Solids* (1974), **35**, No. 4, pp. 479–500.
187. Baryakhtar, V.G., Borovik, A.E., Popov, V.A., Stefanovskii, E.P.: on the Domain Structure of AFM Appearing with a Change of Magnetic Anisotropy Character. *ZhETF* (*Soviet Phys. JETP*) (1970), **59**, No. 10, pp. 1299–1306.
188. Zvezdin, A.K., Matveev, V.M.: On the Theory of Magnetic Properties of Dysprosium Orthoferrite. *ZhETF* (*Soviet Phys. JETP*) (1979), **77**, No. 3, pp. 1076–1086.
189. Belov, K.P., Zvezdin, A.K., Kadomtseva, A.M., Krynetskii, I.B.: On New Orientational Transitions in Orthoferrites Induced by an External Field. *ZhETF* (*Soviet Phys. JETP*) (1974), **67**, No. 5, (11), pp. 1974–1983.
190. Kuleshov, V.S., Popov, V.A.: Degenerate Critical Point of Phase Transitions. *Izv. Akad. Nauk SSSR, Ser. Fiz.* (1977), **44**, No. 7, pp. 1390–1394.
191. Zvezdin, A.K., Kalenkov, S.G.: Orthoferrite Domain Structure near Reorientation Temperature and its Influence on Phase Transition. *Fiz. Tverd. Tela.* (1972), **14**, No. 4, pp. 2835–2839.
192. Belyaeva, A.I., Stelmakhov, Yu.I., Potakova, V.A.: Visual Study of Spin Reorientation in $DyFeO_3$ near T_M. *Fiz. Tverd. Tela.* (1977), **19**, No. 10, pp. 3124–3125.
193. Baryakhtar, V.G., Klepikov, V.F., Stefanovskii, E.P.: On the Possibility of the Existence of an Intermediate State in Rare-Earth Orthoferrites. *Fiz. Metal. i Metallovedenie* (1972), **34**, No. 2, pp. 251–255.
194. Zalesskii, A.V., Savvinov, A.M., Zheludev, I.S., Ivashchenko, A.N.: NMR on Nu-

clei of Fe^{57} and Spin Reorientation in Domains and Domain Walls in $ErFeO_3$ and $DyFeO_3$ Crystals. *ZhETF* (*Soviet Phys. JETP*) (1975), **63**, No. 4, pp. 1449–1453.

195. Farztdinov, M.M., Shamsutdinov, M.A., Khalfina, A.A.: Domain Wall Structure in Orthoferrites. *Fiz. Tverd. Tela.* (1979), **21**, No. 5, pp. 1522–1527.
196. Kooy, C., Enz, H.: Domain Configuration in Layers of $BaFe_6O_{12}$. *Philips Res. Reps.* (1960), **15**, pp. 7–29.
197. Vas'kovskii, V.O., Kandaurova, G.S., Sinitsyn, E.V.: Peculiarities of Domain Structure of Orthoferrite Crystals in Spin Reorientation Region. *Fiz. Tverd. Tela.* (1977), **19**, No. 5, pp. 1245–1251.
198. Mitsek, A.I., Kolmakova, N.P., Gaidanskii, P.E.: Metastable States of Uniaxial Antiferromagnets. *Fiz. Tverd. Tela.* (1969), **11**, No. 5, pp. 1258–1264.
199. Mitsek, A.I., Gaidanskii, P.E., Pushkar, V.H.: Domain Structure of Uniaxial AFM. The Problem of Nucleation. *Phys. Status Solidi* (*b*) (1970), **69**, No. 1, pp. 69–79.
200. Prokhorov, A.S., Rudashevskii, E.G.: Exchange Enhancement and Vibration Quenching of Magnetic Impurity (VMI) in Antiferromagnets (AFM). *Pis'ma v ZhETF* (*Soviet JETP Lett.*) (1975), **22**, No. 4, pp. 214–218.
201. Eremenko, V.V., Novikov, V.P.: Davydov Splitting of the Exciton Line in Antiferromagnetic $RbMnF_3$. *JETP Lett.* (1970), **11**, No. 10, pp. 326–328.
202. Novikov, V.P., Eremenko, V.V., Shapiro, V.V.: Effect of External Factors on the Exciton Line in the Optical Spectrum of the Cubic Antiferromagnet $RbMnF_3$. Soviet *J. Low Temp. Phys.* (1973), **10**, No. 1/2, pp. 95–129.
203. Van der Ziel, J.P.: Optical Spectrum of Antiferromagnetic Cr_2O_3. *Phys. Rev.* (1967), **161**, No. 2, pp. 483–492.
204. Aoyagi, K., Tsushima, K., Sugano, S.: Direct Observation of Davydov Splitting in Antiferromagnetic $YCrO_3$. *Solid State Commun.* (1969), **7**, No. 1, pp. 229–232.
205. Sell, D.D., Greene, R.L., White, R.M.: Optical Exciton–Magnon Absorption in MnF_2. *Phys. Rev.* (1967), **158**, No. 2, pp. 489–510.
206. Loudon, R.: Theory of Infra-Red and Optical Spectra of Antiferromagnets. *Adv. in Phys.* (1968), **17**, No. 66, pp. 243–279.
207. Tanabe, Y., Aoyagi, K.: *Excitons in Magnetic Insulators. Excitons.* Amsterdam: North-Holland, 1982, Vol. 2, pp. 603–664.
208. Eremenko, V.V., Petrov, E.G.: Light Absorption in Antiferromagnets. *Adv. in Phys.* (1977), **26**, No. 1, pp. 31–78.
209. Gaididey, Yu.B., Loktev, V.M., Prikhot'ko, A.F.: Excitons in Magnetically Ordered Crystals. *Fiz. Niz. Temp.* (1977), **3**, No. 5, pp. 549–579.
210. Loginov, A.A., Popov, V.A.: Exciton–Magnon Interaction in Antiferromagnets and its Effect on the Exciton Energy Spectrum. (1981), **7**, No. 1, pp. 88–99.
211. Petrov, E.G.: *Theory of Magnetic Excitons.* Kiev: Naukova Dumka, 1976, 238 pp.
212. Eremenko, V.V.: *Introduction into Optical Spectroscopy of Magnets.* Kiev: Naukova Dumka, 1975, 471 pp.
213. Fujiwara, T., Tanabe, Y.: Temperature Dependence of the Magnon Sideband. *J. Phys. Soc. Japan* (1972), **32**, No. 4, pp. 912–926.
214. Parkinson, J.B., Loudon, R.: Green Function Theory of Magnon Sideband Shapes in Antiferromagnetic Crystals. *J. Phys. C* (1968), **1**, No. 5, pp. 1568–1581.
215. Petrov, E.G.: The Exciton Model of Spin and Optical Excitations of Antiferrodielectrics. *Phys. Status Solidi* (*b*) (1971), **48**, No. 1, pp. 367–379.

216. Dexter, D.L.: Cooperative Optical Absorption in Solids. *Phys. Rev.* (1962), **126**, No. 8, pp. 1962–1984.
217. Tanabe, Y., Moriya, T., Sugano, S.: Magnon–Induced Electric Dipole Transition Moment. *Phys. Rev. Lett.* (1965), **15**, No. 26, pp. 1023–1025.
218. Silvera, J., Halley, J.W.: Infrared Absorption in FeF_2: Phenomenological Theory. *Phys. Rev.* (1966), **149**, No. 2, pp. 415–422.
219. Halley, J.W.: Microscopic Theory of Far-Infrared Two-Magnon Absorption in Antiferromagnets, II. Second-Order Process and Application to MnF_2. *Phys. Rev.* (1967), **154**, No. 2, pp. 458–470.
220. Greenen, R.L., Sell, D.D., Yen, W.M., Schawlow, A.L., White, R.M.: Observation of a Spin-Wave Sideband in the Optical Spectrum of MnF_2. *Phys. Rev. Lett.* (1965), **15**, No. 16, pp. 656–659.
221. Elliott, R.J., Thorpe, M.F., Imbusch, G.F., Loudon, R., Parkinson, J.B.: Magnon–Magnon and Exciton–Magnon Interaction Effects on Antiferromagnetic Spectra. *Phys. Rev. Lett.* (1968), **21**, No. 3, pp. 147–150.
222. Tonegawa, T.: Theory of Magnon Sidebands in MnF_2. *Progr. Theoret. Phys.* (1969), **41**, No. 1, pp. 1–12.
223. Shinagawa, K., Tanabe, Y.: Intensity of Magnon Side-Bands. *J. Phys. Soc. Japan* (1971), **30**, No. 5, pp. 1280–1291.
224. Verdyan, A.I., Eremenko, V.V., Kaner, N.E., Litvinenko, Yu.G., Shapiro, V.V.: Mechanisms of Exciton–Magnon Absorption of Light by the Two-Sublattice Noncollinear Antiferromagnet $CoCO_3$. *Fiz Niz. Temp.* (1980), **6**, No. 5, pp. 644–655.
225. Eremenko, V.V., Litvinenko, Yu.G., Myatlik, V.I.: Magnetic Field Attenuation of Light Absorption by Antiferromagnetic $FeCO_3$. *JETP Lett.* (1970), **12**, No. 2, pp. 47–49.
226. Caird, R.S., Garn, W.B., Fawler, C.M., Thomson, D.B.: Optical Absorption Spectrum of MnF_2 at High Fields. *J. Appl. Phys.* (1971), **42**, No. 4, pp. 1651–1652.
227. Eremenko, V.V., Matyushkin, E.V., Shapiro, V.V., Bron, R.Ya.: Enhancement Mechanism of Antiferromagnet Luminescence in a Magnetic Field, *Soviet J. Low Temp. Phys.* (1977), **3**, No. 1, pp. 61–63.
228. Eremenko, V.V., Novikov, V.P., Petrov, E.G.: Multimagnon Absorption in Optical Spectrum of Antiferromagnetic $RbMnF_3$. *Soviet Phys. JETP* (1974), **39**, No. 6, pp. 1030–1035.
229. Shapiro, V.V., Litvinenko, Yu.G.: Exciton–Magnon Absorption of Light In Crystals Near Magnetic Ordering Temperature. *Fiz Niz. Temp.* (1975), **1**, No. 10, pp. 1295–1298.
230. Petrov, E.G., Loktev, V.M., Gaididei, Yu.B.: On the Theory of Light Absorption by Antiferrodielectrics in the Frequency Region of Double Electronic Excitations of Molecules (Ions). *Phys. Status Solidi* (1970), **41**, No. 1, pp. 117–128.
231. Stokowski, S.E., Sell, D.D.: Exciton–Exciton Transitions in MnF_2. *Phys. Rev. B* (1971), **3**, No. 1, pp. 208–213.
232. Gaididei, Yu.B., Loktev, V.M.: Bound States of Two Frenkel Excitons in Layer Antiferrodielectrics. *Phys. Lett. A* (1973), **46**, No. 1, pp. 67–68.
233. Freeman, S., Hopfield, J.J.: Exciton–Magnon Interaction in Magnetic Insulators. *Phys. Rev. Lett.* (1968), **21**, No. 13, pp. 910–913.
234. Melrzer, R.S., Chen, M.Y., McClure, D.S., Lowe Pariscau, M.: Exciton–Magnon Bound State in MnF_2 and the Exciton Dispersion in MnF_2 and $RbMnF_3$. *Phys. Rev. Lett.* (1968), **21**, No. 13, pp. 913–916.

235. Bakai, A.S., Sergeeva, G.G.: Polarization Effects in Interactions between Intensive Light and Spin Waves. *ZhETF* (*Soviet Phys. JETP*) (1975), **69**, No. 8, pp. 493–498.
236. Araujo de Cid, B., Rezende, S.M.: Coherent Generation of Magnons by Optical Techniques. *J. Appl. Phys.* (1978), **49**, No. 3 (11), pp. 2186–2188.
237. Seidov, Yu.M.: Parametric Excitation of Spin Waves by Optical Pumping. *Dokl. Akad. Nauk SSSR* (1973), **209**, No. 5, pp. 1066–1067.
238. Genkin, G.M.: Parametric Generation of High-Frequency Magnons with Pulses on the Brillouin Zone Boundary. *Pis'ma v ZhETF* (*JETP Lett.*) (1979), **30**, No. 10, pp. 651–654.
239. Holzrichter, J.F., MacFarlane, R.M., Schawlow, A.L.: Magnetization Induced by Optical Pumping in Antiferrmagnetic MnF_2. *Phys. Rev. Lett.* (1971), **26**, No. 11, pp. 652–655.
240. Dokashenko, V.P., Matyushkin, E.V.: To the Question of Magnon Generation in Antiferromagnets upon High Optical Pumping. *Fiz. Niz. Temp.* (1980), **6**, No. 11, pp. 1492–1493.
241. Genkon, G.M., Golubeva, N.G.: Parametric Generation of High Frequency Brillouin Magnons upon Pumping in the Exciton–Magnon–Phonon Absorption Range. *XXI All-Union Conference on Low Temperature Physics* (*Kharkov, September* 23–26, 1980): *Abstracts.* Kharkov, 1980, pp. 238.
242. Wilson, B.A., Hegarty, J., Yen, W.M.: Biexciton Decay in MnF_2. *Phys. Rev. Lett.* (1978), **41**, No. 1, pp. 268–270.
243. Greene, R.L., Sell, D.D., Feigelson, R.S., Imbusch, G.F., Guggenheim, H.J.: Impurity Induced Optical Fluorescence in MnF_2. *Phys. Rev.* (1968), **171**, No. 2, pp. 600–610.
244. Dietz, R.E., Meixner, A.E., Guggenheim, H.J., Misetich, A.: Observation of a Thermodynamic Distinction Between the Brillouin Zone Center and Boundary Exciton States in MnF_2. *Phys. Rev. Lett.* (1968), **21**, No. 15, pp. 1067–1070.
245. Pantei, R., Puthoff, G.: *Fundamentals of Quantum Electronics.* Moscow: Mir, 1972, 384 pp.
246. Fujiwara, T., Gebhardt, W., Petanides, K., Tanabe, Y.: Temperature-Dependent Oscillator Strengths of Optical Absorptions in MnF_2 and $RbMnF_3$. *J. Phys. Soc. Japan* (1972), **33**, No. 1, pp. 39–48.
247. Dokashenko, V.P., Eremenko, V.V., Matyushkin, E.V., Bron, R.Ya.: Two-Exciton Light Absorption in Antiferromagnetic MnF_2 at Four-Photon Excitation. *Soviet J. Low Temp. Phys.* (1977), **3**, No. 11, pp. 708–709.
248. Dokashenko, V.P., Eremenko, V.V., Matyushkin, E.V.: A Change of the Order of Multiquantum Light Absorption in Manganese Fluoride at Varying Excitation Polarization. *Fiz. Niz. Temp.* (1981), **7**, No. 6, pp. 806–808.
249. Loudon, R.: *Light Quantum Theory.* Moscow: Mir, 1976, 488 pp.
250. Arslanbekov, T.U., Delone, N.B., Masakov, A.V., Todirashku, S.S. Fainstein, A.G.: Multiphoton Processes in a Multimode Laser Emission Field. *ZhETF* (*Soviet Phys. JETP*) (1977), **72**, No. 3, pp. 907–916.
251. Bryukner, F., Dneprovskii, V.S., Khattatov, V.U.: Two-Photon Absorption in Cadmium Selenide. *Quantum Electronics* (1974), No. 1, pp. 1360–1364.
252. Dokashenko, V.P., Eremenko, V.V., Matyushkin, E.V.: On the Effect of Statistical Properties of Light on Two-Exciton Multiphoton Absorption Intensity in Antiferromagnetic MnF_2 Crystal. *Soviet J. Low Temp. Phys.* (1978), **4**, No. 7, pp. 426–428.

253. Dukhovnii, A.M., Prilezhaev, D.S.: On a Possibility of Spatial Characteristics Control of Monopulse Laser Emission. *Zh. Techn. Fiz.* (1977), **47**, No. 11, pp. 2440–2441.
254. Eremenko, V.V., Matyushkin, E.V., Petrov, S.V.: Study of Energy Transfer from 3*d* to 4*f* Electron on Antiferromagnetic MnF_2: Eu^{3+} Crystals. *Phys. Stat. Solidi* (1966), **18**, No. 2, pp. 683–686.
255. Matyushkin, E.V., Kukushkin, L.S., Eremenko, V.V.: Peculiarities of the $RbMnF_3 : Nd^{3+}$ Crystals Decay Kinetics due to the Migration of Excitation Energy. *Phys. Stat. Solidi* (1967), **22**, No. 1, pp. 65–69.
256. Matyushkin, E.V., Eremenko, V.V.: Low Temperature Anomalies of Luminescence of Antiferromagnetic $RbMnF_3 : Nd^{3+}$ Crystals. *Ukrain. Fiz. Zh.* (1968), **13**, No. 2, pp. 267–272.
257. Gooen, K., Di Bartolo, B., Alam, M., Powell, R.C., Linz, A.: Thermal Dependence of the Mn^{2+} and Nd^{3+} Fluorescence and of the $Mn^{2+}Nd^{3+}$ Energy Transfer $RbMnF_3$. *Phys. Rev.* (1969), **177**, No. 2, pp. 615–625.
258. Flaherty, J.M., Bartolo, B.: Fluorescence Studies of the $Mn^{2+} - Er^{3+}$ Energy Transfer in MnF_2. *Phys. Rev. B* (1974), **8**, No. 11, pp. 5232–5238.
259. Wilson, B.A., Yen, W.M., Hegarty, J., Imbusch, G.F.: Luminescence from Pure MnF_2 and from MnF_2 Doped with Eu^{3+} and Er^{3+}. *Phys. Rev. B* (1979), **19**, No. 8, pp. 4238–4250.
260. Iverson, M.V., Silbey, W.A.: Optical Properties of $RbMnF_3 : Er^{3+}$. *Phys. Rev. B* (1980), **21**, No. 6, pp. 2522–2532.
261. Holloway, W.W., Prohofsky, E.M., Kestigian, M.: Magnetic Ordering and the Fluorescence of Concentrated Mn Systems. *Phys. Rev.* (1965), **139**, No. 3A, pp. 954–961.
262. Matyushkin, E.V., Eremenko, V.V., Bron, R.Ya.: Luminescence Sensibilization of a Complex Mn^{2+} *F*-center in Magnetoconcentrated Maganese Fluorides. *Ukrain. Fiz. Zh.* (1975), **20**, No. 6, pp. 986–992.
263. Yun, S.J., Koumvakalis, N., Silbey, W.A.: Radiation Damage and Energy Transfer in MnF_2 and $RbMnF_3$ Crystals. *J. Phys. C* (1977), **10**, No. 20, pp. 3987–3997.
264. Moncorge, R., Jacquier, B., Modey, C., Blanchard, H., Brunel, L.: Exciton Dynamics and Energy Transfers in Pure $CsMnF_3$. *J. Phys. (France)* (1982), **43**, pp. 1267–1281.
265. Ueda, K., Tanabe, Y.: Exciton Motion in Antiferromagnets. *J. Phys. Soc. Japan* (1980), **48**, No. 4, pp. 1137–1146.
266. Agranovich, V.M.: *Exciton Theory*. Moscow: Nauka, 1968, 382 pp.
267. Davydov, A.S.: *Theory of Molecular Excitons*. Moscow: Nauka, 1968, 296 pp.
268. MacFarlane, R.M., Luntz, A.C.: Exciton Dynamics and Phase Memory in MnF_2. *Phys. Rev.* (1973), **31**, No. 13, pp. 832–835.
269. Bron, R.Ya., Eremenko, V.V., Matyushkin, E.V.: Exciton Autolocalization in Low-Dimensional Antiferromagnets, 20*th All-Union Conference on Low Temperature Physics* (*Moscow, January* 23–26, 1979). Chernogolovka, 1978, Pt. II, pp. 145–146.
270. Bron, R.Ya., Eremenko, V.V., Matyushkin, E.V.: Localization of Electron Excitation in a Quasi-One-Dimensional Antiferromagnet $CsMnCl_3 \cdot 2H_2O$. *Soviet J. Low Temp. Phys.* (1979), **5**, No. 6, pp. 314–316.
271. Bron, R.Ya., Eremenko, V.V., Matyushkin, E.V.: Exciton Autolocalization in Quasi-Two-Dimensional Antiferromagnet $NaMnCl_3$. *Soviet J. Low Temp. Phys.* (1979), **5**, No. 9, pp. 502–505.

272. McPherson, G., Francis, A.H.: Extensive Energy Transfer in Nearly One-Dimensional Crystals. The Emission Spectrum of $CsMnBr^3$ Doped with Nd^{3+}. *Phys. Rev. Lett*, (1978), **41**, No. 24, pp. 1681–1683.
273. Matyushkin, E.V., Smushkov, V.I.: On Exciton Migration in Antiferromagnetic Manganese Crystals. *Fiz. Nizk. Temper.* (1982), **8**, No. 9, pp. 977–982.
274. McPherson, G.L., Delaney, K.O., Willard, S.C., Francis, A.H.: Energy Transfer in One, Two, and Three Dimensionally Coupled Salts of Divalent Manganese. *Chem. Phys. Lett.* (1979), **68**, No. 1, pp. 9–12.
275. Matyushkin, E.V.: Luminescence from Antiferromagnetic $BaMnF_4$. *Fiz. Kondensir. Sostoyaniya* (1973), No. 26, pp. 126–128.
276. Matyushkin, E.V., Eremenko, V.V., Bron, R.Ya.: Self-Trapping of Excitons in Antiferromagnetic Insulators. *J. Magn. and Magn. Mater.* (1980), **15–18**, pp. 1043–1044.
277. Bazhan, A.N., Fedoseeva, N.V., Spevakova, I.P.: Magnetic Phase Transition from the Antiferromagnetic State to the Paramagnetic One in $NaMnCl_3$. *ZhETF* (*Soviet Phys. JETP*) (1978), **75**, No. 2, pp. 577–584.
278. Rashba, E.I.: Theory of Strong Interactions of Electronic Excitations and Lattice Vibrations in Molecular Crystals. *Opt. i Spektrosk.* (1957), **2**, No. 1, pp. 75–98.
279. Bron, R.Ya., Eremenko, V.V., Matyushkin, E.V., Shapiro, V.V.: Luminescence Enhancement of Quasi-One-Dimensional Antiferromagnet $CsMnCl_3 \cdot 2H_2O$ in a Magnetic Field. *Soviet J. Low Temp Phys.* (1979), **5**, No. 7, pp. 367–370.
280. Eremenko, V.V., Milner, A.A., Popkov, Yu.A., Shapiro, V.V.: New Mechanism of Electrodipole Light Absorption in Antiferromagnets. *Soviet J. Low Temp. Phys.* (1976), **2**, No. 9, pp. 578–580.
281. Dokashenko, V.P., Eremenko, V.V., Matyushkin, E.V., Smushkov, V.I.: Luminescence from Antiferromagnetic Manganese Fluoride upon High Excitation. *Soviety J. Low Temp. Phys.* (1981), **7**, No. 7, pp. 438–441.
282. Strauss, E., Maniscalco, W.J., Yen, W.M., Keller, V.G., Gerhardt, V.: Exciton–Exciton Interaction in $KMnF_3$. *Phys. Rev. Lett.* (1980), **44**, No. 12, pp. 824–827.
283. Yamaguchi, Y., Hirano, M.: Biexciton Decay and Multimagnon Sideband in Emission Spectra of $CsMnF_3$. *Solid State Commun.* (1980), **36**, No. 7, pp. 635–638.
284. Eremenko, V.V., Litvinenko, Yu.G., Garber, T.I.: Magnon Participation in Light Absorption by Antiferromagnetic Oxygen. *JETP Lett.* (1968), **7**, No. 10, pp. 298–301.
285. Verdyan, A.I., Eremenko, V.V., Litvinenko, Yu.G., Shapiro, V.V.: Optical Studies of Spin Wave Spectrum in Weak Ferromagnetic $CoCO_3$. *Soviet J. Low Temp. Phys.* (1979), **5**, No. 1, pp. 36–37.
286. Eremenko, V.V., Litvinenko, Yu.G., Shapiro, V.V., Verdyan, A.I.: Exciton–Magnon Light Absorption in Noncollinear Antiferromagnets at $T \neq 0$. *Soviet J. Low Temp. Phys.* (1977), **3**, No. 9, pp. 573–574.
287. Naumenko, V.M., Eremenko, V.V., Maslennikov, A.I., Kovalenko, L.V.: Long Wavelength IR Spectrum of $CoCO_3$: AFMR High-Frequency Mode and Two-Magnon Absorption. *JETP Lett.* (1978), **27**, No. 1, pp. 17–21.
288. Eremenko, V.V., Mokhir, A.P., Popkov, Yu.A., Sergienko, N.A., Fomin, V.I.: Excitons and Magnons in $CoCO_3$. *Soviet Phys. JETP* (1977), **46**, No. 6, pp. 1231–1237.
289. Belyaeva, A.I., Kuleshov, V.S., Silaev, V.I., Gapon, N.V.: Peculiarities of Spin Wave Spectrum of $CsMnF_3$ and their Manifestation in Light Absorption. *ZhETF* (*Soviet Phys. JETP*) (1971), **61**, No. 4, pp. 1492–1505.

290. Novikov, V.P., Kachur, I.S., Eremenko, V.V.: Peculiarities of Exciton–Magnon Absorption of Light by Antiferromagnetic $CsMnCl_3 \cdot 2H_2O$ Induced by Low Dimensionality of its Magnetic Structure. *Soviet J. Low Temp. Phys.* (1981), **7**, No. 2, pp. 108–112.
291. Robbins, D.J., Day, P.: Temperature Variation of Exciton–Magnon Absorption Bands in Metamagnetic Transition Metal Dihalides. *J. Phys. C* (1976), **9**, No. 5, pp. 867–882.
292. Wood, T.E., Muirhead, A., Day, P.: Optical Study of the Magnetic Phase Diagram of Metamagnetic Ferrous Bromide. *J. Phys. C* (1978), **11**, No. 8, pp. 1619–1633.
293. Day, P., Janke, E., Wood, T.E., Woodwark, D.: Optical Estimation of the Zone-Centre Magnon Gap in $RbCrCl_4$ a Two-Dimensional Easy-Plane Ionic Ferromagnet. *J. Phys. C* (1979), **12**, No. 8, pp. L329–L334.
294. Belyaeva, A.I., Eremenko, V.V., Gapon, N.V., Kotlyarskii, M.M.: Exciton–Magnon Transitions in Absorption Spectra of Solid Solutions of $CsMnF_3$. *Fiz. Tverd. Tela.* (1973), **15**, No. 12, pp. 3532–3534.
295. Milner, A.A., Popkov, Yu.A.: Detection of Magnetic Sublattice Spontaneous Canting in Gexagonal Antiferromagnet. *Pis'ma v ZhETF* (*Soviet Phys. JETP Lett.*) (1977), **25**, No. 5, pp. 244–247.
296. Milner, A.A., Popkov, Yu.A.: Light Absorption Spectrum and Magnetic Behavior of $CsMnF_3$ in a Field up to 300 kOe. *Fiz. Niz. Temp.* (1977), **3**, No. 2, pp. 194–201.
297. Verdyan, A.I., Eremenko, V.V., Litvinenko, Yu.G., Shapiro, V.V.: Light Absorption Dichroism and Magnetic Configuration of Weak Ferromagnetic $CoCO_3$. *Soviet J. Low Temp. Phys.* (1979), **5**, No. 6, pp. 322–325.
298. Petrov, E.G., Ostrovskii, V.S.: Interaction of Light and Gexagonal Antiferromagnets. *Fiz. Niz. Temp.* (1976), **2**, No. 11, pp. 1456–1465.
299. Eremenko, V.V., Milner, A.A., Kharchenko, N.F., Gredeskul, V.M., Popkov, Yu.A., Petrov, V.V.: Weak Ferromagnetism of $CsMnF_3$—Polarization, Spectroscopic and Visual Studies. *Soviet J. Low Temp. Phys.* (1981), **7**, No. 11, pp. 692–696.
300. Dudko, K.L., Eremenko, V.V., Fridman, V.M.: Magnetic Lamination upon Sublattice Flop in Antiferromagnetic Manganese Fluoride. *ZhETF* (*Soviet Phys. JETP*) (1971), **61**, No. 2, pp. 678–688.
301. Milner, A.A., Popkov, Yu.A., Eremenko, V.V.: Spectroscopic Study of Intermediate State in Antiferromagnetic MnF_2. *JETP Lett.* (1973), **18**, No. 1, pp. 20–22.
302. King, A.R., Paquetee, D.: Spin-Flop Domains in MnF_2. *Phys. Rev. Lett.* (1973), **30**, No. 14, pp. 662–666.
303. Eremenko, V.V., Kaner, N.E., Litvinenko, Yu.G., Milner, A.A., Shapiro, V.V: Spectroscopic Determination of Antiferromagnetism Vector Orientation in MnF_2 in Spin-Flop Phase. *Soviet J. Low Temp. Phys.* (1985), **11**, No. 1, pp. 33–36.
304. Eremenko, V.V., Litvinenko, Yu.G., Kazakova, T.I.: Spectral Detection of Magnetic Sublattice Flop in External Magnetic Field in Antiferromagnetic Nickel Tungstate. *Soviet J. Low Temp. Phys.* (1974), **16**, No. 12, pp. 3717–3719.
305. Novikov, V.P., Eremenko, V.V., Kachur, I.S.: Phase Spin-Flop Transition and Intermediate State in Quasi-One-Dimensional Antiferromagnet $CsMnCl_3$ $2H_2O$. *Soviet Phys. JETP* (1982), **55**, No. 2, pp. 327–333.
306. Andlauer, B., Diehl, R., Sholnick, M.S.: Investigation of the Optical Absorption of Fe_3BO_6 after Oxygen Annealing and Under the Influence of Strong Magnetic Fields. *J. Appl. Phys.* (1978), **49**, No. 3 (11), pp. 2200–2202.

307. Eremenko, V.V., Litvinenko, Yu.G., Shapiro, V.V.: Peculiarities of Light Absorption at Magnetic Phase Transition in $FeCO_3$. *Soviet J. Low Temp. Phys.* (1975), **1**, No. 8, pp. 519–522.
308. Kurita, S., Toyokawa, K., Tsushima, K., Sugano, S.: Photoinduced Magnetic Phase Transitionmin Antiferromagnetic $ErCrO_3$. *Solid State Commun.* (1981), **38**, No. 3, pp. 235–239.
309. Golovenchitz, E.I., Sanina, V.A.: Optical Absorption Spectra of $EuCrO_3$. *Fiz. Tverd. Tela.* (1982), **24**, No. 2, pp. 375–383.
310. Eremenko, V.V., Kaner, N.E., Litvinenko, Yu.G., Shapiro, V.V.: Photoinduced One-Ionic Anisotropy in Antiferromagnetic MnF_2. *Soviet Phys. JETP* (1983), **57**, No. 6, pp. 1312–1317.
311. Cowley, R.A., Buyers, W.J.L.: The Properties of Defects in Magnetic Insulators, Rev. Mod. Phys. (1972), **44**, No. 2, pp. 406–450.
312. Weber, R.: Spin Wave Impurity States in Linear Chain Ferromagnets and Antiferromagnets. *Z. Phys.* (1969), **223**, No. 4, pp. 299–337.
313. Izyumov, Yu.A., Medvedev, M.V.: *Theory of Magnetically Ordered Crystals with Impurities.* Moscow: Nauka, 1970, 271 pp.
314. Borovik-Romanov A.S., Meshcheryakov, V.F.: Splitting of Antiferromagnetic Resonance Spectrum in $CoCO_3$. *Pis'ma v ZhETF* (*Soviet Phys. JETP Lett.*) (1968), **8**, No. 8, pp. 425–429.
315. Dumesh, B.S., Egorov, V.M., Meshcheryakov, V.F.: Investigation of the Mn^{2+} and Fe^{2+} Impurities Influence on Antiferromagnetic Resonance Spectrum in $CoCO_3$. *ZhETF* (*Soviet Phys JETP*) (1971), **61**, No. 1, pp. 320–331.
316. Anderson, P.W.: Absence of Diffusion in Certain Random Lattices. *Phys. Rev.* (1958), **109**, No. 5, pp. 1492–1505.
317. Lifshitz, I.M.: On Energy Spectrum Structure and Quantum States of Disordered Condensed Systems. *Uspekhi Fiz. Nauk* (1964), **83**, No. 4, pp. 617–663.
318. Ivanov, M.A.: Quasi-local Vibration Dynamics at High Concentration of Impurity Centers. *Fiz. Tverd. Tela.* (1970), **12**, No. 7, pp. 1895–1905.
319. Ivanov, M.A., Pogorelov, Yu.G., Botvinko, M.N.: Elastic Vibration Spectrum of Doped Crystal in the Presence of Local States Near the Zone Edge. *ZhETF* (*Soviet Phys. JETP*) (1976), **70**, No. 2, pp. 610–620.
320. Ivanov, M.A., Pogorelov, Yu.G.: Electronic States in Crystals with Doped Levels Near the Band Edge. *ZhETF* (*Soviet Phys. JETP*) (1977), **72**, No. 6, pp. 2198–2209.
321. Ivanov, M.A., Pogorelov, Yu.G.: Crystal Electronic Spectrum in the Presence of Large Radius Impurity States. *ZhETF* (*Soviet Phys. JETP*) (1979), **76**, No. 3, pp. 1012–1022.
322. Ivanov, M.A., Pogorelov, Yu.G., Loktev, V.M., Kocharyan, K.N., Prokhorov, A.S., Rudashevsky, E.B.: Collective Rearrangement of a Spin Excitation Spectrum of an Antiferromagnet with Magnetic Impurities in an External Magnetic Field. *Solid State Commun.* (1980), **33**, No. 6, pp. 623–626.
323. Ivanov, M.A., Pogorelov, Yu.G.: Infrared Absorption on Impurity Excitations Near the Upper Edge of Spin Wave Band of the Antiferromagnet. *Solid State Commun.* (1980), **34**, No. 8, pp. 629–633.
324. Ivanov, M.A., Loktev, V.M., Pogorelov, Yu.G.: Rearrangement of the Spin Excitation Spectrum of an Anisotropic Antiferromagnet with Magnetic Impurities in an External Field. Kiev, 1980, 42 pp. (Preprint/Acad. Sci. Ukrain SSR Inst. Theoret. Phys.; ITP-80-40 pp.)

325. Invanov, M.A., Loktev, V.M., Pogorelov, Yu.G.: Rearrangement of the Spin Excitation Spectrum of an Anisotropic Antiferromagnet with Magnetic Impurities in an External Field. *Soviet J. Low Temp. Phys.* (1981), **7**, No. 11, pp. 1401–1415.
326. Ivanov, M.A., Loktev, V.M., Pogorelov, Yu.G.: Low-Frequency Spin Excitations of Transition Element Impurities in Antiferromagnets of the Type of "Easy" Plane. *Fiz. Tverd. Tela.* (1983), **25**, No. 6, pp. 1644–1649.
327. Ivanov, M.A., Loktev, V.M., Pogorelov, Yu.G.: Spin Excitation Spectrum of Antiferromagnetic $CoCO_3$ with Fe^{2+} Impurities. Kiev, 1983, 31 pp. (Preprint/ Acad. Sci. Ukrain. SSR Inst. Theoret. Phys.; ITP-83-52 pp.)
328. Lifshitz, I.M., Gredeskul, S.A., Pastur, L.A.: *Introduction in the Disordered System Theory.* Moscow: Nauka, 1982, 358 pp.
329. Buyers, W.J.L., Pepper, D.E., Elliott, R.J.: Theory of Spin Waves in Disordered Antiferromagnets, I. Application to (Mn, Co)F_2 and K(Mn, Co)F_3. *J. Phys. C* (1972), **5**, No. 18, pp. 2611–2628.
330. Buyers, W.J.L., Cowley, R.A., Holden, T.M., Stevenson, R.W.H.: Observation of a Localized Magnon in Co-Doped MnF_2. *J. Appl. Phys.* (1968), **39**, No. 2, pp. 1118–1119.
331. Dürr, U., Uwira, B.: Influence of Localization on Properties of the Impurity-Induced Magnon in $FeF_2 : Co^{2+}$. *J. Phys. C* (1979), **12**, No. 20, pp. L793–L796.
332. Rashba, E.I.: Theory of Doped Light Absorption in Molecular Crystals. *Opt. i Spektrosk.* (1957), **2**, No. 5, pp. 568–577.
333. Rashba, E.I., Gurgenishvili, G.E.: On the Theory of Edge Absorption in Semiconductors. *Fiz. Tverd. Tela.* (1962), **4**, No. 4, pp. 1029–1031.
334. Rashba, E.I.: Theory of Doped Absorption near Exciton Bands at Isotopic Substitution. *Fiz. Tverd. Tela.* (1962), **4**, No. 11, pp. 3301–3320.
335. Broude, V.L.: Spectral Studies of Benzene. *Uspekhi Fiz. Nauk* (1961), **74**, No. 4, pp. 577–608.
336. Broude, V.L., Rashba, E.I., Sheka, E.F.: Anomalous Impurity Absorption near Exciton Bands in Molecular Crystals. *Dokl. Akad. Nauk SSSR* (1961), **139**, No. 5, pp. 1085–1088.
337. Broude, V.L., Onoprienko, M.I.: Absorption Spectra of Normal and Deuterated Benzene at 20 K. *Opt. i Spektrosk.* (1961), **10**, No. 5, pp. 634–639.
338. Andreev, A.F., Lifshitz, I.M.: Quantum Theory of Defects in Crystals. *ZhETF* (*Soviet Phys. JETP*) (1969), **56**, No. 6, pp. 2057–2068.
339. Lifshitz, I.M.: On Degenerate Regular Perturbations, I. Discrete Spectrum. *ZhETF* (*Soviet Phys. JETP*) (1947), **17**, No. 5, pp. 1017–1025.
340. Lifshitz, I.M.: On Degenerate Regular Perturbations, II. Quasi-Continuous and Continuous Spectra. *ZhETF* (*Soviet Phys. JETP*) (1947), **17**, No. 2, pp. 1076–1089.
341. Callaway, J.: Theory of Scattering in Solids. *J. Math. Phys.* (1964), **5**, No. 6, pp. 783–798.
342. Koster, G.F., Slater, J.C.: Wave Functions for Impurity Levels. *Phys. Rev.* (1954), **95**, No. 5, pp. 1167–1176.
343. Koster, G.F., Slater, J.C.: Simplified Impurity Calculation. *Phys. Rev.* (1954), **96**, No. 5, pp. 1208–1223.
344. Takeno, S.: Impurity States in a Heisenberg Ferromagnet. *Progr. Theoret. Phys.* (1963), **30**, No. 4, pp. 565–566.
345. Takeno, S.: Spin-Wave Impurity Levels in a Heisenberg Ferromagnet. *Progr. Theoret. Phys.* (1963), **30**, No. 6, pp. 731–742.

346. Tonegawa, T.: Theory of Spin-Wave Impurity States in an Antiferromagnet. *Progr. Theoret. Phys.* (1968), **40**, No. 6, pp. 1195–1226.
347. Lovesey, S.W.: Spin-Wave Theory of Impurity States in a Heisenberg Antiferromagnet, I. Positive Impurity-Host Exchange Coupling. *J. Phys. C* (1968), **1**, No. 1, pp. 102–124.
348. Dietz, R.E., Parisot, G., Meixner, A.E., Guggenheim, H.J.: Impurity Magnons in MnF_2 : Ni. *J. Appl. Phys.* (1970), **41**, No. 3, pp. 888–889.
349. Moch, P., Parisot, G., Dietz, R.E., Guggenheim, H.J.: Observation of Localized Magnons by Raman Scattering in Ni-Doped MnF_2. *Phys. Rev. Lett.* (1968), **21**, No. 23, pp. 1596–1599.
350. Thorpe, M.F.: Theory of Localized Magnons in Ni^{++} Doped Manganese Salts. *Phys. Rev. B* (1970), **2**, No. 7, pp. 2690–2702.
351. Svensson, E.S., Holden, T.M., Buyers, W.J.L., Cowley, R.A., Stevenson, R.W.H.: On the Resonant Perturbation of Spin Waves by Impurities. *Solid State Commun.* (1969), **7**, No. 23, pp. 1693–1696.
352. Misetich, A., Dietz, R.E., Guggenheim, H.J.: *Localized Excitations in Solids.* New York: Plenum Press, 1968, 379 pp.
353. Naumenko, V.M., Pishko, V.V.: Rearrangement of Magnetic Excitation Spectrum of Weakly Doped $CoCO_3$ in an External Field. *Soviet J. Low Temp. Phys.* (1985), **11**, No. 5, pp. 496–500.
354. Sanders, R.W., Rezends, S.M., Motokawa, M., Belanger, R.M., Jacarino, V.: FIR Study of AFMR and the Local Magnon Mode in Mn : FeF_2. *J. Magn. and Magn. Mater.* (1980), **15–18**, pp. 725–726.
355. Sanders, R.W., Belanger, R.M., Motokawa, M., Jaccarino, V., Rezende, S.M.: Far-Infrared Laser Study of Magnetic Polaritons in FeF_2 and Mn Impurity Mode in FeF_2 : Mn. *Phys. Rev. B* (1981), **23**, No. 3, pp. 1190–1204.
356. Thayamballi, P., Hone, D.: Effect of S_z-Nonconserving Interactions on Local-Mode Dynamics in FeF_2 : Mn. *Phys. Rev. B* (1983), **27**, No. 5, pp. 2924–2931.
357. Loktev, V.M.: Theoretical Study of Static, Resonant and Optical Properties of Anisotropic Magnets. Authorized Abstract of Dissertation of Doctor of Physics and Mathematics, Kiev, 1983.
358. Eremenko, V.V., Naumenko, V.M., Petrov, S.V., Pishko, V.V.: Rearrangement of the Magnetic Excitation Spectrum of Antiferromagnetic CoF_2 Doped with MnF_2 of Low Concentration, *Soviet Phys. JETP* (1982), **55**, No. 3, pp. 481–489.
359. Wiecko, C., Hone, D.: Theory of Impurity Banding and Magnetic Resonance in Heisenberg Antiferromagnets of $T = 0$, *J. Phys. C* (1980), **13**, No. 20, pp. 3883–3894.
360. Hone, D., Wiecko, C.: Theory of Impurity Band Resonance in Heisenberg Magnets at $T = 0$. *J. Magn. and Magn. Mater.* (1980), **15–18**, pp. 723–724.
361. Pogorelov, Yu.G., Loktev, V.M.: Impurity Spin States in Anisotropic Antiferromagnets, I. One-Impurity Approximation. *Soviet J. Low Temp. Phys.* (1979), **5**, No. 5, pp. 483–490.
362. Rezende, S.M.: Theory of the Resonance of Local Magnon Modes in Antiferromagnets. *Phys. Rev. B* (1983), **27**, No. 5, pp. 3032–3042.
363. Johnson, K.C., Sievers, A.J.: High Pressure Study of the AFMR in FeF_2. *Abstracts of the 17th Conference on Magnetism and Magnetic Materials, Chicago,* 1971. New York: AIP, 1972, 3D-1.
364. Blewitt, R., Weber, R.: High-Magnetic-Field Spectroscopy of MnF_2 and MnF_2 : Fe^{2+} in the Far-Infrared, *J. Appl. Phys.* (1970), **41**, No. 3, pp. 884–885.

365. Shklovskii, B.I., Efros, A.L.: *Electronic Properties of Doped Semiconductors*. Moscow: Nauka, 1979, 416 pp.
366. Thouless, D.J.: Electrons in Disordered Systems and the Theory of Localization. *Phys. Rep.* (1974), **13**, No. 3, pp. 93–142.
367. Knor, K.E., Smith, P.V.: Localization of Electron States in Two-Dimensional Disordered Potential Arrays. *J. Phys. C* (1971), **4**, No. 14, pp. 2029–2040.
368. Kikuchi, M.: Numerical Studies of Localization in Structurally Disordered Systems. *J. Phys. Soc. Japan* (1974), **37**, No. 4, pp. 904–911.
369. Buyers, W.J.L., Holden, T.M., Svensson, E.C., Cowley, R.A., Hutchings, M.T.: Excitations in $KCoF_3$, II. Theoretical, *J. Phys. C* (1971), **4**, No. 14, pp. 2139–2159.
370. Buyers, W.J.L., Holden, T.M., Svensson, E.3., Cowley R.A., Stevenson, R.W.H.: Character of Excitations in Substitutionally Disordered Antiferromagnets. *Phys. Rev. Lett.* (1971), **27**, No. 21, pp. 1442–1445.
371. Anderson, P.W.: Localized Magnetic States in Metals. *Phys. Rev.* (1961), **124**, No. 1, pp. 41–53.
372. Richards, P.L.: Far-Infrared Magnetic Resonance in CoF_2, NiF_2, $KNiF_3$ and YbIG. *J. Appl. Phys.* (1963), **34**, No. 4, pp. 1237–1238.
373. Martel, P., Cowley, R.A., Stevenson, R.W.H.: Experimental Studies of the Magnetic and Phonon Excitations in Cobalt Fluoride, *Canad. J. Phys.* (1968), **46**, No. 11, pp. 1355–1370.
374. Brünner, H., Renk, K.F.: Antiferromagnetic Resonance in CoF_2 at 10^{12} Hz. *J. Appl. Phys.* (1970), **41**, No. 5, pp. 2250–2251.
375. Enders, B., Richards, P.L., Tennant, W.E., Catalano, E.: Antiferromagnetic Resonance Modes in $(Co, Mn)F_2$ and $(Fe, Mn)F_2$. *Abstracts of the 18th Annual Conference on Magnetism and Magnetic Materials, Denver*, 1972. New York: AIP, 1972, 8F-4.
376. Naumenko, V.M., Eremenko, V.V., Bandura, V.M., Pishko, V.V.: Coherent Rearrangement of the Spin Wave Spectrum of Antiferromagnetic Cobalt Fluoride Doped with Manganese ($CoF_2 + 4 \cdot 10^{-3}\ Mn^{2+}$). *Pis'ma v ZhETF (Soviet Phys. JETP Lett.)* (1980), **32**, No. 6, pp. 436–439.
377. Naumenko, V.M., Fomin, V.I., Eremenko, V.V.: Vacuum Set-Up for Magneto-Optical Studies in the Far-Infrared Spectrum. *Pribor. i Tekhn. Experim.* (1967), No. 5, pp. 223–224.
378. Gredeskul, V.M., Gredeskul, S.A., Eremenko, V.V., Naumenko, V.M.: Magnetization and Resonance in Orthorhombic Antiferromagnets with the Dzyaloshinsky Interaction ($CoWO_4$). *J. Phys. Chem. Soc.* (1972), **33**, No. 4, pp. 859–880.
379. Naumenko, V.M., Eremenko, V.V., Klochko, A.V.: Millimeter and Submillimeter Range Pulse Spectrometer. *Pribor. i Tekhn. Experim.* (1981), No. 4, pp. 159–162.
380. Allen, S.J., Guggenheim, H.J.: Magnetic Excitations in Antiferromagnetic CoF_2, I. Spin–Optical–Phonon Interaction. *Phys. Rev. B* (1971), **4**, No. 3, pp. 937–949.
381. Prinz, G.A., Forester, D.W., Lewis, J.L.: Analysis of Far-Infrared Spectra of Antiferromagnetic $FeCO_3$. *Phys. Rev.* B (1973), **8**, No. 5 (I), pp. 2155–2165.

Subject Index